NUMERICAL SOLUTION OF HIGHLY NONLINEAR PROBLEMS

Fixed Point Algorithms and Complementarity Problems

Collection of articles based on the lectures presented at the Symposium on Fixed Point Algorithms and Complementarity, held at the University of Southampton, 3-5 July, 1979.

edited by

Walter FORSTER

Faculty of Mathematical Studies
The University of Southampton, England

1980

NORTH-HOLLAND PUBLISHING COMPANY – AMSTERDAM • NEW YORK • OXFORD

Repl. QA371
N82

7212-4878

MATH.-STAT.

© North-Holland Publishing Company, 1980

All rights reserved. No part of this publication may be reproduced, stored in a retrieval system, or transmitted, in any form or by any means, electronic, mechanical, photocopying, recording or otherwise, without the prior permission of the copyright owner.

Publishers:

NORTH-HOLLAND PUBLISHING COMPANY
AMSTERDAM • NEW YORK • OXFORD

Sole Distributors for the U.S.A. and Canada:

ELSEVIER NORTH-HOLLAND INC.
52 VANDERBILT AVENUE, NEW YORK, N.Y. 10017

Library of Congress Cataloging in Publication Data

Main entry under title:

Numerical solution of highly nonlinear problems.

"Collection of articles based on the lectures presented at the Symposium on Fixed Point Algorithm and Complementarity, held at the University of Southampton, 3-5 July, 1979."

1. Differential equations, Nonlinear--Numerical solutions--Addresses, essays, lectures. 2. Fixed point theory--Addresses, essays, lectures. 3. Linear complementarity problem--Addresses, essays, lectures. I. Forster, Walter, 1940- II. Symposium on Fixed Point Algorithm and Complementarity, University of Southampton, 1979.
QA371.N82 515'.35 79-25252

ISBN 0-444-85427-4

PRINTED IN THE NETHERLANDS

QA371
N82
MATH

PREFACE

The SYMPOSIUM ON FIXED POINT ALGORITHMS AND COMPLEMENTARITY was held at the University of Southampton, England, from 3rd to 5th July 1979. The Organizing Committee consisted of R.W. Cottle (Stanford), W. Forster (Southampton) and H.J. Lüthi (Zurich). The Symposium was attended by 60 participants from 20 different countries and was sponsored by U.N.E.S.C.O., European Research Office (London), Department of Mathematics (University of Southampton), I.B.M. U.K. Ltd., Lloyds Bank Ltd., Office of Naval Research (London).

Fixed point algorithms can be regarded as the culmination of 70 years of mathematical research starting with Brouwer's nonconstructive fixed point theorem. Fixed point theorems have long been used to establish existence for highly nonlinear problems. Poincaré seems to have been the first one to suggest this approach. The great breakthrough came in 1967 when Scarf succeeded in giving the first constructive proof of Brouwer's fixed point theorem. This was the starting point for the development of a large number of constructive methods for finding fixed points. Together with the capabilities of computers such methods have become powerful tools for finding solutions to general systems of nonlinear equations, price equilibria, etc. etc.

A number of eminent mathematicians whose names are inseparably linked with the development of fixed point algorithms and many young researchers contributed to the success of the Symposium. The papers collected in this volume are either extensions of papers presented at the Symposium or have been specially prepared for this volume. All papers have been retyped, but because of the time factor involved only a limited amount of editing could be done. The papers cover algorithms, applications, theory and historical background. Together with the volumes originating from the meetings in Clemson (1974), Bonn (1978) and Madison (1979) this volume will play its role in the further advancement of methods for the numerical solution of highly nonlinear problems.

Walter Forster
Faculty of Mathematical Studies
The University
Southampton, England

October 1979.

CONTENTS

LECTURES

on fixed point algorithms (pivoting algorithms and related topics) and complementarity problems.

E.L. Allgower (Fort Collins):

HOMOTOPIES FOR SEVERAL SOLUTIONS.

S.J. Byrne and R.W.H. Sargent (London):

A VERSION OF LEMKE'S ALGORITHM USING ONLY ELEMENTARY PRINCIPAL PIVOTS.

D.I.A. Cohen (New York):

A COMBINATORIAL APPROACH TO THE KAKUTANI FIXED POINT THEOREM.

R.W. Cottle (Stanford):

A THEOREM ON THE PARTITIONING OF SIMPLOTOPES AND ITS APPLICATION TO A PROPERTY OF Q-MATRICES.

B.C. Eaves (Stanford) and C.E. Lemke (New York):

AN EQUIVALENCE BETWEEN SYSTEMS OF PIECEWISE LINEAR EQUATIONS AND THE LINEAR COMPLEMENTARITY PROBLEM.

L. Filus (Warsaw):

FIXED POINT ALGORITHMS AND SOME GRAPH PROBLEMS.

W. Forster (Southampton):

A COMPUTATIONAL FIXED POINT THEOREM FOR COMPLEXES.

C.B. Garcia (Chicago):

THE FLEX SIMPLICIAL ALGORITHM.

K. Georg (Bonn):

AN APPLICATION OF SIMPLICIAL ALGORITHMS TO VARIATIONAL INEQUALITIES.

F.J. Gould (Chicago):

APPLICATION OF THE NEWTON HOMOTOPY TO COMPLEMENTARITY PROBLEMS.

H.W. Kuhn (Princeton):

ALGORITHMS FOR COMPUTING TRAFFIC EQUILIBRIA.

G. van der Laan (Amsterdam):

A CLASS OF RESTART FIXED POINT ALGORITHMS WITHOUT AN EXTRA DIMENSION.

Tien-Yien Li (Wisconsin):

PIECEWISE SMOOTH CONTINUATION AND OPTIMIZATION.

H.J. Lüthi (Zurich):

A SPERNER-TYPE PROOF OF KRONECKER'S THEOREM.

L. Mathiesen and T. Hansen (Bergen):

THE COMPUTATION OF EQUILIBRIUM PRICES FOR AN ECONOMY WITH INSTITUTIONAL CONSTRAINTS ON PRICES.

Jong-Shi Pang (Pittsburgh):

A PARAMETRIC LINEAR COMPLEMENTARITY TECHNIQUE FOR THE COMPUTATION OF EQUILIBRIUM PRICES IN AN ECONOMIC SPATIAL MODEL.

H.O. Peitgen (Bremen):

TOPOLOGICAL PERTURBATIONS IN THE NUMERICAL STUDY OF NONLINEAR EIGENVALUE PROBLEMS: ARTIFICIAL BIFURCATIONS.

M. Prüfer (Bremen):

A NEW LABELLING WITH APPLICATIONS IN THE LERAY-SCHAUDER CONTINUATION METHOD AND GLOBAL BIFURCATIONS.

R. Saigal (Evanston):

ON SOLVING LARGE STRUCTURAL FIXED POINT PROBLEMS.

H.E. Scarf (Yale University):

THE APPLICATION OF FIXED POINT TECHNIQUES TO INTEGER PROGRAMMING.

R. Schramm (Tel-Aviv):

AN UPPER BOUND ON THE NUMBER OF SOLUTIONS OF PIECEWISE LINEAR EQUATIONS.

H.W. Siegberg (Bonn):

BROUWER DEGREE: HISTORY AND NUMERICAL COMPUTATION.

E. Sperner (Hamburg):

FIFTY YEARS OF FURTHER DEVELOPMENT OF A COMBINATORIAL LEMMA.

M. Stynes (Cork):

AN N-DIMENSIONAL BISECTION METHOD FOR SOLVING SYSTEMS OF N EQUATIONS IN N UNKNOWNS.

A.J.J. Talman (Amsterdam):

COMPUTATION OF FIXED POINTS ON THE PRODUCT SPACE OF UNIT SIMPLICES AND AN APPLICATION TO NON-COOPERATIVE N PERSON GAMES.

M.J. Todd (New York):

ON RECENT VARIABLE DIMENSION SIMPLICIAL ALGORITHMS.

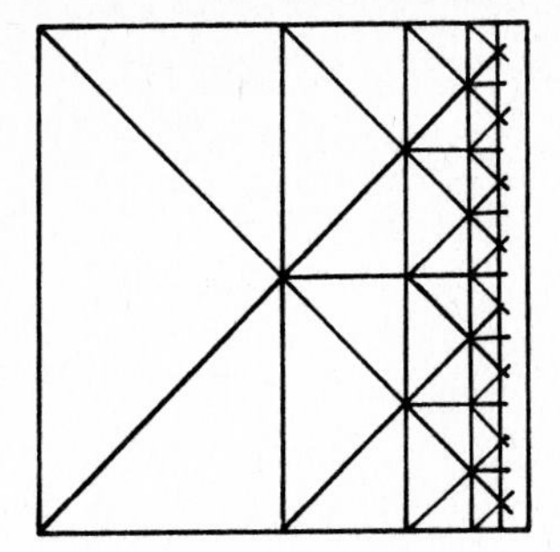

The papers in this section contain algorithms.

Numerical Solution of Highly Nonlinear Problems
W. Forster (ed.)
© North-Holland Publishing Company, 1980

CONVERGENCE AND PROPERTIES OF RECENT VARIABLE DIMENSION ALGORITHMS

G. van der Laan and A.J.J. Talman

Free University, Amsterdam,
The Netherlands

In this paper algorithms recently developed by the authors are discussed. The applications on some special bounded sets are described and some geometrical interpretations are given. Moreover, it is proved that if the affine hull of the unit simplex is triangulated as proposed by the authors in an earlier paper, Merrill's condition guarantees convergence of the basic algorithm on this set. A similar result can be proved on R^n, viz. if the (α^*, β^*)-triangulation is used. It is argued that by using vector labelling the convergence of a fixed point algorithm depends on the initiated system of linear equations and the underlying triangulation. Finally, the basic algorithm on R^n has been generalized to a class of algorithms. An extreme case is treated and it is proved that Merrill's condition is again sufficient.

1. INTRODUCTION

After the basic algorithm of Scarf [15] and [16] to compute fixed points, several more efficient algorithms based on simplicial subdivision have been developed. One method, due to Merrill [12], Kuhn and MacKinnon [4] and also Lüthi [11], is a restart algorithm and will be referred to as the Sandwich method; another method, introduced by Eaves [1] and Eaves and Saigal [2], involves an automatically refining triangulation and is called the homotopy algorithm. The Sandwich method has the disadvantage that a layer of artificial (labelled) points is needed influencing the path of generated simplices, whereas in the homotopy algorithm the factor of incrementation must be less than or equal to 2. Moreover, both methods operate with an extra dimension and guarantee a path of full-dimensional adjacent simplices.

In their papers [5] and [6] Van der Laan and Talman presented a variable dimension restart algorithm which does not need an extra dimension or artificial points. Furthermore, their algorithm differs from the other ones by the fact that not a path of full-dimensional simplices is generated but a unique path of lower-dimensional simplices. To be more precise, the method starts with one point only, a zero-dimensional simplex, which can be arbitrarily chosen, e.g. from prior information, and generates a path of adjacent simplices of variable dimension, and terminates with a full-dimensional simplex which is a good approximation to a fixed point, and provides the starting point for the next application of the algorithm with a finer grid.

In section 2. a concise description of the algorithm of Van der Laan and Talman is given to compute a fixed point of a function (or mapping) on the (n-1)-dimensional unit simplex and its affine hull T^n. Also conditions to guarantee that the algorithm converges on T^n are discussed. Therefore we need a triangulation of T^n which was for the first time introduced also by Van der Laan and Talman [7].

Section 3. describes the variable dimension algorithm on some special bounded sets of R^n and on R^n itself. Again convergence conditions are given. Some properties and interpretations of the method are discussed in section 4. Based on these features Van der Laan and Talman [10] introduced a triangulation appropriate to a homotopy algorithm which allows for an arbitrarily chosen sequence of factors of incrementation. Moreover, they [9] developed a class of algorithms on R^n , which allows for terminating with lower-dimensional simplices. An extreme case of this class is discussed in section 5. Another application is the computation of fixed points on the product space of simplices (see [8]).

2. VARIABLE DIMENSION ALGORITHM ON S^n AND T^n

Assume we want to compute a fixed point of a continuous function f from

$$S^n = \{x \in R^n_+ \mid \sum_{i=1}^{n} x_i = 1\}$$

into itself. Therefore S^n is triangulated in the standard way (see Kuhn [3] and Scarf [16]) with grid size m , i.e. the triangulation is the collection of (n-1)-dimensional simplices $\sigma(y^1, \pi)$ with vertices $y^1, \ldots, y^n$ being grid points in S^n such that π is a permutation of $n-1$ elements of $I_n = \{1, \ldots, n\}$ and $y^{i+1} = y^i + m^{-1} q(\pi_i)$, $i = 1, \ldots, n-1$ where $q(i)$ is the i-th column of the $n \times n$ triangulation matrix Q defined by

$$Q = \begin{bmatrix} -1 & 0 & . & . & . & 0 & 1 \\ 1 & . & & & & & 0 \\ . & . & . & & & & . \\ . & & . & . & & & . \\ . & & & . & . & & . \\ . & & & & . & -1 & 0 \\ 0 & . & . & . & 0 & 1 & -1 \end{bmatrix}$$

Observe that every simplex of the triangulation has n different representations since $\Sigma^n_{i=1} q(i)$ is the zero vector. Let T be a proper subset of t elements of I_n , then for a grid point y^1 and a permutation $\pi(T) = (\pi_1, \ldots, \pi_t)$ of the elements of T , the t-simplex with vertices $y^1, \ldots, y^{t+1}$ where $y^{i+1} = y^i + m^{-1} q(\pi_i)$, $i = 1, \ldots, t$ will be denoted as $\sigma(y^1, \pi(T))$.

Every point x of S^n is labelled by

$$\ell(x) = i \quad \text{if} \quad i = \min\{j \mid f_j(x) - x_j \le f_h(x) - x_h , h = 1, \ldots, n , x_j > 0\} .$$

A simplex $\sigma(y^1, \ldots, y^n)$ is called completely labelled if its vertices carry all the labels of the set I_n. A completely labelled simplex is a good approximation of a fixed point of f (Saigal [14]). Moreover a $(t-1)$-simplex $\sigma(y^1, \ldots, y^t)$, $1 \le t \le n-1$, is called T-complete if the vertices of σ carry all the labels of the set T. Note that every zero-dimensional simplex $\{y\}$ is $\{\ell(y)\}$-complete.

Let v be an arbitrarily chosen grid point, called starting point. Then the variable dimension algorithm proceeds as follows, where $e(i)$ is the i-th unit vector.

Step 0. Set $t = 0$, $y^1 = v$, $T = \emptyset$, $\pi(T) = \emptyset$, $\sigma = \{y^1\}$, $\bar{y} = y^1$ and $R_i = 0$, $i = 1, \ldots, n$.

Step 1. Calculate $\ell(\bar{y})$. If $\ell(\bar{y})$ is not an element of T go to step 3. Otherwise $\ell(\bar{y}) = \ell(y^s)$ for exactly one vertex $y^s \neq \bar{y}$.

Step 2. If $s = t+1$ and $R_{\pi_t} = 0$, go to step 4. Otherwise σ and R are adapted according to table 1. by replacing y^s. Return to step 1 with $\bar{y}$ equal to the new vertex of σ.

Step 3. If $t = n-1$, a completely labelled simplex is found and the algorithm terminates. Otherwise, a $(T \cup \{\ell(\bar{y})\})$-complete simplex is found and T becomes $T \cup \{\ell(\bar{y})\}$, $\pi(T)$ becomes $(\pi_1, \ldots, \pi_t, \ell(\bar{y}))$, σ becomes $\sigma(y^1, \pi(T))$ and t becomes $t+1$. Return to step 1 with $\bar{y}$ equal to y^{t+1}.

Step 4. Let for some s', $s' \le t$, $y^{s'}$ be the vertex of σ with label π_t. Then T becomes $T \backslash \{\pi_t\}$, $\pi(T)$ becomes $(\pi_1, \ldots, \pi_{t-1})$, σ becomes $\sigma(y^1, \pi(T))$, t becomes $t-1$ and return to step 2 with $s = s'$.

	y^1 becomes	$\pi(T)$ becomes	R becomes
$s = 1$	$y^1 + m^{-1}\ q(\pi_1)$	$(\pi_2 , \ldots , \pi_t , \pi_1)$	$R + e(\pi_1)$
$2 \leq s \leq t$	y^1	$(\pi_1 , \ldots , \pi_{s-1} , \pi_s , \pi_{s+2} , \ldots , \pi_t)$	R
$s = t+1$	$y^1 - m^{-1}\ q(\pi_t)$	$(\pi_t , \pi_1 , \ldots , \pi_{t-1})$	$R - e(\pi_t)$

Table 1. Pivot rules (s is the index of the vertex which must be replaced).

The algorithm generates a path of adjacent simplices starting with the zero-dimensional simplex $\{v\}$. First a search is made in the direction $q(\ell(v))$, i.e. the $\ell(v)$-th component is decreased to find the other labels. One-dimensional simplices $\sigma(y^1, y^2)$ are generated with $y^1 = v + R_{\ell(v)} q(\ell(v))/m$ and $y^2 = y^1 + m^{-1} q(\ell(v))$. The algorithm proceeds with two-dimensional simplices as soon as a second label is found, which must occur since $\ell(x) \neq i$ if $x_i = 0$. In general, if in step 1 a new label is found, say k, we have that for certain t, subset T of I_n, permutation $\pi(T)$, simplex $\sigma(y^1, \ldots, y^{t+1})$ and vector R holds

(i) $y^1 = v + \sum_{j=1}^{n} R_j q(j)/m$,

(ii) $R_j = 0$ for $j \notin T$ and $R_j \geq 0$ for $j \in T$,

(iii) $y^{i+1} = y^i + m^{-1} q(\pi_i)$, $i = 1, \ldots, t$,

(iv) σ is $(T \cup \{k\})$-complete.

Then the algorithm terminates with a completely labelled simplex $\sigma(y^1, \ldots, y^n)$ if $t = n-1$, whereas otherwise a search is made with the labels j, in all directions $q(j)$, $j \in T \cup \{k\}$, to find the other labels. To do so, the algorithm generates $(t+1)$-simplices starting with $\sigma(y^1, \ldots, y^{t+2})$, where $y^{t+2} = y^{t+1} + m^{-1} q(k)$, until either again a new label is found or in step 2, $s = t+1$ and R_{π_t} negative, and from (i) it is clear that the method searches in the direction $-q(\pi_t)$ while label π_t already has been found. To avoid this, the replacement step is not performed, but in step 4 the vertex y^{t+1} is deleted, together with label π_t by removing the vertex of σ having label π_t, and the method proceeds with $(t-1)$-simplices, making a search in the direction $q(j)$, $j \in T \setminus \{\pi_t\}$.

In Van der Laan and Talman [5] it is proved that the algorithm can never return to a simplex visited already before and that all steps are feasible. Therefore, since the number of t-simplices, $0 \leq t \leq n-1$, is finite, the algorithm always terminates with a completely labelled subsimplex being a good approximation to a fixed point. Clearly, using this approximation to choose the starting point, the algorithm can be restarted with a finer grid until a given accuracy is obtained.

To apply the algorithm on

$$T^n = \{x \in R^n \mid \sum_{i=1}^{n} x_i = 1\} \quad ,$$

this set is triangulated according to the T-triangulation with grid size m , i.e. the triangulation is the collection of simplices $\sigma(y^1, \pi)$ with vertices $y^1, \dots, y^n$ such that

$$y^1 = z + \sum_{i=1}^{n-1} \lambda_i \, t(i)$$

where λ_i , $i = 1, \dots, n-1$, is integer and z an arbitrarily chosen point in T^n , and $y^{i+1} = y^i + m^{-1} t(\pi_i)$, where $\pi = (\pi_1, \dots, \pi_{n-1})$ is a permutation of $n-1$ elements on I_n . The vector $t(i)$ is the i-th column of the $n \times n$ triangulation matrix T defined by

$$T = \begin{bmatrix} -(n-1) & 1 & . & . & . & . & 1 \\ 1 & . & & & & & . \\ . & & . & & & & . \\ . & & & . & & & . \\ . & & & & . & & . \\ . & & & & & . & 1 \\ 1 & . & . & . & . & 1 & -(n-1) \end{bmatrix} .$$

It is argued by Van der Laan and Talman [7] that this triangulation is very appropriate to be used in fixed point algorithms on T^n .
Every point of T^n is labelled by

$$\ell(x) = i \quad \text{where} \quad i = \min \{j \mid f_j(x) - x_j \le f_h(x) - x_h \; , \; h = 1, \dots, n\} .$$

Again a completely labelled simplex is a good approximation to a fixed point. To generate such a simplex, if there exists one, the algorithm starts with an arbitrary point v of T^n . T^n is triangulated such that v is a grid point. The algorithm follows now exactly the same steps as described above for S^n except that in table 1. $q(i)$ has to be changed to $t(i)$, $i = 1, \dots, n$. Using the same arguments either a path of adjacent simplices of variable dimension is generated terminating with a completely labelled n-simplex, or the algorithm diverges along a path to infinity. A condition to guarantee that the algorithm terminates is given in the following theorem.

Theorem 2.1: Let f be a continuous function on T^n and let v be the starting point. Assume there exist $\mu < \min v_i$ and $\varepsilon > 0$ such that for all $x \in B(\mu) = \{y \in T^n \mid \min y_i = \mu\}$ there is an index $i \in I_n$ with $x_i < v_i$ and $f_j(x) - x_j < f_i(x) - x_i - \varepsilon$ for at least one index $j \in I_n$. Then the algorithm terminates if the triangulation is small enough.

Proof: Let the grid of the triangulation be so fine that

$$\sup_{x,y \in \sigma} |\max (f_i(x) - x_i - f_i(y) + y_i| < \varepsilon$$

for every simplex σ meeting $B(\mu)$. We will prove that no simplex having points in common with $B(\mu)$ can be generated. Let x be such a point, then there is an index i such that $x_i < v_i$ and $f_j(x) - x_j < f_i(x) - x_i - \varepsilon$ for some $j \in I_n$. If σ is generated we must have that $i \in T$ since otherwise the i-th component was not decreased. Moreover $f_j(y^k) - y^k_j < f_i(y^k) - y^k_i$, $k = 1, \ldots, t+1$. Hence $\ell(y^k) \neq i$ for all vertices y^k, $k = 1, \ldots, t+1$, which contradicts $i \in T$. Therefore the algorithm can only generate simplices in the interior of the convex hull of $B(\mu)$. Since $B(\mu)$ is bounded, the number of generated simplices must be finite.

□

Note that this condition depends on the starting point, so it is not sure that the algorithm converges if a restart is made with a finer grid. Stronger conditions guaranteeing that the algorithm converges for every starting point can be easily found.

To compute a fixed point of a mapping Φ from S^n (or T^n) into itself, each point x is labelled with a vector label $\ell(x) \in R^n$. Let f be a linear approximation to Φ with respect to the triangulation. Then $\ell(x) = x - f(x) + e$, where e is the vector with $e_i = 1$, $i = 1, \ldots, n$. Let $e(i)$ be the i-th unit vector. We call a $(t-1)$-simplex $\sigma(y^1, \ldots, y^t)$ $1 \leq t \leq n-1$, T-complete if the set of linear equations

$$\sum_{i=1}^{t} \lambda_i \, \ell(y^i) + \sum_{j \notin T} \mu_j^* \, e(j) = e$$

has a non-negative solution λ_i^* $i = 1, \ldots, t$ and μ_j^* $j \in I_n \backslash T$. By summing over all equations, it follows immediately that for every (feasible) solution holds

$$\sum_{i=1}^{t} n \lambda_i^* + \sum_{j \notin T} \mu_j^* = n .$$

A t-simplex $\sigma(y^1, \ldots, y^{t+1})$ is called completely labelled if $\mu_j^* = 0$ for all j . If σ is completely labelled we have in general that $t = n-1$, but in contrast to integer labelling it is possible that t is less than $n-1$. Let $\sigma(y^1, \ldots, y^{t+1})$ be completely labelled, then

$$y^* = \sum_{i=1}^{t+1} \lambda_i^* y^i$$

is a good approximation to a fixed point (Saigal [14]) . The algorithm starts with an arbitrary grid point v of S^n (or T^n) and with the set of linear equations $I\mu = e$ and proceeds as follows.

Step 0. Set $t = 0$, $y^1 = v$, $T = \emptyset$, $\pi(T) = \emptyset$, $\sigma = \{y^1\}$, $s = 1$, $\bar{y} = y^1$, $R_i = 0$, $i = 1, \ldots, n$ and $\mu_h = 1$, $h = 1, \ldots, n$.

Step 1. Calculate $\ell(\bar{y})$. Make a pivot step with $\ell(\bar{y})$ in the system of linear equations

$$\sum_{\substack{i=1 \\ i \neq s}}^{t+1} \lambda_i \, \ell(y^i) + \sum_{h \notin T} \mu_j \, e(h) = e .$$

If $\mu_j = 0$ for all j , the algorithm terminates. If not, either for some j $e(j)$ is eliminated and go to step 3 , or $\ell(y^{s'})$ is eliminated for just one vertex $y^{s'} \neq \bar{y}$.

Step 2. If $s' = t+1$ and $R_{\pi_t} = 0$, go to step 4. Otherwise $\sigma(y^1, \pi(T))$ and R are adapted according to table 1. by replacing $y^{s'}$. Return to step 1 with $\bar{y}$ equal to the new vertex of σ and s to the index of $\bar{y}$.

Step 3. A $(T \cup \{j\})$-complete simplex is found and T becomes $T \cup \{j\}$, $\pi(T)$ becomes $(\pi_1, \ldots, \pi_t, j)$, σ becomes $\sigma(y^1, \pi(T))$ and t becomes $t+1$. Return to step 1 with $\bar{y}$ equal to y^{t+1} and $s = t+1$.

Step 4. The vertex y^{t+1} is deleted. A pivot step with $e(\pi_t)$ is made in the current set of linear equations

$$\sum_{i=1}^{t} \lambda_i \, \ell(y^i) + \sum_{h \notin T} \mu_j \, e(h) = e \, ,$$

and T becomes $T \backslash \{\pi_t\}$, $\pi(T)$ becomes $(\pi_1, \ldots, \pi_{t-1})$, σ becomes $\sigma(y^1, \pi(T))$ and t becomes $t-1$.
If for some j, $e(j)$ is eliminated, return to step 3.
Otherwise $\ell(y^{s'})$ is eliminated for a unique s', $s' \leq t+1$.
Return to step 2.

Assuming non-degeneracy, the algorithm generates a path of adjacent simplices. Clearly, on T^n the algorithm terminates with a completely labelled simplex or the path goes to infinity. On S^n the algorithm terminates with a good approximation if all replacement-steps and the extension to a higher dimensional simplex are feasible. To prove the first, let $\sigma(y^1, \pi(T)) = \sigma(y^1, \ldots, y^{t+1})$ be a simplex generated by the algorithm and $\tau(y^1, \ldots, y^{s-1}, y^{s+1}, \ldots, y^{t+1})$ be a facet of σ on the j-th side of S^n, i.e. $y^i_j = 0$, $i \neq s$.
Then we have the following lemma.

<u>Lemma 2.2</u>: If $v_j \neq 0$ either τ is completely labelled or the set of linear equations

$$\sum_{i \neq s} \lambda_i \, \ell(y^i) + \textstyle\sum_h \mu_h \, e(h) = e$$

has no feasible solution.

<u>Proof</u>: If there is a feasible solution λ^*_i, $i \neq s$ and μ^*_h, $h \notin T$, then we have

$$n \sum_{i \neq s} \lambda^*_i + \textstyle\sum_h \mu^*_h = n$$

implying that either

$$\sum_{i \neq s} \lambda_i^* = 1$$

and τ is completely labelled or $\sum_{i \neq s} \lambda_i^* < 1$.

In the latter case j must be an element of T (since $v_j \neq 0$) and hence the j-th equation is

$$\sum_{i \neq s} \lambda_i^* \ell_j(y^i) = 1 \quad .$$

Furthermore, $\ell_j(y^i) = y_j^i - f_j(y^i) + 1 \leq 1$, since $y_j^i = 0$.
Therefore

$$\sum_{i \neq s} \lambda_i^* \geq 1 \quad ,$$

which gives a contradiction.

□

The lemma means that y^s cannot be replaced if $v_j \neq 0$. If $v_j = 0$, then either $j \in T$, which is identical to the case just mentioned, or $j \notin T$, which implies that step 4 has to be performed. Secondly, the extension to a higher dimensional simplex is always feasible except when $y_j^{t+1} = 0$ and $e(j)$ is just eliminated. Then, following the proof of lemma 2.2, $\sigma(y^1 , \ldots , y^{t+1})$ must be completely labelled. Consequently all steps are feasible and the algorithm must terminate.

Finally we will prove that Merrill's condition is sufficient for the convergence on T^n .

Theorem 2.3: Let Φ be an upper semi continuous mapping on T^n . For some $\delta > 0$ let there exist $w \in T^n$ and $\rho < 1/n$ such that for all $x, z \in T^n$ with

$$\min_{i=1,\ldots,n} x_i < \rho \quad \text{and} \quad \max_{i=1,\ldots,n} |z_i - x_i| < \delta$$

holds $(f(x) - x)'(w-z) > 0$ for all $f(x) \in \Phi(x)$.
Then the algorithm converges if the mesh of the triangulation is small enough.

Proof: Let the mesh of the triangulation be less than δ and let v be an arbitrarily chosen starting point. Define ρ' by

$$\rho' = \min\{\rho - \delta \, , \, \min_i v_i + \min_i n(v_i - w_i)\}$$

Let $\sigma(y^1, \pi(T))$ be a t-dimensional simplex such that

$$\min_{i=1,\ldots,n} y_i < \rho'$$

for some $y \in \sigma$. We will prove that σ has no T-complete facet $\tau(y^1, \ldots, y^{s-1}, y^{s+1}, \ldots, y^{t+1})$, $s = 1, \ldots, t+1$. Define $\lambda_s = 0$. Then by multiplying the system of linear equations with $(w - y)$ we obtain

$$\left(\sum_{i=1}^{t+1} \lambda_i \{-f(y^i) + y^i\} + \sum_{h \notin T} \mu_h e(h)\right)'(w - y) = e'(w - y) = 0 \, ,$$

since

$$\sum_{i=1}^{n} w_i = \sum_{i=1}^{n} y_i = 1 \, .$$

From the condition it follows that $(-f(y^i) + y^i)'(w - y)$ is negative for all i. So it remains to prove that $w_h - y_h$ is negative for all $h \notin T$ since at least one λ_i or μ_h is positive.
Let for some k,

$$y_k = \min_i y_i < \rho' < \min_i v_i + \min_j n(v_j - w_j) \, .$$

Hence, $y_k < v_k + n(v_h - w_h)$ for all $h \notin T$.
Because of the structure of the triangulation matrix T

$$y_h \geq v_h + n^{-1}(v_k - y_k) \qquad \text{for all } h \notin T \, .$$

Both inequalities together give for all $h \notin T$

$$y_h > v_h + n^{-1} v_k - n^{-1} v_k - v_h + w_h = w_h \, .$$

Therefore no simplex can be generated outside the convex compact set $\{x \in T^n \mid \min x_i \leq \rho'\}$, which proves the theorem.

□

Observe that the proof is not valid if the standard triangulation is taken. Finally, we remark that the condition "for some δ " can be strengthened to "for all δ " . In that case the algorithm converges for every grid size.

3. VARIABLE DIMENSION FIXED POINT ALGORITHM ON (SPECIAL BOUNDED SETS OF) R^n

First the algorithm described in the previous section is modified to compute a fixed point of a function f (or mapping Φ) from the set $V^n(\alpha\ ;\ a)$ into itself where

$$V^n(\alpha\ ;\ a) = \{x \in R^n \mid \sum_{i=1}^{n} x_i \geq \alpha \quad \text{and} \quad x_i \leq a_i \ , \ i = 1, \dots, n\}$$

where α and a_i are given numbers such that

$$\sum_{i=1}^{n} a_i > \alpha \ .$$

Clearly, the H-triangulation of R^n with grid size m restricted to $V^n(\alpha\ ;\ a)$ triangulates $V^n(\alpha\ ;\ a)$ if the vertex a is a gridpoint [illegible] m is a multiple of

$$\sum_{i=1}^{n} a_i - \alpha \ ,$$

i.e. the triangulation is the collection of n-simplices $\sigma(y^1, \pi)$ in $V^n(\alpha\ ;\ a)$ such that π is a permutation of n elements of I_{n+1} ,

$$y^1 = a - \sum_{i=1}^{n} \lambda_i\, h(i)/m$$

for non negative integers λ_i , $i = 1, \dots, n$ and $y^{i+1} = y^i + h(\pi_i)/m$, $i = 1, \dots, n$, where $h(j)$ is the j-th column of the $n \times (n+1)$ matrix H defined by

$$H = \begin{bmatrix} 1 & -1 & . & . & . & . & 0 \\ 0 & 1 & -1 & & & & . \\ . & & . & . & & & . \\ . & & & . & . & & . \\ . & & & & . & -1 & . \\ 0 & . & . & . & . & 1 & -1 \end{bmatrix}$$

Each point x of $V^n(\alpha\ ;\ a)$ is labelled as follows. In case of integer labelling let J be the set of indices j such that $f_j(x) - x_j \geq f_h(x) - x_h$ for all $h = 1, \ldots, n$.
If $f_j(x) - x_j \leq 0$, $j \in J$ and

$$\sum_{i=1}^{n} x_i > \alpha ,$$

then $\ell(x) = n+1$. Otherwise $\ell(x) = i$ where i is the lowest index in J such that $x_i < a_i$. Note that this labelling is proper in the sense that $\ell(x) \neq i$ if $x_i = a_i$, $i = 1, \ldots, n$, and $\ell(x) \neq n+1$ if

$$\sum_{j=1}^{n} x_j = \alpha .$$

In case of vector labelling a point x of $V^n(\alpha\ ;\ a)$ receives the $(n+1)$-vector label $\ell(x)$ defined by $\ell_i(x) = f_i(x) - x_i + 1$, $i = 1, \ldots, n$, and $\ell_{n+1}(x) = 1$, where f is a linear approximation to Φ with respect to the triangulation. Again a completely labelled simplex must be a good approximation to a fixed point.

Starting with an arbitrarily chosen gridpoint v the algorithm described in the previous section, modified for the dimension and the triangulation matrix, generates a path of adjacent simplices of variable dimension such that the common faces are *T*-complete, and terminates with a completely labelled simplex, using the standard argument of section 2.

In the same way the algorithm can be modified to compute a fixed point of a function or mapping from the unit cube $C^n = \{x \in R^n_+ \mid x_i \leq 1,\ i = 1, \ldots, n\}$ into itself. This set is triangulated by the K-triangulation with gridsize m of R^n restricted to C^n such that m is an integer and the zero-vector is a gridpoint.

The triangulation matrix is the $n \times (n+1)$-matrix defined by $u(i) = e(i)$, $i = 1, \dots, n$, and $u(n+1) = -e$. In case of integer labelling let J be the set of indices j such that $f_j(x) - x_j \geq f_h(x) - x_h$ for all $h = 1, \dots, n$. If $f_j(x) - x_j \leq 0$, $j \in J$, and $x_i > 0$ for all $i \in I_n$, then $\ell(x) = n+1$. Otherwise $\ell(x) = i$ where i is the lowest index in J such that $x_i < 1$. Again this labelling rule is proper, i.e. $\ell(x) \neq i$ if $x_i = 1$, and $\ell(x) \neq n+1$ if $x_j = 0$ for at least one index j.

In case of vector labelling the labelling rule is identical to that of $V^n(\alpha ; a)$. Clearly, applying the algorithm with the necessary modifications, gives a completely labelled simplex. To apply the algorithm in R^n, R^n is triangulated with gridsize m according to the $n \times (n+1)$-triangulation matrix A^* defined by

$$A^* = \begin{bmatrix} n+\sqrt{n+1} & -1 & \cdot & \cdot & \cdot & -1 & -1-\sqrt{n+1} \\ -1 & \cdot & & & & \cdot & \cdot \\ \cdot & & \cdot & & & \cdot & \cdot \\ \cdot & & & \cdot & & \cdot & \cdot \\ \cdot & & & & \cdot & -1 & \cdot \\ -1 & -1 & \cdot & \cdot & -1 & n+\sqrt{n+1} & -1-\sqrt{n+1} \end{bmatrix} .$$

The i-th column of A^* will be denoted by $a^*(i)$. As argued by Van der Laan and Talman [7] this triangulation is very appropriate to be used in fixed point algorithms on R^n since the simplices are as "round" as possible. This is caused by the fact that each column of A^* has the same length. The application can now easily be done by labelling a point $x \in R^n$ with $n+1$ if $f_i(x) - x_i \leq 0$ for all i, and with i if $i = \min\{j \mid f_j(x) - x_j \geq f_h(x) - x_h, \; h = 1, \dots, n\}$ and $f_i(x) - x_i > 0$ in case of integer labelling, and with the $n+1$ vector label

$$\ell(x) = \begin{bmatrix} f(x) - x \\ 1 \end{bmatrix}$$

in case of vector labelling.

In the following theorems conditions are stated to guarantee that the path of generated simplices is finite.

Theorem 3.1: (Integer labelling)
Let f be a continuous function on R^n and v the starting point. Assume there exist $\mu > \max_i |v_i|$ and $\varepsilon > 0$ such that for all

$x \in B(\mu) = \{y \in R^n \mid \max_i |y_i| = \mu\}$ we have at least one of the

following cases:

1. $e'x < e'v$ and for at least one index j, $f_j(x) - x_j > \varepsilon$.
2. There is an index i with $x_i > v_i$, and
   $f_j(x) - x_j > f_i(x) - x_i + \varepsilon$ for at least one index j or
   $f_j(x) - x_j < -\varepsilon$ for all $j \in I_n$.

Then the algorithm terminates if the triangulation is fine enough.

Proof: Let the mesh be so small that

$$\sup_{x,y \in \sigma} (\max_i |f_i(x) - x_i - f_i(y) - y_i|) < \varepsilon$$

for every simplex σ meeting $B(\mu)$.
Let $\sigma(y^1, \pi(T))$ be a t-simplex such that $\sigma \cap B(\mu) \neq \emptyset$. then for a point $x \in \sigma \cap B(\mu)$ we have that (1) or (2) holds.
If (1) is satisfied then $n+1 \in T$ since $e'x < e'v$, and $f_j(y^i) - y^i_j > 0$ for all $i = 1, \ldots, t+1$, since $f_j(x) - x_j > \varepsilon$.
Hence $\ell(y^i) \neq n+1$ for all i which gives a contradiction.
If (2) is satisfied then $i \in T$ since $x_i > v_i$, and for at least one j, $f_j(y^k) - y^k_j > f_i(y^k) - y^k_i$ for all k, or for all j,

$f_j(y^k) - y^k_j < 0$ also for all k.
Hence $\ell(y^k) \neq i$, $k = 1, \ldots, t+1$ and again a contradiction is obtained. Therefore the algorithm can generate only simplices in the interior of the convex hull of $B(\mu)$, which proves the theorem.

□

Theorem 3.2: (Vector labelling)
For some $\delta > 0$, let there exist $w \in R^n$ and $\rho > 0$ such that for all $x, z \in R^n$ with $\max |x_i| > \rho$ and $\max_i |z_i - x_i| < \delta$ holds

$(f(x) - x)'(w - z) > 0$ (Merrill's condition) for all $f(x) \in \Phi(x)$.
Then the algorithm converges if the mesh of the triangulation is small enough.

Proof: Let the mesh of the triangulation be less than δ and let v be the starting point. Define ρ' by

$$\rho' = \max [\rho + \delta , \max_i |v_i| + \max \{\max_i (n + \sqrt{n+1}) |v_i - w_i| , (n + \sqrt{n+1})(1 + \sqrt{n+1})^{-1} | \sum_i (w_i - v_i) | \}] .$$

Let $\sigma(y^1 , \pi(T))$ be a t-simplex such that for some $y \in \sigma$, $\max_i |y_i| > \rho'$. We will prove that σ has no T-complete facet

$\tau(y^1 , \ldots , y^{s-1} , y^{s+1} , \ldots , y^{t+1})$, $s = 1, \ldots , t+1$.
By defining $\mu_{n+1} = 0$ if $n+1 \in T$, and $\lambda_s = 0$, we obtain after multiplying the system of linear equations with the $(n+1)$-vector $((w - y)',0)'$

$$\sum_{i=1}^{t+1} \lambda_i (f(y^i) - y^i)'(w - y) + \sum_{h \in I_n \setminus T} \mu_h (w_h - y_h) - \mu_{n+1} \sum_i (w_i - y_i) = 0 ,$$

since

$$\sum_{i=1}^{t+1} \lambda_i = 1 - \mu_{n+1} .$$

It follows immediately from Merrill's condition that $(f(y^i) - y^i)'(w - y) > 0$ since $\max_i |y_i| > \rho' \geq \rho + \delta$ and

therefore $\max_i |y_i^k| > \rho$ for all vertices y^k , $k = 1, \ldots , t+1$.

Furthermore we will prove that $w_h - y_h > 0$ for all $h \in I_n \setminus T$ and that

$$\sum_i (w_i - y_i) < 0$$

if $n+1 \notin T$. If so, then the system has no feasible solution since at least one of the λ_i's or μ_{n+1} is positive. To prove that $w_h - y_h$ is positive for all $h \in I_n \backslash T$, we first assume that for some k, $y_k = \max_i |y_i|$. Hence,

$$y_k > v_k + (n+\sqrt{n+1})(v_h - w_h) \qquad h \in I_n \backslash T .$$

By the structure of the triangulation matrix A^*

$$y_h \leq v_h - (n+\sqrt{n+1})^{-1}(y_k - v_k) \qquad h \in I_n \backslash T .$$

Combining both inequalities,

$$y_h < v_h - (v_h - w_h) = w_h \qquad h \in I_n \backslash T .$$

If for some k, $-y_k = \max_i |y_i|$, again by the structure of A^*

$$y_h = v_h + (y_h - v_h) \leq v_h + (y_k - v_k) \qquad h \in I_n \backslash T .$$

Together with

$$-y_k > \rho' \geq -v_k + (n+\sqrt{n+1})\, |v_h - w_h| \geq -v_k + v_h - w_h ,$$

this again implies $y_h < w_h$.
It remains to prove that

$$\sum_i (w_i - y_i) < 0 \qquad \text{if} \quad n+1 \notin T .$$

Again assume that for some k, $y_k = \max_i |y_i|$. Since $n+1 \notin T$, we have from A^* that

$$\sum_{i \neq k} y_i \geq \sum_{i \neq k} v_i - (n-1)(n+\sqrt{n+1})^{-1}(y_k - v_k) .$$

Hence,

$$\sum_{i=1}^{n} (w_i - y_i) \leq \sum_{i=1}^{n} (w_i - v_i) - (1+\sqrt{n+1})(n+\sqrt{n+1})^{-1}(y_k - v_k)$$

which is negative, since

$$y_k > \rho' \geq v_k + (n+\sqrt{n+1})(1+\sqrt{n+1})^{-1} \sum_{i=1}^{n} (w_i - v_i) \quad .$$

Finally, if for some k , $-y_k = \max_i |y_i|$, then

$$-y_k > \rho' \geq -v_k + (n+\sqrt{n+1})(1+\sqrt{n+1})^{-1} \left| \sum_{i=1}^{n} (w_i - v_i) \right| \geq$$

$$\geq -v_k + (1+\sqrt{n+1})^{-1} \sum_{i=1}^{n} (w_i - v_i) \quad .$$

Since $(n+1) \notin T$ and $v_k - y_k > 0$ we have again from A* that

$$\sum_{i \neq k} (y_i - v_i) \geq (2+\sqrt{n+1})(v_k - y_k) \quad .$$

Therefore

$$\sum_{i=1}^{n} (w_i - y_i) = \sum_{i=1}^{n} (w_i - v_i) + \sum_{i \neq k} (v_i - y_i) + (v_k - y_k) < 0$$

which completes the proof of the theorem since only simplices can be generated in the bounded set $\{x \in R^n \mid \max_i |x_i| < \rho' \}$.

□

This theorem is also valid for other $(\alpha , -1)$-triangulations of R^n with $\alpha > n-1$ (see [7]). The proof is the same as above except that we have to take ρ' equal to

$$\rho' = \max \left[\rho + \delta , \max_i |v_i| + \max \{\max \alpha |v_i - w_i|, \alpha(\alpha-(n-1)^{-1}) \left| \sum_{i=1}^{n} (w_i - v_i) \right| \} \right] .$$

Note that the proof is not valid for the K-triangulation, since $\rho' \to \infty$ if $\alpha \to \infty$.

4. INTERPRETATION OF THE ALGORITHM

Let the variable dimension algorithm be applied on S^n. Define for T being a proper subset of I_n the regions $A(T)$ by

$$A(T) = \{x \in S^n \mid x = v + \sum_{j \in T} \lambda_j q(j) \quad \text{for non-negative } \lambda_j, \ j \in T\} .$$

Observe that for all T $A(T)$ is triangulated by the triangulation of S^n restricted to $A(T)$. Let the set of proper points of $A(T)$, denoted by $\mathring{A}(T)$, be defined by

$$\mathring{A}(T) = \{x \in A(T) \mid x = v + \sum_{j \in T} \lambda_j q(j) \quad \text{for positive } \lambda_j, \ j \in T\} .$$

Then the collection of all $\mathring{A}(T)$ partitions S^n and if $x \in A(T)$ there are unique λ_j, $j \in T$, such that

$$x = v + \sum_{j \in T} \lambda_j q(j) .$$

Note that $A(\emptyset) = \mathring{A}(\emptyset) = \{v\}$.

For other sets in the previous sections, the regions $A(T)$ can be defined in the same way. They are illustrated in the figures 1a - 1e. In all figures the sets are triangulated as in the previous sections.

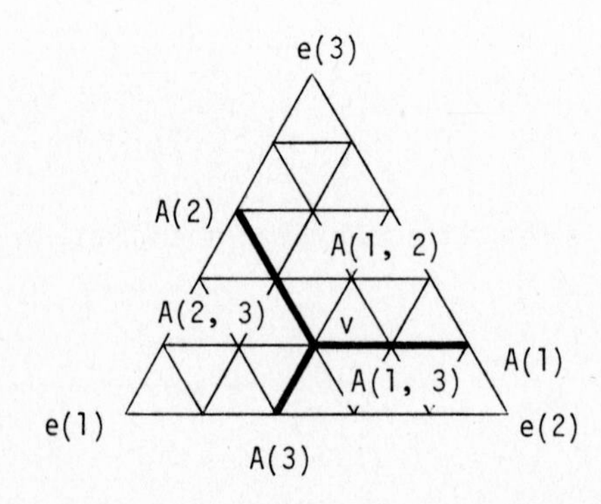

Figure 1a. S^n, $n = 3$, $m = 5$, $v = (\frac{2}{5}, \frac{2}{5}, \frac{1}{5})$

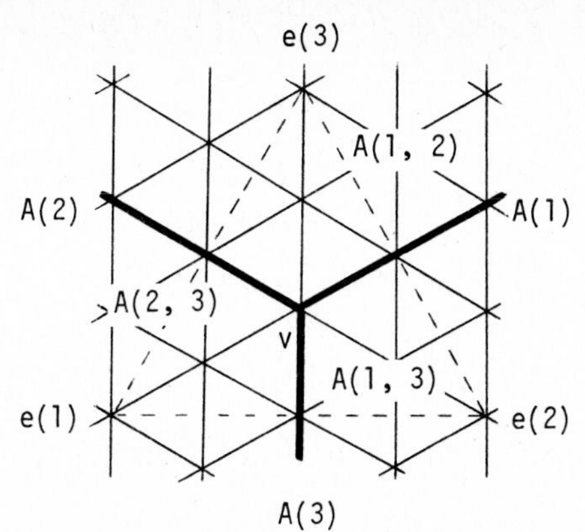

Figure 1b. T^n , $n = 3$, $m = 6$, $v = (\frac{1}{3}, \frac{1}{3}, \frac{1}{3})$

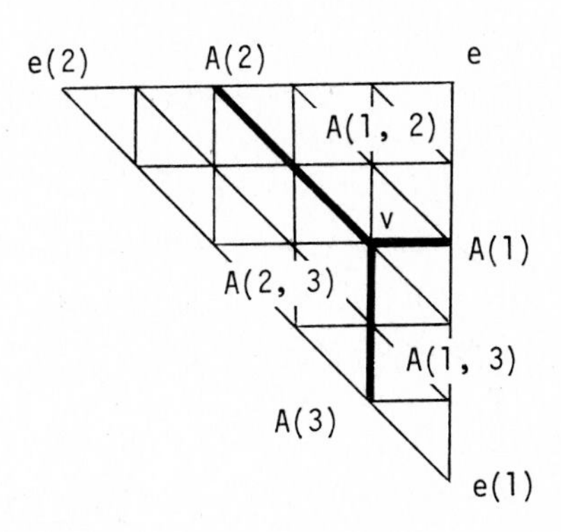

Figure 1c. $V^n(1, e)$, $n = 2$, $m = 5$, $v = (\frac{4}{5}, \frac{3}{5})$

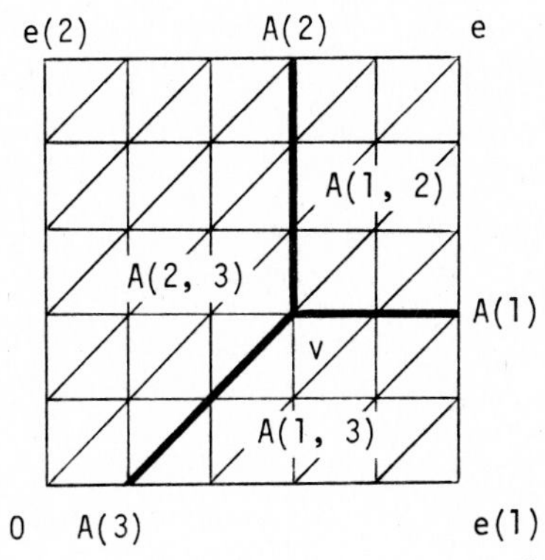

Figure 1d. C^n , $n = 2$, $m = 5$, $v = (\frac{3}{5}, \frac{2}{5})$

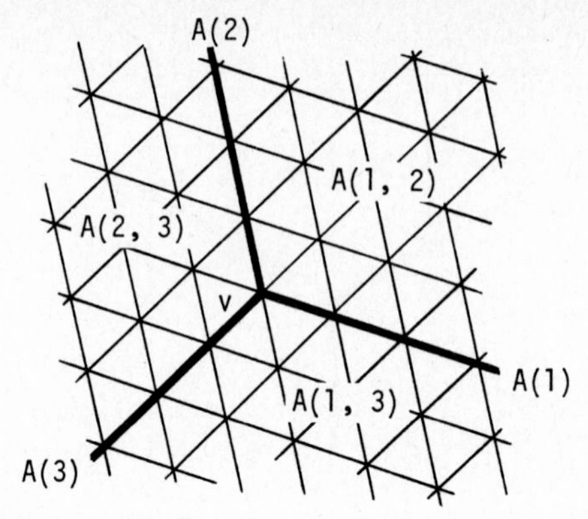

Figure 1e. R^n , $n = 2$, $m = 5$, $v = (0, 0)$

Let $\sigma(y^1, \pi(T))$ be a t-simplex generated by the algorithm. Then T is the set of found labels or eliminated unit vectors, and σ has the property that it is a simplex in $A(T)$ and y^{t+1} is a proper point of $A(T)$. So a search is made in the region $A(T)$ to find a $T \cup \{k\}$ - complete t-simplex for some $k \notin T$. As soon as such a simplex is found, the algorithm proceeds in $A(T \cup \{k\})$, generating a path of (t+1)-simplices with $(T \cup \{k\})$-complete common facets. If, however, a replacement step would imply a change from $A(T)$ to an adjacent region $A(T \cup \{k\} \setminus \{j\})$ for some $j \in T$, $k \notin T$, then the replacement step is not performed but the algorithm continues in $A(T \setminus \{j\})$, generating a path of (t-1)-simplices with $(T \setminus \{j\})$-complete common facets. So, in $A(T)$ the algorithm generates a path of t-simplices with T-complete common facets, starting either in step 3 with a simplex having a T-complete facet in $A(T \setminus \{j\})$ for some $j \in T$, or in step 4 with a $(T \cup \{j\})$-complete simplex with $j \notin T$, and terminates either with a $T \cup \{k\}$ - complete simplex for some $k \notin T$ or with a simplex having a T-complete facet in $A(T \setminus \{k\})$ for some $k \in T$. Starting with the zero-dimensional simplex $\{v\}$ in $A(\emptyset)$ a sequence of such paths is generated, which terminates with a completely labelled simplex (or the path goes to infinity).

Based on these fundamental properties, a generalization of the algorithm is presented in Van der Laan and Talman (see [8] and [9]). We will discuss this matter in the following section.

The regions $A(T)$ also allow for an interpretation with artificially labelled points. For S^n, we obtain a triangulation of $S^n \times [0, 1]$, where $S^n \times \{1\}$ denotes the natural level, by connecting each grid point $x \in \mathring{A}(T) \times \{1\}$ with the $n-t$ vertices $\hat{e}(i) = (e(i)', 0)'$, $i \notin T$. So there are n points on the artificial level, viz. $\hat{e}(1), \ldots, \hat{e}(n)$. In case of integer labelling $\ell(\hat{e}(i)) = i$ whereas in case of vector labelling $\ell(\hat{e}(i)) = e(i)$. Starting in the classic way with the n-simplex $\sigma(\hat{e}(1), \ldots, \hat{e}(n), (v', 1)')$, having $\tau(\hat{e}(1), \ldots, \hat{e}(n))$ as a completely labelled boundary facet, a path on n-simplices is generated, having completely labelled common facets, which terminates with a simplex having a completely labelled boundary facet on $S^n \times \{1\}$. The intersection of this path with $S^n \times \{1\}$ is the sequence of simplices of the variable dimension algorithm. The triangulation of $S^n \times [0, 1]$ is illustrated in figure 2.

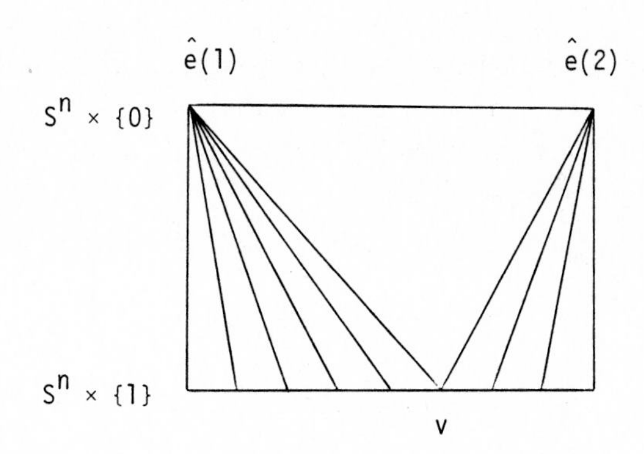

Figure 2. Triangulation of $S^n \times [0, 1]$,

$n = 2$, $m = 8$, $v = (\frac{3}{8}, \frac{5}{8})$

The algorithm can be interpreted in the same way for other sets by defining appropriately artificial points $\hat{w}(i) = (w(i)', 0)$ with labelling $\ell(\hat{w}(i)) = i$ respectively $\ell(\hat{w}(i)) = e(i)$. The points and the set which is triangulated are given in table 2, where conv(A) denotes the convex hull of the set A.

Note that in all cases the artificial points $w(i)$, $i = 1, \dots, n, (n+1)$, are points in the region $A(I_n \backslash \{i\})$ respectively $A(I_{n+1} \backslash \{i\})$. Moreover, they are proper points of this region, except in the case where the sets are bounded and the starting point v is on the boundary.

We conclude this section with some remarks. Firstly, a homotopy algorithm which allows for a sequence of factors of incrementation of two or more has been developed by Van der Laan and Talman [10], based on the triangulation of $S^n \times [0, 1]$ discussed just above. Secondly, in the previous section it was argued that using the K-triangulation, Merrill's condition is not sufficient to guarantee the convergence of the variable dimension algorithm. However, Todd (1978) showed that Merrill's condition is sufficient if the artificial labelling of points in the zero level is chosen in an appropriate way, i.e. the system of linear equations $A\mu = e$ is not initiated with $A = I$. So, in fact the convergence depends on the initiated system of linear equations and on the underlying triangulation.

5. THE GENERALIZED VARIABLE DIMENSION ALGORITHM

A class of fixed point algorithms on R^n was developed by Van der Laan and Talman [9] based on the features of their basic algorithm. One of the two extreme cases of the class is this algorithm. The other one is treated in this section. Therefore we need a new integer labelling rule.

Definition 5.1: (Integer labelling)
The integer labelling $\ell(x)$ of a point $x \in R^n$ is as follows. Let

$$i = \min_h \{h \mid |f_h(x) - x_h| \geq |f_k(x) - x_k| , \quad k = 1, \dots, n\} .$$

Then $\ell(x) = i$ if $f_i(x) - x_i \geq 0$, and $\ell(x) = n+i$ otherwise.

	Artificial points	Triangulated set
S^n	$w(i) = e(i)$, $i = 1, \ldots, n$	$S^n \times [0, 1]$
T^n	$w(i) = v - t(i)$, $i = 1, \ldots, n$	conv $(T^n \times \{1\} \cup \{\sigma(\hat{w}(1), \ldots, \hat{w}(n))\})$
V^n	$w(i) = a - (\sum_{j=1}^{n} a_j - \alpha)\, e(i)$, $i = 1, \ldots, n$ $w(n+1) = a$	$V^n \times [0, 1]$
C^n	$w(i) = e - e(i)$, $i = 1, \ldots, n$ $w(n+1) = e$	conv $(C^n \times \{1\} \cup \{\sigma(\hat{w}(1), \ldots, \hat{w}(n+1))\})$
R^n	$w(i) = v - a^*(i)$, $i = 1, \ldots, n+1$	conv $(R^n \times \{1\} \cup \{\sigma(\hat{w}(1), \ldots, \hat{w}(n+1))\})$

Table 2. Artificial points and triangulated sets.

Observe that the total number of labels is $2n$ instead of $n+1$. However, let a t-simplex σ of a triangulation of R^n be such that two different vertices of σ are labelled, for some j, with j and $j+n$. Then σ is a good approximation of a fixed point as stated in the following lemma, and is called a j-stopping simplex.

Lemma 5.2: Let the mesh of the triangulation be less than δ and let

$$\max_i |x_i - y_i| < \delta \quad \text{imply} \quad \max_i |f_i(x) - f_i(y)| < \varepsilon .$$

Then for any x^* in σ

$$\max_i |f_i(x^*) - x_i^*| \leq 2(\varepsilon + \delta) .$$

The proof of the lemma can be found in [9].

To generate a j-stopping simplex, R^n is subdivided in regions $A(T)$ as follows. Let Z^1 be the collection of subsets T of I_{2n} such that $j \leq n$ and $j \in T$ implies $n+j \notin T$, and $j > n$ and $j \in T$ implies $j-n \notin T$. Moreover, let E^* be the $(n \times 2n)$-matrix $E^* = [I \ \ -I]$ and z an arbitrarily chosen point in R^n, then for $T \in Z^1$, define

$$A(T) = \{x \in R^n \mid x = z + \sum_{j \in T} \lambda_j e^*(j) \quad \text{for non-negative } \lambda_j , j \in T\}.$$

Clearly, the collection of $\mathring{A}(T)$, $T \in Z^1$, partition R^n. Note also that the λ_j's, $j \in T$, corresponding with a point x, $x \in A(T)$, are unique.

Now $A(T)$, $T \in Z^1$, is triangulated with grid size m, $m > 0$, in the standard way, i.e. the triangulation is the collection of simplices $\sigma(y^1, \pi(T))$ in $A(T)$ with vertices $y^1, \ldots, y^{t+1}$ such that

$$y^1 = z + \sum_{j \in T} \mu_j e^*(j)$$

for non-negative integers and $y^{i+1} = y^i + q(\pi_i)/m$, $i = 1, \ldots, t$, where $|T| = t$.

Then the union of the triangulations of $A(T)$ for $T \in Z^2$, where Z^2 is the subset of Z^1 such that either $j \in T$ or $n+j \in T$ if $T \in Z^2$, triangulates R^n since for all $T_1, T_2 \in Z^2$ the triangulation of $A(T_1) \cap A(T_2)$, induced by the ones of $A(T_1)$ and $A(T_2)$, is the triangulation of $A(T_1 \cap T_2)$ defined above. For $n = 2$ the triangulation of R^n is illustrated in figure 3.

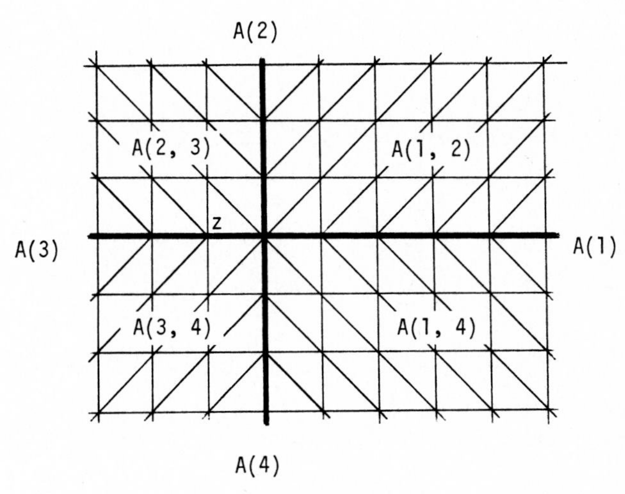

Figure 3. Triangulation of R^n, $n = 2$.

Observe that the triangulation obtained in this way is the K'-triangulation proposed by Todd [17]. The application of the variable dimension algorithm is straightforward. Let v be an arbitrarily chosen starting point, then we set z equal to v. The algorithm starts with the zero-dimensional simplex $\{v\}$ and generates a path of adjacent t-simplices until a stopping simplex is found (or the path goes to infinity) following the steps of section 2. Again, if T is the set of current labels, simplices of the triangulation of $A(T)$ are generated such that they have T-complete common facets until either for some k a $T \cup \{k\}$-complete simplex is found, or a simplex is generated having a T-complete facet in $A(T\backslash\{h\})$ for some $h \in T$. In the latter case label h is deleted. Note that this occurs if and only if $y_j^t = v_j^t$ and y^{t+1} has to be removed.

An integer labelling method closely related to this extreme case and based on the K-triangulation was developed by Reiser [13]. However, one may expect fewer iterations if the K'-triangulation is used since it takes advantage of the starting point, viz. movements from this point in a given direction can be made very rapidly (see also Todd [17]). Conditions to guarantee that the algorithm converges can be found in [9] and also in [13].

Next we describe the method for vector labelling.

Definition 5.3.1: (Vector labelling)
A point $x \in R^n$ receives the vector label $\ell(x) \in R^{2n}$ where

$$\ell_i(x) = f_i(x) - x_i + 1 \qquad i = 1, \ldots, n$$

and

$$\ell_{i+n}(x) = -f_i(x) + x_i + 1 \qquad i = 1, \ldots, n$$

where $f(x)$ is defined as in section 2.

For $T \in I_{2n}$ the $(t-1)$-simplex $\sigma(y^1, \ldots, y^t)$ is called T-complete if the system of linear equations

$$\sum_{i=1}^{t} \lambda_i \ell(y^i) + \sum_{j \in I_{2n} \backslash T} \mu_j e(j) = e$$

has a non-negative solution. If $\sigma(y^1, \ldots, y^t)$ is T-complete with solution λ_i^*, $i = 1, \ldots, t$, and μ_j^*, $j \in I_{2n} \backslash T$, then we have the following lemma.

Lemma 5.4: Define $\mu_h^* = 0$ if $h \in T$, then

$$2 \sum_{i=1}^{t} \lambda_i^* + \mu_h^* + \mu_{h+n}^* = 2 \qquad h = 1, \ldots, n.$$

The lemma follows immediately by summing the h-th and $(n+h)$-th equation.
A T-complete simplex $\sigma(y^1, \ldots, y^t)$ is called completely labelled if for all $h \in I_{2n} \backslash T$, $\mu_h^* = 0$. Obviously,

$$y^* = \sum_{i=1}^{t} \lambda_i^* y^i$$

is a good approximation of a fixed point.
Moreover, we have from lemma 5.4 that if $\sigma(y^1, \dots, y^t)$ is T-complete such that j and $j+n \in T$ for some $j \in I_n$, then σ is completely labelled. We call such a simplex again a j-stopping simplex. With minor modifications, the algorithm proceeds now as described in section 2, starting with an arbitrarily chosen point v and the system of $2n$ linear equations $I\mu = e$. Setting z equal to v, a path of adjacent simplices of the K'-triangulation of variable dimension is generated sucht that the algorithm operates with simplices in $A(T)$ having T-complete common facets where T is the set of eliminated vectors.
The algorithm terminates as soon as a completely labelled simplex is found.
In the following theorem we prove that Merrill's condition is sufficient to guarantee convergence (see section 3).

Theorem 5.5: (Vector labelling)
If Merrill's condition is satisfied, the algorithm converges if the mesh of the triangulation is small enough.

Proof: Let the mesh of the triangulation be less than δ and let v be the starting point. Define ρ' by

$$\rho' = \max \{\rho + \delta , \max_i |v_i| + \sum_{j=1}^{n} |w_j - v_j| \} .$$

Let $\sigma(y^1, \pi(T))$ be a t-simplex such that for some $y \in \sigma$, $\max_i |y_i| > \rho'$. We will prove that σ has no T-complete facet $\tau(y^1, \dots, y^{s-1}, y^{s+1}, \dots, y^{t+1})$, $s = 1, \dots, t+1$. By multiplicating the system of linear equations with the 2n-vector $((w - y)',(y - w)')'$ we obtain, if we set $\lambda_s = 0$ and $\mu_h = 0$ for $h \in T$,

$$\sum_{i=1}^{t+1} \lambda_i (f(y^i) - y^i + e)'(w - y) + \sum_{i=1}^{t+1} \lambda_i (-f(y^i) + y^i + e)'(y - w) +$$

$$\sum_{j=1}^{n} \mu_j (w - y)_j + \sum_{j=1}^{n} \mu_{j+n} (y - w)_j = 0$$

or

$$2 \sum_{i=1}^{t+1} \lambda_i (f(y^i) - y^i)'(w - y) + \sum_{j=1}^{n} (\mu_j - \mu_{j+n})(w_j - y_j) = 0 .$$

By Merrill's condition $(f(y^i) - y^i)'(w - y)$ is positive. So, to prove that the system has no feasible solution, it remains to show that

$$\sum_{j=1}^{n} (\mu_j - \mu_{j+n})(w_j - y_j)$$

is positive. Note that if

$$\sum_{i=1}^{t+1} \lambda_i = 1 \quad ,$$

μ_j and μ_{j+n} are zero for all $j = 1, \ldots, n$ and hence there is no feasible solution. So let

$$\sum_i \lambda_i < 1 \quad .$$

If $\mu_j > 0$, then $j \notin T$ and therefore $y_j \leq v_j$. If for some i, $\mu_{i+n} > 0$ then $i+n \notin T$ and hence $y_i \geq v_i$. Since

$$\mu_j + \mu_{j+n} = 2 - 2 \sum_i \lambda_i > 0 \quad ,$$

μ_j or $\mu_{j+n} \notin T$ or both. In the latter case $y_j = v_j$. Consequently, $\mu_j (v_j - y_j) \geq 0$ and $\mu_{j+n} (y_j - v_j) \geq 0$, for all $j \in I_n$.

Next we assume that for some k, $y_k = \max_i |y_i|$.
Then

$$y_k > \rho' \geq v_k + \sum_{j=1}^{n} |w_j - v_j| \quad .$$

So $k \in T$ and hence form lemma 5.4

$$\mu_{n+k} = 2 - 2 \sum_i \lambda_i > 0 \quad .$$

Therefore

$$\sum_{j=1}^{n} (\mu_j - \mu_{j+n})(w_j - y_j)$$

$$\geq \sum_{j=1}^{n} \mu_j (w_j - v_j) + \mu_{k+n} (y_k - v_k) + \sum_{j=1}^{n} \mu_{j+n} (v_j - w_j)$$

$$> \mu_{k+n} (\mu_{k+n}^{-1} \sum_{j=1}^{n} \mu_j (w_j - v_j) + \mu_{k+n}^{-1} \sum_{j=1}^{n} \mu_{j+n} (v_j - w_j)$$

$$+ \sum_{j=1}^{n} |w_j - v_j|)$$

$$\geq \mu_{k+n} (-\mu_{k+n}^{-1} \sum_{j=1}^{n} (\mu_j + \mu_{j+n}) |w_j - v_j|$$

$$+ \sum_{j=1}^{n} |w_j - v_j|) = 0$$

since in the last step

$$\mu_j + \mu_{j+n} = \mu_{k+n} = 2 - 2 \sum_{i=1}^{t+1} \lambda_i$$

which follows again from lemma 5.4.

If for some k , $-y_k = \max_i |y_i|$ then

$$-y_k > \rho' \geq -v_k + \sum_{j=1}^{n} |w_j - v_j|$$

and $k+n \in T$ which implies that

$$\mu_k = 2 - 2 \sum_i \lambda_i > 0 .$$

Following the same arguments as just above we obtain again that

$$\sum_{j=1}^{n} (\mu_j - \mu_{j+n}) (w_j - y_j)$$

must be positive.

Combining all of this together we have that no simplices outside the set $\{x \in R^n \mid \max_i |x_i| > \rho'\}$ can be generated. Since the number of simplices within this bounded set is finite, the algorithm must terminate with a stopping simplex.

□

Finally we mention some applications of the algorithm. Firstly, a closely related algorithm on the product space of unit simplices was used by Van der Laan and Talman [8] to compute the equilibrium strategies of a non-cooperative n-person game. The product space of n unit simplices is then the strategy space. Another interesting application of the algorithm was given by Todd [19] to compute a zero point of a continuous function from the set $\{y \in R^n \mid \max |y_i| \leq 1\}$ to R^n such that $f(-y) = -f(y)$ if $\max_i |y_i| = 1$ (Borsuk-Ulam Theorem).

REFERENCES

1. Eaves, B.C., Homotopies for Computation of Fixed Points, Mathematical Programming 3, 1-22 (1972).

2. Eaves, B.C.; Saigal, R., Homotopies for Computation of Fixed Points on Unbounded Regions, Mathematical Programming 3, 225-237 (1972).

3. Kuhn, H.W., Some Combinatorial Lemmas in Topology, IBM Journal of Research and Development 4, 518-524 (1960).

4. Kuhn, H.W.; MacKinnon, J.G., Sandwich Method for Finding Fixed Points, Journal of Optimization Theory and Applications 17, 189-204 (1975).

5. Laan, G. van der; Talman, A.J.J., A Restart Algorithm for Computing Fixed Points without an Extra Dimension, to appear in Mathematical Programming.

6. Laan, G. van der; Talman, A.J.J., A Restart Algorithm without an Artificial Level for Computing Fixed Points on Unbounded Regions, to appear in Proceedings of the Conference on Functional Differential Equations and Approximations of Fixed Points, Bonn, July 1978.

7. Laan, G. van der; Talman, A.J.J., An Improvement of Fixed Point Algorithms by Using a Good Triangulation, submitted to Mathematical Programming.

8. Laan, G. van der; Talman, A.J.J., On the Computation of Fixed Points in the Product Space of Unit Simplices and an Application to Non-Cooperative N- Person Games, to appear in Mathematics of Operations Research.

9. Laan, G. van der; Talman, A.J.J., A Class of Simplicial Subdivisions for Restart Fixed Point Algorithms without an Extra Dimension, submitted to Mathematical Programming.

10. Laan, G. van der; Talman, A.J.J., A New Subdivision for Computing Fixed Points with a Homotopy Algorithm, submitted to Mathematical Programming.

11. Lüthi, H.J., A Simplicial Approximation of a Solution for the Nonlinear Complementarity Problem, Mathematical Programming 9, 278-293 (1975).

12. Merrill, O.H., Applications and Extensions of an Algorithm that Computes Fixed Points of Certain Upper Semi-Continuous Point to Set Mappings, Ph.D. Thesis, University of Michigan, 1972.

13. Reiser, P.M., A Modified Integer Labelling for Complementarity Algorithms, Institut für Operations Research, Universität Zürich, June 1978.

14. Saigal, R., Investigations into the Efficiency of Fixed Point Algorithms, in: Fixed Points: Algorithms and Applications, S. Karamardian (Ed.), Academic Press, New York 1977, pp. 203-223.

15. Scarf, H.E., The Approximation of Fixed Points of a Continuous Mapping, SIAM J. Appl. Math. 15, 1328-1343 (1967).

16. Scarf, H.E., The Computation of Economic Equilibria, (with collaboration of T. Hansen), Yale University Press, New Haven (Connecticut) 1973.

17. Todd, M.J., Improving the Convergence of Fixed Point Algorithms, Mathematical Programming Study 7, 151-169 (1978).

18. Todd, M.J., Fixed-Point Algorithms that Allow Restarting Without an Extra Dimension, College of Engineering, Cornell University, Ithaca, Technical Report No. 379, May 1978.

19. Todd, M.J., A Variable-Dimension Simplicial Algorithm for Antipodal Fixed-Point Theorems, College of Engineering, Cornell University, Ithaca, Technical Report No. 417, April 1979.

Numerical Solution of Highly Nonlinear Problems
W. Forster (ed.)
© North-Holland Publishing Company, 1980

A NOTE ON "A NEW ALGORITHM FOR COMPUTING FIXED POINTS" BY VAN DER LAAN AND TALMAN

Masakazu Kojima

Tokyo Institute of Technology,
Tokyo, Japan

The purpose of this short note is to explain the fundamental structure of the method used by van der Laan and Talman [5, 6] (see also [4]) for computing fixed points in terms of the unified framework given by Eaves and Scarf [2] for the fixed point and complementarity theory. Van der Laan's and Talman's method was originally developed for computing fixed points of continuous maps from the (n-1)-dimensional unit simplex

$$S^n = \{x = (x_1, \ldots, x_n) \in R^n : \sum_{i=1}^{n} x_i = 1 \text{ and } x_i \geq 0 \ (1 \leq i \leq n)\}$$

into itself [5], and then generalized for maps from R^n into itself [6]. Todd [4] modified the method to compute a solution of a system of equations

$$f(x) = 0 \quad , \tag{1}$$

where $f: R^n \to R^n$ is continuous. Obviously, the problem of finding a fixed point of a continuous map $g: R^n \to R^n$ is equivalent to the system of equations $g(x) - x = 0$. So we deal with the system (1) instead of the fixed point problem in R^n . In the papers [4] and [6] integer and vector labellings were given to trace a path of completely labelled simplices. We will show that if $f: R^n \to R^n$ is continuously differentiable, the differential equations approach (see e.g. [1]) can be utilized for tracing the same path.

To explain the fundamental structure of the method, we need to introduce two subdivisions by $(n+1)$ polyhedral cones. Let p^i, q^i $(0 \le i \le n)$ be $2(n+1)$ vectors in R^n. Define

$$P(i) = \{ \sum_{j \neq i} \lambda_j p^j + x^0 \ : \lambda_j \ge 0 \quad (j \neq i) \}$$

and

$$Q(i) = \{ \sum_{j \neq i} \mu_j q^j \ : \mu_j \ge 0 \quad (j \neq i) \} \quad ,$$

where x^0 is a point in R^n from which we will start the method. We assume:

(2) For each i the set of n vectors p^j $(j \neq i)$ $(q^j$ $(j \neq i))$ is linearly independent.

(3) The collection $J = \{P(i) : 0 \le i \le n\}$ $(K = \{Q(i) : 0 \le i \le n\})$ is a subdivision of R^n.

In the papers [4] and [6], $-\sum_{i=1}^{n} e^i$, e^1, ..., e^n were taken as p^0, p^1, ..., p^n. The choice of the $2(n+1)$ vectors p^i, q^i $(0 \le i \le n)$ affects the path which will be generated by the method. Let J^* be a subdivision of R^n such that each piece of J^* is contained in some cone in J. We assume that either

(4) f is piecewise linear on J^*

or

(5) f is continuously differentiable on R^n.

For each $\alpha \in R$, let

$$\alpha^+ = \max\{0 , \alpha\} \quad \text{and} \quad \alpha^- = \min\{0 , \alpha\} \quad .$$

Now we define the map $h: R^{1+n} \to R^n$ as follows:

$$h(y) = f(\sum_{j=0}^{n} y_j^+ p^j + x^0) + \sum_{j=0}^{n} y_j^- q^j$$

for every $y = (y_0 , y_1 , \ldots , y_n) \in R^{1+n}$.

Consider the system of equations

(6) $$h(y) = 0 .$$

If f satisfies (4) (or (5)), then $h: R^{1+n} \to R^n$ is piecewise linear (or piecewise continuously differentiable). Hence if, in addition, the system (6) satisfies an appropriate regularity condition then the solution set, denoted by Y, to (6) consists of disjoint piecewise linear (or piecewise smooth) one-manifolds, each of which can be traced by complementary pivoting (or solving differential equations ([1])).

In the remainder of the paper, we assume that

(7) each connected component is either a one-manifold which is homeomorphic to the unit circle or a one-manifold which is unbounded in both directions.

We will list some fundamental properties of the solution set Y to (6).

Proposition 1. (The initial ray). Let

$$L_0 = \{y \in R^{1+n} : f(x^0) = \sum_{j=0}^{n} (-y_j)\, q^j \ , \quad y \le 0\} \quad .$$

Then L_0 is a half line contained in Y.

Proof. $L_0 \subset Y$ is obvious. Hence it suffices to show that L_0 is a half line. From the assumption (3), we can find an i, such that $f(x^0) \in Q(i)$, i.e.

$$f(x^0) = \sum_{j=0}^{n} \bar{\mu}_j\, q^j \qquad \text{and} \quad \bar{\mu}_i = 0$$

for some $\bar{\mu} = (\bar{\mu}_0 , \bar{\mu}_1 , \dots , \bar{\mu}_n) \ge 0$. Also $-q^i$ must belong to to some $Q(k)$. If k was distinct from i then we would have

$$\sum_{j \ne k} \nu_j\, q^j = - q^i$$

for some $\nu_j \ge 0$ $(j \ne k)$, a contradiction to the assumption (2). Hence $k = i$, and there exist $\bar{\nu} = (\bar{\nu}_0 , \bar{\nu}_1 , \dots , \bar{\nu}_n) \ge 0$ such that

$$\sum_{j \ne i} \bar{\nu}_j\, q^j = - \bar{\nu}_i\, q^i \qquad \text{and} \quad \bar{\nu}_i = 1 \quad .$$

By construction, we obtain

$$L_0 = \{-\bar{\mu} - t\bar{\nu} : t \ge 0\} \quad .$$

Q.E.D.

We can similarly prove:

Proposition 2. (Solution rays). Let x^* be a solution to (1), and let

$$L = \{y \in R^{1+n} : x^* = \sum_{j=0}^{n} y_j p^j + x^0 , \quad y \geq 0\} \ .$$

Then L is a half line contained in Y .

Proposition 3. Let S_0 be a connected component of Y which contains L_0 . Suppose that the set

$$\{ \sum_{j=0}^{n} y_j^+ p^j + x^0 : y \in S_0\}$$

is bounded. Then it contains a solution to (1).

Proof. By assumption (7) and Proposition 1, $S_0 \backslash L_0$ is unbounded. Hence some component y_{j*} of y tends to infinity along $S_0 \backslash L_0$. Assume that $y_{j*} \to +\infty$ along $S_0 \backslash L_0$. If some component $y_k \leq 0$ and $y_{j*} \to +\infty$ along a sequence in $S_0 \backslash L_0$, then

$$x = \sum_{j=0}^{n} y_j^+ p^j + x^0$$

would tend to infinity along the sequence and we would have a contradiction. Hence $y_j > 0$ $(0 \leq j \leq n)$ for some $y \in S_0 \backslash L_0$, which implies that

$$x = \sum_{j=0}^{n} y_j^+ p^j + x^0$$

is a solution to (1).

Now we deal with the case where the set $\{y^+ : y \in S_0 \backslash L_0\}$ is bounded and some component $y_{j*} \to -\infty$ along $S_0 \backslash L_0$. Since the set

$$\{f(\sum_{j=0}^{n} y_j^+ p^j + x^0) : y \in S_0 \backslash L_0\}$$

is bounded, the set

$$A = \{ \sum_{j=0}^{n} y_j^- q^j : y \in S_0 \backslash L_0\}$$

must be bounded. If $y_j < 0$ $(0 \leq j \leq n)$ for some $y \in S_0 \backslash L_0$, then

$$f(x^0) + \sum_{i=0}^{n} y_i^- q^i = 0 \quad ,$$

i.e. $y \in L_0$ and we have a contradiction. Hence some component $y_j \geq 0$ and $y_{j*} \to -\infty$ along a sequence in $S_0 \backslash L_0$. In this case the set A is not bounded because of the assumption (2), and again we have a contradiction.

Q.E.D.

Todd [4] showed that a modification of the algorithm given by van der Laan and Talman [5] succeeds in approximating a solution to (1) under Merrill's condition. If we take p^j, q^j $(0 \leq j \leq n)$ such that $(p^i)^T q^j > 0$ $(i \neq j)$ as in [4], Merrill's condition ensures that the set

$$\{ \sum_{j=0}^{n} y_j^+ p^j + x^0 : y \in S_0 \}$$

is bounded (see Theorem 4.3 and its proof in [4]).

Finally we point out that the system of equations (6) involves parametric complementarity problems. If $y \geq 0$ is a solution to (6), then

$$x = \sum_{j=0}^{n} y_j^+ p^j + x^0$$

is a solution to (1) (see also Proposition 1). So we restrict our attention to solutions y of (6) satisfying $y_i \leq 0$ for some i. For simplicity of notation, we assume that $y_0 \leq 0$ and consider the set B of solutions to $h(y) = 0$ satisfying $y_0 \leq 0$. Let $M = [q^1 , \ldots , q^n]$, $d = +M^{-1} q^0$ and

$$g(x) = M^{-1} f(\sum_{j=1}^{n} x_i p^i + x^0)$$

for every $x = (x_1 , \ldots , x_n) \geq 0$.

Then $(y_0 , z) \in R^{1+n}$, where $y_0 \leq 0$ and $z \in R^n$, belongs to B if and only if

$$z^- + g(z^+) + y_0 d = 0 \quad .$$

Furthermore (y_0 , z) satisfies the above equality if and only if $x = z^+$ and $w = -z^-$ satisfy

(8) $0 \leq x , \quad w = g(x) + y_0 d \geq 0 , \quad x^T w = 0$.

For each fixed y_0, the problem of finding $(x, w) \in R^{2n}$ satisfying (8) is called a complementarity problem. Thus the system (6) involves a complementarity problem with a real nonpositive parameter y_0. See Sections 5 and 6 of [3].

REFERENCES

1. Chow, S.N.; Mallet-Paret, J.; Yorke, J.A., Finding Zeros of Maps: Homotopy Methods that are Constructive with Probability one, to appear in Journal of Nonlinear Analysis - Theory, Methods and Applications.

2. Eaves, B.C.; Scarf, H., The Solution of Systems of Piecewise Linear Equations, Mathematics of Operations Research 1, 1-27 (1976).

3. Kojima, M., Studies of PL Approximations of Piecewise-C^1 Mappings in Fixed Points and Complementarity Theory, Mathematics of Operations Research 3, 17-36 (1978).

4. Todd, M.J., Fixed-Point Algorithms that allow Restarting without an extra Dimension, Tech. Rept. No. 379, Cornell University, Ithaca, New York 14853, May 1978.

5. van der Laan, G.; Talman, J.J., A new Algorithm for Computing Fixed Points, Free University, Amsterdam, March 1978.

6. van der Laan, G.; Talman, J.J., A Restart Algorithm without an artificial Level for computing Fixed Points on unbounded Regions, Free University, Amsterdam, 1978.

Numerical Solution of Highly Nonlinear Problems
W. Forster (ed.)
© North-Holland Publishing Company, 1980

GLOBAL AND LOCAL CONVERGENCE AND MONOTONICITY RESULTS FOR A RECENT VARIABLE-DIMENSION SIMPLICIAL ALGORITHM

Michael J. Todd

Cornell University,
Ithaca, New York

Van der Laan and Talman have proposed an algorithm that produces simplices of various dimensions lying in coordinate subspaces in computing an approximate zero of a continuous function on R^n . Here we give global and local convergence results and some "monotonicity" results when solving a convex optimization problem. The latter theorems apply also to Merrill's restart algorithm and give strong results concerning the paths generated.

This research was supported in part by National Science Foundation Grant No. ENG 76-08749.

1. INTRODUCTION

We are concerned with the recent algorithm of van der Laan and Talman for computing zeros of a continuous function on R^n [7] . This algorithm generates a sequence of simplices of R^n of varying dimension: at any iteration a j-simplex in a j-dimensional coordinate subspace of R^n is produced, with j between 0 and n . The method is related to an earlier algorithm of Reiser [10] using integer labelling for the complementarity problem.

The van der Laan-Talman algorithm can be viewed as following the path of zeros of a piecewise-linear function from a subset R^{n+1} to R^n . The corresponding subdivision of the subset of R^{n+1} is neither simplicial nor locally finite, however. This geometrical interpretation is due to Wright, see [18]. In section 2 we present the algorithm and show that it can be viewed this way. We also give linear systems that can be used to generate the sequence of simplices. While van der Laan and Talman used a basis of size $2n$ (we consider only their version with $k = n$, which seems the most attractive), we show that a basis of size $n+1$ suffices. However, we require a more complicated set of inequalities than they, and the resulting pivoting rules are slightly more complicated. In fact, the linear systems and associated inequalities are similar to those that arise in Merrill's restart algorithm [8] when the linearity of the artificial function is exploited [16], [17]. Section 2 places the algorithm of van der Laan and Talman in the framework of other piecewise-linear algorithms; for general references see Allgower and Georg [1] and Eaves [3].

Section 3 addresses the global convergence properties of this algorithm. While van der Laan and Talman gave a criterion for its success, this condition was rather obscure. We show that Merrill's well-known condition [8] suffices for finite convergence from any starting point. Merrill's condition encompasses problems arising from convex optimization applications.

In section 4 we consider such optimization problems in depth. We give results here both for the van der Laan-Talman algorithm and for that of Merrill [8]. Roughly, the results are as follows. If the convex optimization problem has a point which is "very feasible" (the ball around the point of radius equal to the mesh size lies in the feasible region) then any simplex further from the starting point than such a point contains a feasible point, and any terminal simplex contains a feasible point. If the terminal simplex contains a feasible point, it contains a feasible point that is "optimal" to a related optimization problem. Finally, the algorithm continually produces points that are "optimal" to a related optimization problem with an additional constraint. Hence the algorithm can be viewed as tracing such "optimal" solutions as the auxiliary constraint is progressively relaxed. This viewpoint is reminiscent of Eaves [2] and [4] - see also Kojima [5]. The only difference between Merrill's algorithm and that of van der Laan and Talman in this context is that different norms are appropriate: the ℓ_2-norm for the former and the ℓ_1-norm for the latter.

Section 5 gives local convergence results for the van der Laan and Talman algorithm. We assume the smoothness and regularity conditions of Saigal [11] and Saigal and Todd [12]. Then we give a restarting implementation of the algorithm that attains quadratic convergence and the property that, asymptotically, $n+1$ function evaluations suffice in each restart. The method of proof, first considering an idealized situation and then bounding perturbations from this situation, seems applicable to many such algorithms. We generally use the

$$\ell_\infty\text{-norm}: \quad ||x|| = \max_i |x_i| \quad ,$$

and the induced operator norm $||A|| = \max\{||Ax|| \mid ||x|| = 1\}$. Occasionally we use also the

$$\ell_2\text{-norm} \quad ||x||_2 = (x^T x)^{1/2}$$

and the

$$\ell_1\text{-norm} \quad ||x||_1 = \sum_i |x_i| \quad ,$$

at such times we write $||.||_\infty$ for $||.||$ for clarity or emphasis.

2. THE ALGORITHM OF VAN DER LAAN AND TALMAN

Here we describe briefly the algorithm proposed by van der Laan and Talman [7] in the case where (in their notation) $k = n$. We show how it can be implemented using a basis of size only $(n+1)\times(n+1)$, with several unit columns, whereas they use a basis of size $2n\times 2n$, again with unit columns.

First we introduce some notation enabling us to work with orthants of coordinate subspaces easily. For each $x \in R^n$, $\mathrm{sgn}(x)$ denotes the vector whose i-th component is 0 , $+1$ or -1 according as x_i is zero, positive or negative; any such $(0, \pm 1)$-vector is called a sign vector. For each sign vector $s \in R^n$, we denote by $C(s)$ the "orthant" $\{x \in R^n \mid x_i = \lambda_i s_i$, some $\lambda_i \geq 0$, for all $i\}$; actually $C(s)$ is an orthant of the subspace with x_j zero whenever s_j is.

Let T be any triangulation of R^n with the property that any simplex of T meeting an orthant $C(s)$ lies completely in that orthant. Examples are K_1 and J_1 ; see e.g., [13]. For each $\sigma \in T$, $\mathrm{sgn}(x)$ is the same for each x in the relative interior of σ ; we denote this sign vector by $\mathrm{sgn}(\sigma)$.

In R^n , $u^1, \ldots, u^n$ denote the n unit vectors and u the vector of ones $\sum_i u^i$.

Now let $f : R^n \to R^n$ be a continuous function. (Point-to-set mappings will be considered in section 3 and 4; the extension is trivial. We use a function solely for notational convenience.) Let σ be a j-simplex of T in $C(s)$ where s has exactly j nonzero components. Let $\{i_{j+1}, \ldots, i_p\} = I_+ = \{i \mid s_i \geq 0\}$ and $\{i_{p+1}, \ldots, i_{2n}\} = I_- = \{i \mid s_i \leq 0\}$. Suppose $\tau = [v^1, \ldots, v^j]$ is a facet of σ and denote by f^i $f(v^i)$ for $1 \leq i \leq j$. Then, following van der Laan and Talman, we say τ is a complete facet of σ iff there is a solution x to

$$\sum_{i=1}^{j} \begin{bmatrix} f^i + u \\ \\ -f^i - u \end{bmatrix} x_i + \sum_{k=j+1}^{p} \begin{bmatrix} u^{i_k} \\ \\ 0 \end{bmatrix} x_k + \sum_{k=p+1}^{2n} \begin{bmatrix} 0 \\ \\ u^{i_k} \end{bmatrix} x_k = \begin{bmatrix} u \\ \\ u \end{bmatrix}; \quad (1)$$

$$x \geq 0 \ .$$

The system (1) is of order $2n$ and has j non-unit columns. We show below that we can equivalently work with a system of size $n+1$ with $j+1$ non-unit columns:

Theorem 2.1: Let $\{h \mid 1 \leq h \leq n\ ,\ s_h = 0\} = \{h_1, \ldots, h_{n-j}\}$. Then τ is a complete facet of σ iff there is a solution to

$$\begin{bmatrix} 1 & \ldots & 1 & 0 & \ldots & 0 & 1 \\ f^1 & \ldots & f^j & u^{h_1} & \ldots & u^{h_{n-j}} & -s \end{bmatrix} \begin{bmatrix} w_1 \\ \\ w_{n+1} \end{bmatrix} = \begin{bmatrix} 1 \\ \\ 0 \end{bmatrix} ; \quad (2)$$

$w_i \geq 0\ ,\ 1 \leq i \leq j\ ;\ w_{n+1} \geq w_k\ ,\ w_{n+1} \geq -w_k\ ,\ j < k \leq n\ ;$ $w_{n+1} \geq 0$.

Proof: Suppose first that (1) has a solution x . By adding all the equations we obtain

$$\sum_{i=1}^{j} x_i + \sum_{k=j+1}^{2n} x_k/2n = 1 . \quad (3)$$

Subtracting this from all the other equations yields

$$\begin{aligned} &\sum_{i=1}^{j} f^i x_i + \sum_{k=j+1}^{p} u^{i_k} x_k - \left[\sum_{k=j+1}^{2n} x_k/2n \right] u = 0 ; \\ &\sum_{i=1}^{j} f^i x_i - \sum_{k=p+1}^{2n} u^{i_k} x_k + \left[\sum_{k=j+1}^{2n} x_k/2n \right] u = 0 . \end{aligned} \quad (4)$$

Now define w as follows. For $1 \leq i \leq j$, set $w_i = x_i$; also set

$$w_{n+1} = \sum_{k=j+1}^{2n} x_k/2n .$$

Then the first equation of (2) is satisfied and these components of w are negative. Now consider h_q . It lies in I_+ and I_- ; suppose $h_q = i_{k_1} = i_{k_2}$, $j < k_1 \leq p < k_2 \leq 2n$. Set

$$w_{j+q} = (-\sum_{i=1}^{j} f^i x_i)_{h_q} = x_{k_1} - w_{n+1} = - x_{k_2} + w_{n+1} .$$

Then the remaining equations of (2) are satisfied and $w_{n+1} \geq w_{j+q} \geq -w_{n+1}$. Hence we have a solution to (2) .

Conversely suppose w solves (2) . Set $x_i = w_i$ for $1 \leq i \leq j$. If $i_{k_1} = i_{k_2} = h_q$, set $x_{k_1} = w_{n+1} + w_{j+q}$ and

$x_{k_2} = w_{n+1} - w_{j+q}$. Finally, if $s_{i_k} > 0$ set

$x_k = 2\,w_{n+1} = (2 \sum_{i=1}^{j} f^i x_i)_{i_k}$. The resulting vector x is

clearly nonnegative. It is not hard to establish that x also satisfies (3) and (4) and hence (1) .

□

Note that if σ is the 0-simplex $[0]$ whose vertex is the origin and if τ is its facet the empty simplex then τ is a complete facet of σ since the matrix of (2) is a permuted identity and $w = (0, \ldots, 0, 1)$ solves (2) .

Suppose now $\sigma = [v^0, v^1, \ldots, v^j]$ with $\mathrm{sgn}(\sigma) = s$ and $\{h \mid s_h = 0\}$. Let $f^i = f(v^i)$, $0 \leq i \leq j$. We say σ is complete if

$$\begin{bmatrix} 1 & \ldots & 1 & 0 & \ldots & 0 & 1 \\ f^0 & \ldots & f^j & u^{h_1} & \ldots & u^{h_{n-j}} & -s \end{bmatrix} \begin{bmatrix} w_0 \\ \\ w_{n+1} \end{bmatrix} = \begin{bmatrix} 1 \\ 0 \end{bmatrix} ; \qquad (5)$$

$w_i \geq 0$, $0 \leq i \leq j$; $w_{n+1} \geq w_k$, $w_{n+1} \geq -w_k$, $j < k \leq n$; $w_{n+1} \geq 0$

has a solution. To avoid problems of degeneracy, we say σ is very complete if there is an $(n + 2) \times (n + 2)$ matrix W with rows $w^0, \ldots, w^{n\ 1}$ such that

$$\begin{bmatrix} 1 & \ldots & 1 & 0 & \ldots & 0 & 1 \\ f^0 & \ldots & f^j & u^{h_1} & \ldots & u^{h_{n\ j}} & -s \end{bmatrix} W = I ; \qquad (6)$$

$w^i \geq_\ell 0$, $0 \leq i \leq j$; $w^{n+1} \geq_\ell w^k$ and $w^{n+1} \geq_\ell -w^k$, $j < k \leq n$; $w^{n+1} \geq_\ell 0$.

Here $w \geq_\ell w'$ means w is lexicographically greater than w'. Finally, σ is a terminal simplex if (5) has a solution with w_{n+1} (and hence w_k, $j < k \leq n$) zero.

The 0-simplex [0] is easily seen to be very complete.

For any very complete simplex σ, there are two "basic" solutions to (6) with one of the inequalities holding as an equality. Moving from one of these solutions to the other is similar to a linear programming pivot step; see [16] and [17] for details in handling the special inequalities.

If one of these "basic" solutions has $w^i = 0$ for some $0 \leq i \leq j$, then we see that the facet $\tau = [v^0, \ldots, v^{i-1}, v^{i+1}, \ldots, v^j]$ is a complete facet of σ. If $\mathrm{sgn}(\tau) = s$, then the simplex σ' in $C(s)$, σ' containing τ, $\sigma' \neq \sigma$, can also be seen to be very complete. If $\mathrm{sgn}(\tau) \neq s$, let $t_h = 0 \neq s_h$; by expressing

$$\begin{bmatrix} 1 \\ \\ -s \end{bmatrix} \quad \text{as} \quad \begin{bmatrix} 1 \\ \\ -t \end{bmatrix} + (-s_h) \begin{bmatrix} 0 \\ \\ u^k \end{bmatrix} \qquad \text{we see that}$$

τ is itself very complete.

If one of the "basic solutions" of σ has $\varepsilon w^k = w^{n+1}$ for some $j < k \leq n+1$, $\varepsilon = \pm 1$, let $h = h_{k-j}$, and $r_h = -\varepsilon$, $r_i = s_i$ otherwise. Then σ is a facet of a unique simplex ρ in $C(r)$, and we check that σ is a complete facet of ρ and that ρ is very complete.

The algorithm proceeds from the starting 0-simplex [0] generating very complete simplices by proceeding from "basic" solution to "basic" solution and reinterpreting as described above. For details we refer to [7] or to [18], where a detailed description of the algorithm is given for an application to the Borsuk-Ulam theorem. We give a geometrical interpretation of the algorithm below.

The linear systems (2), (5) and (6) are similar to systems that arise in Merrill's homotopy restart algorithm [8] when exploiting the linearity of the artificial function as in [16]. These systems are associated with a certain subdivision of $R^n \times [0, 1]$ involving cubical pieces in $R^n \times \{0\}$. A similar interpretation of van der Laan and Talman's algorithm is possible, with just one cubical piece. This geometrical interpretation is due to Wright, see [18]. First we state

Proposition 2.2: The simplex $\sigma = [v^0, \ldots, v^j]$ with $\mathrm{sgn}(\sigma) = s$ is complete iff there is some z with $||z|| \leq 1$, $z_i = -s_i$ if $s_i \neq 0$, and

$$\sum_{i=0}^{j} \lambda_i f(v^i) + \nu z = 0 , \quad \sum_{i=0}^{j} \lambda_i + \nu = 1 ,$$

$\lambda_i \geq 0$, $\nu \geq 0$ having a solution.

Proof: The result follows directly from (5) ; if $w_{n+1} > 0$ set

$$z = -s + \sum_{q=1}^{n-j} u^{h_q} w_{j+q} .$$

□

Now consider T as a triangulation of $R^n \times \{1\}$ and let B^n be the n-cube $\{x \in R^n \mid ||x|| \leq 1\}$. We define the subdivision L of conv $(B^n \times \{0\} \cup R^n \times \{1\})$ by characterizing the (n+1)-dimensional pieces as conv $(B^n(s) \times \{0\} \cup \sigma \times \{1\})$ where σ is a j-simplex of T with $\mathrm{sgn}(\sigma) = s$, s having j nonzero components, and $B^n(s)$ the (n-j)-dimensional face $\{x \in B^n \mid x_i = s_i \text{ if } s_i \neq 0\}$ of the cube. We suggest that the reader draw some of these pieces when $n = 2$ to understand the subdivision L. Define the piecewise-linear function G by setting $G(x, 0) = -x$, $G(v, 1) = f(v)$ for v a vertex of T and extending linearly on each piece of L. In the nondegenerate case, proposition 2.2 shows that the algorithm generates simplices that correspond to pieces of L containing zeros of G. See section 3 of [18]. Thus the algorithm can be viewed as tracing the zeros of a piecewise-linear homotopy G, just as van der Laan's and Talman's "conical" algorithm [6] can [15].

3. GLOBAL CONVERGENCE

Van der Laan and Talman have given a condition that guarantees finite convergence of their algorithm [7]. Here we show that Merrill's condition [8] also assures convergence. The advantage is that Merrill's condition has been shown to be satisfied by optimization problems under reasonable restrictions and by problems of obtaining a fixed point of a function $g : R^n \to R^n$ with $g(R^n)$ bounded [8].

Suppose we are given an upper semicontinuous point-to-set mapping $F : R^n \to R^{n*}$. (R^{n*} is the set of compact convex nonempty subsets of R^n.) The algorithm of the previous section can be applied; for each vertex v of T we set $f(v)$ to be an arbitrary member of $F(v)$. Consider the following

Condition (M): There exist $x^0 \in R^n$, $\delta > 0$ and $\mu > 0$ such that whenever $||x - x^0|| > \mu$, $||z - x|| \leq \delta$ and $f \in F(x)$, $f^T (z - x^0) < 0$.

Merrill [8] proved that this condition was sufficient to guarantee global convergence of his algorithm when the artificial function was $r(x) = c - x$ and the triangulation used had (projected) mesh at most δ. We show that the same holds for the van der Laan-Talman algorithm.

Theorem 3.1: Suppose T has mesh at most δ and that condition (M) holds with this value of δ and some $x^0 \in R^n$, $\mu > 0$. Then the algorithm of van der Laan and Talman computes a zero of a piecewise-linear approximation to F in a finite number of steps. Furthermore, each such zero lies in $\{x \in R^n \mid ||x - x^0|| \leq \mu + \delta\}$.

Proof: Suppose at some stage the algorithm generates the complete simplex $\sigma = [v^0, \ldots, v^j]$ with $\text{sgn}(\sigma) = s$. Then proposition 2.2 states that there is a vector $z \in R^n$ with $||z|| \leq 1$ and $z_i = -s_i$ whenever $s_i \neq 0$ such that

$$\sum_{i=1}^{j} \lambda_i f(v^i) + \nu z = 0 ;$$

$$\sum_{i=0}^{j} \lambda_i + \nu = 1 ; \tag{7}$$

$$\lambda_i \geq 0 , \quad \nu \geq 0 .$$

Suppose now that $||v^0 - x^0|| > \max \{\mu + \delta , (n+1) ||x^0||\}$. We aim for a contradiction. Each v^i has $||v^i - x^0|| > \mu$, so by (M) we have

$$f(v^i)^T (v^0 - x^0) < 0 . \tag{8}$$

Now v^0 lies in $C(s)$, so $z^T v^0 = \sum_{s_i \neq 0} z_i v_i^0 \leq -||v^0||$. Also $z^T x^0 = \sum_i z_i x_i^0 \leq \sum |x_i^0| \leq n ||x^0||$. Since $||v^0|| > (n+1) ||x^0|| - ||x^0|| = n ||x^0||$, we obtain

$$z^T (v^0 - x^0) < 0 . \tag{9}$$

But (8) and (9) contradict (7). It follows that all simplices generated by the algorithm lie within $\max \{\mu + \delta , (n+1) ||x^0||\}$ of $||x^0||$. There are only a finite number of such simplices and thus finite convergence is assured.

For the terminal simplex the argument above only needs $||v^0 - x^0|| > \mu + \delta$ since $\nu = 0$. Hence the terminal simplex lies within $\mu + \delta$ of x^0, completing the proof.

□

This result implies global convergence of the restarted algorithm under condition (M) as in Merrill [8]. There is however a slight subtlety involved. When the algorithm is restarted we shift the origin to the linear approximate zero of F just found. In this new coordinatization, x^0 is represented differently. However all such linear approximate zeros lie within $\mu + \delta$ of the fixed x^0, so that they lie in a compact set and a convergent subsequence can be obtained. After the first iteration x^0 is within $\mu + \delta$ of the new origin, so there is a uniform bound of $(n+1)(\mu + \delta)$ on the distance from the current origin at which a simplex can be generated.

4. MONOTONICITY IN OPTIMIZATION PROBLEMS

Suppose we are trying to minimize $\Theta(x)$ subject to $\Psi(x) \leq 0$ where Θ and Ψ are convex functions on R^n. (In the convex case no loss of generality is involved in considering only one constraint; if there are several inequality constraints let Ψ be the pointwise supremum of the constraint functions.) We then search for a zero of the point-to-set mapping F defined by

$$F(x) = \begin{cases} -\partial\Theta(x) & \text{if } \Psi(x) < 0 ; \\ -\partial\Psi(x) & \text{if } \Psi(x) > 0 ; \\ -\text{conv}\{\partial\Theta(x) \cup \partial\Psi(x)\} & \text{if } \Psi(x) = 0 . \end{cases} \quad (10)$$

See [8] or [13]. Here $\partial\Theta$ is the subgradient of Θ ; if Θ is differentiable at x , $\partial\Theta(x) = \{\nabla\Theta(x)\}$. Note that by defining Ψ as the constant function -1 , we have included unconstrained optimization also, with the standard mapping.

Merrill [8] obtained several global convergence conditions for his algorithm applied to this problem; see also [13]. In addition he demonstrated certain bounds that can be obtained when a zero of a piecewise-linear approximation to F is generated.

Here we provide results concerning the points generated throughout the algorithm, rather than just at its conclusion. Our results apply both to Merrill's algorithm and that of van der Laan and Talman as discussed in section 2 . The flavour of our conclusions is similar to that of Eaves' results in [2] and [4]; artificial constraints are being relaxed until they are inessential. See also Kojima [5]. However, we deal explicitly with the inaccuracies caused by simplicial approximation.

Let $[v^0 , \ldots , v^j]$ be a simplex of diameter at most δ . Suppose $H = \{h \mid \Psi(v^h) \leq 0\}$ and $I = \{i \mid \Psi(v^i) > 0\}$ partition $\{0 , 1 , \ldots , j\}$ and suppose $f(v^h) = t^h \in -\partial\Theta(v^h)$ for $h \in H$ and $f(v^i) = p^i \in -\partial\Psi(v^i)$ for $i \in I$. Suppose that for some $z \in R^n$,

$$\begin{aligned} &\sum_H \lambda_h + \sum_I \lambda_i + \nu = 1 , \\ &\sum_H \lambda_h t^h + \sum_I \lambda_i p^i + \nu z = 0 , \qquad (11) \\ &\lambda_h \geq 0 , \quad \lambda_i \geq 0 , \quad \nu \geq 0 . \end{aligned}$$

Define $\lambda_H = \sum_H \lambda_h$ and $\lambda_I = \sum_I \lambda_i$. If $\lambda_H > 0$, define $v^H = \sum_H \lambda_h v^h/\lambda_H$, otherwise let v^H be arbitrary. Define v^I similarly. Finally denote by

$\tilde{\Theta}(x) \quad \sup\{\Theta(y) \mid \|y - x\| \le \delta\}$ and by

$\tilde{\Psi}(x) \quad \sup\{\Psi(y) \mid \|y - x\| \le \delta\}$.

Lemma 4.1: Under the hypotheses and notations above, for each $y \in R^n$ we have

$$\lambda_H \, \Theta(v^H) \le \lambda_H \, \tilde{\Theta}(y) + \lambda_I \, \tilde{\Psi}(y) + \nu \, z^T(v - y) \tag{12}$$

for any v with $\|v - v^i\| \le \delta$ all i .

Proof: Choose any v as above. For each $h \in H$ we have

$$\begin{aligned}(t^h)^T(y - v) = (t^h)^T((y - v + v^h) - v^h) &\ge \Theta(v^h) - \Theta(y - v + v^h) \\ &\ge \Theta(v^h) - \tilde{\Theta}(y) \quad .\end{aligned}$$

Similarly,

$(p^i)^T(y - v) \ge \Psi(v^i) - \tilde{\Psi}(y)$ for each $i \in I$. Hence taking the inner product of (11) with $y - v$ yields

$$\sum_H \lambda_h \, (\Theta(v^h) - \tilde{\Theta}(y)) + \sum_I \lambda_i \, (-\tilde{\Psi}(y)) + \nu \, z^T(y - v) \le 0 \quad .$$

The desired result follows since the convexity of Θ implies

$$\lambda_H \, \Theta(v^H) \le \sum_H \lambda_h \, \Theta(v^h) \; .$$

□

Let us now consider Merrill's algorithm. Suppose we use the artificial function $c - x$ and a triangulation with (projected) mesh at most δ . Let the simplex $[(v^0 , 1), \ldots , (v^j , 1), (v^{j+1} , 0), \ldots , (v^{n+1} , 0)]$ be generated at some stage. Let H and I be defined as above and let $K = \{j+1 , \ldots , n+1\}$. We have

$$\begin{aligned}&\sum_H t^h \, \lambda_h + \sum_I p^i \, \lambda_i + \sum_K (c - v^k) \, \lambda_k = 0 \\ &\sum_H \lambda_h + \sum_I \lambda_i + \sum_K \lambda_k = 1 \\ &\lambda_h \ge 0 \;\; , \;\; \lambda_i \ge 0 \;\; , \;\; \lambda_k \ge 0 \;\; .\end{aligned} \tag{13}$$

Define v^H and v^I as above; also let $\nu = \sum_K \lambda_k$ and if $\nu > 0$ let $v^K = \sum_K \lambda_k v^k/\nu$; otherwise v^K is arbitrary. Then with $z = c - v^K$ we obtain (11) and applying Lemma 4.1 we deduce

$$\lambda_H \, \Theta(v^H) \leq \lambda_H \tilde{\Theta}(y) + \lambda_I \tilde{\Psi}(y) + \nu(c - v^K)^T(v^K - y) \qquad (14)$$

for all $y \in R^n$. We then easily obtain

Theorem 4.2:

a) Suppose for some $y \in R^n$, $\tilde{\Psi}(y) < 0$. Then if $\nu = 0$ or $||v^K - c||_2 > ||y - c||_2$, λ_H is positive and $\Theta(v^H) \leq \tilde{\Theta}(y)$.

b) If $\nu = 0$ and $\lambda_H > 0$, then $\Theta(v^H) \leq \tilde{\Theta}(x)$ for any $x \in R^n$ with $\tilde{\Psi}(x) \leq 0$.

c) If $\nu > 0$ and $\lambda_H > 0$, then $\Theta(v^H) \leq \tilde{\Theta}(x)$ for all $x \in R^n$ with $\tilde{\Psi}(x) \leq 0$ and $||x - c||_2 \leq ||v^K - c||_2$.

Proof: Part (a) with $\nu = 0$ follows directly from (14), as does part (b). For the remaining parts, note that

$$||v^K - c||_2 > ||y - c||_2 \Rightarrow$$

$$\Rightarrow (v^K - c)^T(v^K - c) > (v^K - c)^T(y - c)$$

$$\Rightarrow (c - v^K)^T(v^K - y) < 0$$

and similarly with weak inequalities.

□

Part (a) assures the convergence of Merrill's algorithm if $\tilde{\Psi}(y) < 0$ for some y and Θ and Ψ have no common direction of recession. It also shows that when $\tilde{\Psi}(y) < 0$ for some y and finite convergence is achieved $(\nu = 0)$, a feasible solution results. These results can be found in Merrill [8] (see also [13]) but here we have more information about the path of the algorithm.

Part (b) shows in what sense the zero of a piecewise-linear approximation to F is an optimal solution. This result complements the bounds on Θ obtained by Merrill [8]. In the unconstrained case $(\Psi \equiv -1)$ Merrill had this result precisely.

The most interesting part is (c). It demonstrates that in some sense the algorithm generates "optimal" solutions to the problem with the auxiliary constraint $||x - c||_2 \leq \alpha$ as α is gradually increased. This is in the spirit of Eaves' methods in [2] and [4].

We next turn to the van der Laan-Talman algorithm as applied to F with a triangulation of mesh at most δ. Suppose the complete simplex $\sigma = [v^0, \ldots, v^j]$ with $\text{sgn}(\sigma) = s$ is generated. Then, using proposition 2.2 and lemma 4.1, we obtain, with notation as above,

$$\lambda_H \, \Theta(v^H) \leq \lambda_H \, \tilde{\Theta}(y) + \lambda_I \, \tilde{\Psi}(y) + \nu \, z^T(v^0 - y) \tag{15}$$

for all $y \in R^n$, where $||z||_\infty \leq 1$ and $z_i = -s_i$ whenever $s_i \neq 0$. We then deduce

Theorem 4.3:

a) Suppose for some $y \in R^n$, $\tilde{\Psi}(y) < 0$. Then if $\nu = 0$ or $||v^0||_1 > ||y||_1$, λ_H is positive and $\Theta(v^H) \leq \tilde{\Theta}(y)$.

b) If $\nu = 0$ and $\lambda_H > 0$, then $\Theta(v^H) \leq \tilde{\Theta}(x)$ for any x with $\tilde{\Psi}(x) \leq 0$.

c) If $\nu > 0$ and $\lambda_H > 0$, then $\Theta(v^H) \leq \tilde{\Theta}(y)$ for any x with $\tilde{\Psi}(x) \leq 0$ and $||x||_1 \leq ||v^0||_1$.

Proof: The proof is similar to that for the previous theorem. We merely note that $||v^0||_1 = -z^T v^0$ since $v_i^0 \neq 0$ implies $\text{sgn}(v_i^0) = s_i$ and that $||y||_1 \geq ||y||_1 \, ||z||_\infty \geq -z^T y$, so $||v^0|| > ||y||_1$ implies $z^T(v^0 - y) < 0$ and similarly with weak inequalities.

□

Theorem 4.3 has a similar interpretation to that of theorem 4.2. Now the algorithm generates "optimal" solutions to the problem with the auxiliary constraint $||x||_1 \leq \alpha$ as α is gradually increased.

Suppose that $\tilde{\Psi}(y) < 0$ for some $y \in R^n$. Part (a) of each theorem then implies that we generate a feasible solution v^H with each algorithm. Let us suppose that in fact $\tilde{\Psi}(v^H) < 0$. We may now restart each algorithm, with $c = v^H$ in Merrill's algorithm and the origin translated to v^H in the van der Laan-Talman algorithm. The parts (a) then show that feasibility is maintained and that $\Theta(\bar{v}^H) \leq \tilde{\Theta}(v^H)$ with $\bar{v}^H$ the next solution. Suppose that such a sequence of restarts generates $x^1, x^2, \ldots$. We see that

$\tilde{\Psi}(x^i) < 0 \Rightarrow \Psi(x^{i+1}) \leq 0$, and that if $\tilde{\Psi}(x^i) < 0$ for all i , $\Theta(x^{i+1}) \leq \tilde{\Theta}(x^i)$ for all i . This is the monotonicity property referred to above.

5. LOCAL CONVERGENCE

In this section we consider the application of the van der Laan and Talman algorithm to compute a zero of a function $f : R^n \to R^n$ satisfying the following regularity condition.

Condition (R): f is continuously differentiable and its derivative Df satisfies the Lipschitz condition

$$||Df(x) - Df(y)|| \leq \kappa||x - y|| \quad \text{for all} \quad x, y \in R^n \ .$$

There is a uniform bound μ with Df nonsingular and $||Df^{-1}|| \leq \mu$ at every zero of f .

We suppose the algorithm is implemented and restarted as follows. At stage k we have an approximate zero x^{k-1} of f , a matrix D_{k-1}^{-1} where D_{k-1} is an approximation to $Df(x^{k-1})$, and a grid size ε_k . Initially, x^0 is a guess for a zero of f and ε_1 measures our estimate of its accuracy. D_0 is chosen to promote finite convergence; if condition (M) holds, choose $D_0 = -I$, the negative identity. We then apply the algorithm of section 2 to the function f^k defined by $f^k(x) = -D_{k-1}^{-1} f(x^{k-1} - \varepsilon_k w + x)$, where

$w = (n+1)^{-1} (n , n-1 , \ldots , 1)^T$. We use one of the triangulations K_1 or J_1 scaled by ε_k .

Suppose an approximate zero y^k of f^k is found. Then $\bar{x}^k = x^{k-1} - \varepsilon_k w + y^k$ is an approximate zero of f . We restart as follows. Since we are using lexicographic rules the terminal simplex has dimension n . Hence the algorithm also generates an approximate inverse derivative $\tilde{D}_k^{-1}$ of f^k at y^k (see, for instance, [11], [14]). Then $\bar{D}_k^{-1} = \tilde{D}_k^{-1} D_{k-1}^{-1}$ is an approximate inverse derivative of f at $\bar{x}^k$.

Next, we evaluate a single discrete Newton step $s^k = -\bar{D}_k^{-1} f(\bar{x}^k)$ at $\bar{x}^k$. If $||s^k|| \geq \varepsilon_k/2$ or $||f(\bar{x}^k + s^k)|| \geq ||f(\bar{x}^k)||$ we set

$x^k = \bar{x}^k$, $D_k^{-1} = D_{k-1}^{-1}$ and $\varepsilon_{k+1} = \varepsilon_k/2$. Otherwise we set

$x^k = \bar{x}^k + s^k$, $D_k^{-1} = \bar{D}_k^{-1}$ and $\varepsilon_{k+1} = ||s^k||$. This description is similar to that for the accelerated fixed-point algorithm with parameter $m = 0$ of [12].

(The version described above is most convenient for analysis. In practice one would avoid multiplying each function value $f(x^{k-1} - \varepsilon_k w + v)$ by $-D_{k-1}^{-1}$. Instead we would replace the columns

$(0, u^{h_q})$ and $(1, -s)$ in (6) by $(0, -D_{k-1} u^{h_q})$ and $(1, D_{k-1} s)$ respectively. Assuming nondegeneracy the algorithms are identical.)

We suppose that this algorithm is well-defined, i.e., that each y^k is obtained from x^{k-1} in a finite number of iterations. We also assume that all x^k's lie in a fixed bounded set. We aim to prove that, with these hypotheses and condition (R), for sufficiently large k, $n+1$ function evaluations suffice to obtain y^k from x^{k-1} (the $(n+1)$-step property). In addition we wish to show that the $\bar{x}^k$'s converge quadratically to a zero x^* of f: $||\bar{x}^{k+1} - x^*|| \leq \eta||\bar{x}^k - x^*||^2$ for some η.

First we note the following bounds:

Lemma 5.1: (Saigal [11]). In the notation above,

$$||f(\bar{x}^k)|| \leq \frac{1}{2} \kappa \varepsilon_k^2 \quad \text{and}$$

$$||\bar{D}_k - Df(\bar{x}^k)|| \leq \kappa n \varepsilon_k .$$

□

Since $\varepsilon_k \to 0$ and $||x^k - \bar{x}^k|| \leq \varepsilon_k/2$, the $\bar{x}^k$'s remain in a bounded set. The lemma then implies that every limit point of $\{\bar{x}^k\}$, and hence of $\{x^k\}$, is a zero of f. In order to be able to say something about the matrices $\bar{D}_k$, we need the following result (which follows easily from the Neumann lemma, see p. 45 of [9]) :

Lemma 5.2: Let A and E be real $n \times n$ matrices and assume A is nonsingular with $||A^{-1}|| \leq \alpha$. If $||E|| \leq \beta$ with $\alpha\beta < 1$ then $A + E$ is nonsingular and

$$||(A + E)^{-1} - A^{-1}|| \leq \alpha^2 \beta .$$

□

We then obtain directly

Lemma 5.3: For any $\delta > 0$, there is an index k_0 such that each x^k, $k \geq k_0$, lies within δ of some zero of f. Further k_0 can be chosen so that $||\bar{D}_k^{-1}|| \leq 2\mu$ and

$$||\bar{D}_k^{-1} f(\bar{x}^k)|| \leq \mu \kappa \varepsilon_k^2 < \varepsilon_k/2 \quad \text{for } k \geq k_0 .$$

Proof: If the first part failed, we would be able to extract a convergent subsequence from those x^k that did not lie within δ of a zero. But the limit of such a subsequence is necessarily a zero of f, a contradiction. For the second part, choose k_0 so that $n\, \varepsilon_{k_0} < (2 \mu \kappa)^{-1}$ and also all $\bar{x}^k$'s, $k \geq k_0$, lie within $(2 \mu \kappa)^{-1}$ of a zero of f. Then for $k \geq k_0$, $\bar{D}_k$ lies within $(2 \mu)^{-1}$ of $Df(\bar{x}^k)$ which lies within $(2 \mu)^{-1}$ of $Df(\hat{x}^k)$ with $\hat{x}^k$ a zero of f. Now apply lemma 5.2 with

$A = Df(\hat{x}^k)$, $\alpha = \mu$ and $E = \bar{D}_k - A$. Thus $||\bar{D}_k^{-1}|| \leq 2\mu$. Now

$$||\bar{D}_k^{-1} f(\bar{x}^k)|| \leq ||\bar{D}_k^{-1}||\ ||f(\bar{x}^k)|| \leq 2\mu \cdot \tfrac{1}{2} \kappa \varepsilon_k^2 < \varepsilon_k/2n .$$

The result follows.

□

Thus for sufficiently large k we have $||s^k|| < \varepsilon_k/2$. To show that the algorithm accelerates for sufficiently large k, we must prove that

$||f(\bar{x}^k + s^k)|| < ||f(\bar{x}^k)||$. But 3.2.12 of [9] yields

$$||f(\bar{x}^k + s^k) - f(\bar{x}^k) - Df(\bar{x}^k)\, s^k|| \leq \tfrac{1}{2} \kappa\, ||s^k||^2 ,$$

whence we obtain, for $k \geq k_0$,

$$\begin{aligned}||f(\bar{x}^k + s^k)|| &\leq ||f(\bar{x}^k) + Df(\bar{x}^k)\, s^k|| + \tfrac{1}{2}\kappa\, ||\bar{D}_k^{-1} f(\bar{x}^k)||^2 \\ &\leq ||(\bar{D}_k - Df(\bar{x}^k))\, \bar{D}_k^{-1} f(\bar{x}^k)|| + \tfrac{1}{2}\kappa\, ||\bar{D}_k^{-1}||^2\, ||f(\bar{x}^k)||^2 \\ &\leq [\kappa\, n\, \varepsilon_k \,.\, 2\mu + \tfrac{1}{2}\kappa \,.\, 4\mu^2 \,.\, \tfrac{1}{2}\kappa\, \varepsilon_k^2]\, ||f(\bar{x}^k)||\end{aligned}$$

by Lemmas 5.1 and 5.3. Now take k sufficiently large that the quantity in brackets is strictly less than one, e.g., so that $\varepsilon_k < (4\, n\, \mu\, \kappa)^{-1}$; then $||f(\bar{x}^k + s^k)|| < ||f(\bar{x}^k)||$. We have therefore shown that for k sufficiently large the algorithm always accelerates. We have also seen that in this case we have $\varepsilon_{k+1} \leq \mu\, \kappa\, \varepsilon_k^2$, so that the grid size is shrinking quadratically. This implies that $||f(\bar{x}^k)||$ is shrinking quadratically. However, we do not yet know that the $\bar{x}^k$'s are converging to a single zero x^*. This and the quadratic convergence will follow from the (n+1)-step property.

The strategy we use is the following. We first consider the idealized situation with f^k given by $f^k(x) = \varepsilon_k w - x$; this is the case if f is affine, x^{k-1} is a zero of f and D_{k-1}^{-1} is the exact inverse derivative of f. For this simple function we exhibit the simplices generated and the corresponding matrices as in (2) and their inverses. Hence the (n+1)-step property is verified in the idealized situation. Next we show that the actual matrix corresponding to such a simplex is close to that in the simple situation. Finally an application of Lemma 5.2 shows that the actual matrix inverse is close enough to the idealized matrix inverse that the required inequalities hold. This demonstrates that exactly the same simplices are generated so that again the (n+1)-step property holds.

To use the bounds most effectively we first remove the effect of the scale factor ε_k. Hence we require

Lemma 5.4: A simplex is very complete with respect to the function f iff it is very complete with respect to the function $\alpha^{-1} f$ for any positive α.

Proof: Suppose σ is very complete with respect to f . The the system (6) has a "basic" solution. Without loss of generality we can assume that we have a matrix

$$\begin{bmatrix} e^T & 0 & 1 \\ A & 0 & -s \\ B & I & 0 \end{bmatrix} \tag{16}$$

whose inverse

$$\begin{bmatrix} p & C & 0 \\ -Bp & -BC & I \\ \rho & d^T & 0 \end{bmatrix}$$

satisfies $(p , C) \geq_\ell 0$; $(\rho , d^T) \geq_\ell 0$; and for each row (Φ , g^T) of $(-Bp , -BC)$, $(\rho , d^T) >_\ell (\Phi , g^T)$ and $(\rho , d^T) >_\ell -(\Phi , g^T)$. When f is replaced by $\alpha^{-1} f$, the matrix in (16) and its inverse are replaced by

$$\begin{bmatrix} e^T & 0 & 1 \\ \alpha^{-1}A & 0 & -s \\ \alpha^{-1}B & I & 0 \end{bmatrix}$$

and

$$\begin{bmatrix} \alpha\gamma p & \alpha C - \alpha(1 - \alpha)\gamma p d^T & 0 \\ -\gamma B p & -BC + (1 - \alpha)\gamma B p d^T & I \\ \gamma\rho & d^T - (1 - \alpha)\gamma\rho d^T & 0 \end{bmatrix}$$

respectively, where $\gamma = \alpha + \rho - \alpha\rho$. It is easy to check that this matrix satisfies the same set of inequalities, showing that σ is very complete with respect to the function $\alpha^{-1} f$. We have, of course, also proved the converse, since $f = (\alpha^{-1})^{-1}(\alpha^{-1} f)$.

□

Let us now consider the idealized situation where we use $f^k(x) = \varepsilon_k w - x$ and the triangulation $\varepsilon_k K_1$, say. By the lemma above, we could alternatively use $\tilde{f}^k(x) = w - x/\varepsilon_k$. But clearly this is equivalent to using $\bar{f}(x) = w - x$ with the triangulation K_1 .

Define $v^0 = 0$, $v^1 = u^1$, ... , $v^n = u^1 + u^2 + \ldots + u^n$. Then $[v^0, \ldots, v^n]$ is a simplex of K_1 , and it is also a simplex of J_1 . We show below that the algorithm produces the sequences of simplices $[v^0]$, $[v^0, v^1]$, $[v^0, \ldots, v^n]$.

Lemma 5.5: For each $j = 0, 1, \ldots, n$, $\tau = [v^0, \ldots, v^{j-1}]$ is a complete facet of $\sigma = [v^0, \ldots, v^j]$, and σ is very complete, with the function $\bar{f}$ above.

Proof: First note that $\text{sgn}(\sigma) = v^j$. Hence the matrix in (2) corresponding to τ and σ is

$$\begin{bmatrix} 1 & \ldots & 1 & 0 & \ldots & 0 & 1 \\ \bar{f}(v^0) & \ldots & \bar{f}(v^{j-1}) & u^{j+1} & \ldots & u^n & -v^j \end{bmatrix}$$

or

$$\left[\begin{array}{cccccc|cc|c}
1 & 1 & \cdot & \cdot & \cdot & 1 & 0 & & 0 \\
\frac{n}{n+1} & -\frac{1}{n+1} & \cdot & \cdot & \cdot & -\frac{1}{n+1} & & & -1 \\
\frac{n-1}{n+1} & \frac{n-1}{n+1} & -\frac{2}{n+1} & \cdots & & -\frac{2}{n+1} & & & \cdot \\
\cdot & \cdot & \cdot & \cdot & & & & & \cdot \\
\cdot & \cdot & \cdot & \cdot & & & \bigcirc & & \cdot \\
\cdot & \cdot & & \cdot & \cdot & & & & \cdot \\
\cdot & \cdot & & & \cdot & -\frac{j-1}{n+1} & & & \cdot \\
\frac{n+1-j}{n+1} & \frac{n+1-j}{n+1} & \cdot & \cdot & \cdot & \frac{n+1-j}{n+1} & & & -1 \\
\hline
\frac{n-j}{n+1} & \frac{n-j}{n+1} & \cdot & \cdot & \cdot & \frac{n-j}{n+1} & 1 & \bigcirc & 0 \\
\cdot & \cdot & & & & \cdot & \cdot & & \cdot \\
\cdot & \cdot & & & & \cdot & & \cdot & \cdot \\
\cdot & \cdot & & & & \cdot & & \cdot & \cdot \\
\frac{1}{n+1} & \frac{1}{n+1} & \cdot & \cdot & \cdot & \frac{1}{n+1} & \bigcirc & 1 & 0
\end{array}\right] \tag{17}$$

The inverse of this matrix is

$$
W = \left[\begin{array}{cccccc|c}
\frac{n+2-j}{2n+2-j} & 1 & & & & \frac{-n}{2n+2-j} & \\
\frac{1}{2n+2-j} & -1 & 1 & & \bigcirc & \frac{1}{2n+2-j} & \\
\cdot & \cdot & \cdot & & & \cdot & \\
\cdot & & \cdot & \cdot & & \cdot & \bigcirc \\
\cdot & & & \cdot & \cdot & \cdot & \\
\frac{1}{2n+2-j} & \bigcirc & & -1 & 1 & \frac{1}{2n+2-j} & \\
\hline
-\frac{n-j}{2n+2-j} & & & & & -\frac{n-j}{2n+2-j} & \\
\cdot & & & & & \cdot & \\
\cdot & & \bigcirc & & & \cdot & I \\
\cdot & & & & & \cdot & \\
-\frac{1}{2n+2-j} & & & & & -\frac{1}{2n+2-j} & \\
\hline
\frac{n+1-j}{2n+2-j} & & 0 & & & -\frac{n+1}{2n+2-j} & 0
\end{array}\right] . \tag{18}
$$

Note that, if the rows of W are denoted $w^1, \ldots, w^{n+1}$, we have $w^1 \geq_\ell 0, \ldots, w^j \geq_\ell 0$, $w^{n+1} \geq_\ell w^k$ and $w^{n+1} \geq_\ell -w^k$ for $j < k \leq n$. Hence τ is a complete facet of σ and σ is very complete.

□

Corollary 5.6: The (n+1)-step property holds for the function $\bar{f}$ when the triangulation K_1 or J_1 is employed.

□

We now consider the actual situation. Here we are using f^k defined by $f^k(x) = -D_{k-1}^{-1} f(x^{k-1} - \varepsilon_k w + x)$. We claim that this is close to $\varepsilon_k w - x$. In fact, we have

$$f(x^{k-1} - \varepsilon_k w + x) = f(x^{k-1}) + DF(x^{k-1})(x - \varepsilon_k w) + e(x) \quad , \text{ with}$$

$$||e(x)|| \leq \frac{1}{2} \kappa ||x||^2 \quad \text{(see 3.2.12 of [9]). Hence}$$

$$f^k(x) - (\varepsilon_k w - x) = -D_{k-1}^{-1} f(x^{k-1}) + D_{k-1}^{-1}(D_{k-1} - Df(x^{k-1}))(x - \varepsilon_k w) + e(x). \tag{19}$$

Lemma 5.7: For any $\gamma > 0$, there is a k_1 sufficiently large such that, for $k \geq k_1$, $||f^k(\varepsilon_k v^i) - (\varepsilon_k w - \varepsilon_k v^i)|| < \gamma \varepsilon_k$ for $i = 0, 1, \ldots, n$.

Proof: For $k > k_0$ as given in lemma 5.3, $D_k = \bar{D}_k$ of lemma 5.1 holds with D_k instead of $\bar{D}_k$. Thus $||D_{k-1} - Df(x^{k-1})|| \to 0$. Since $||D_{k-1}^{-1}|| \leq 2\mu$ for $k > k_0$ and $||\varepsilon_k v^i - \varepsilon_k w|| \leq \varepsilon_k$, the second term in (19) can be made less than $\gamma\varepsilon_k/3$ for k sufficiently large.
The third term is easily dealt with since $||e(\varepsilon_k v^i)|| \leq \frac{1}{2} \kappa \varepsilon_k^2 < \gamma \varepsilon_k/3$ for k sufficiently large. Now for $k > k_0$, $\varepsilon_k = ||D_{k-1}^{-1} f(\bar{x}^{k-1})||$; lemma 3.5 of [12] shows that

$||D_{k-1}^{-1} f(x^{k-1})||/||D_{k-1}^{-1} f(\bar{x}^{k-1})|| \to 0$ as $k \to \infty$, so that the first term is less than $\gamma \varepsilon_k/3$ for k sufficiently large. The lemma is now proved.

□

Theorem 5.8: Under the assumptions above, the (n+1)-step property holds for k sufficiently large.

Proof: We show that when k is sufficiently large, each simplex $[\varepsilon_k v^0, \varepsilon_k v^1, \ldots, \varepsilon_k v^j]$ is very complete for f^k. By lemma 5.4, this is equivalent to showing that $[v^0, v^1, \ldots, v^j]$ is very complete for $\hat{f}^k$ defined by $\hat{f}^k(x) = \varepsilon_k^{-1} f^k(x)$. Lemma 5.7 shows that for any $\gamma > 0$ there is some sufficiently large k so that the difference between $\hat{f}^k(x)$ and $w - v^i$ has norm at most γ. Now consider the matrix

$$\begin{bmatrix} 1 & \ldots & 1 & 0 & \ldots & 0 & 1 \\ \hat{f}^k(v^0) & \ldots & \hat{f}^k(v^{j-1}) & u^{j+1} & \ldots & u^n & -v^j \end{bmatrix} \tag{20}$$

Let A be the matrix in (17) and let $A + E$ be the matrix in (20). Then

$$||A^{-1}|| \le \frac{2n+6}{n+2} < 3 \quad \text{and}$$

$$||E|| \le (n+1) \max_i ||\hat{f}^k(v^i) - (w - v^i)|| \le (n+1)\gamma \ .$$

Now apply lemma 5.2. If γ was chosen so that $3(n+1)\gamma \le 1$, $A + E$ is invertible. Furthermore, $(A + E)^{-1}$ is within $9(n+1)\gamma$ of A^{-1} in norm. But $A^{-1} = W$ is given explicitly in (18). We note that if no entry in W changes by $1/2(2n+2-j) \ge 1/(4n+4)$, then the required inequalities still hold. Thus if we choose $\gamma \le 1/36(n+1)^2$, the simplex is very complete for f^k. The theorem follows.

□

The argument is straightforward from here. When the (n+1)-step property holds, $||\bar{x}^k - x^{k-1}|| \le \varepsilon_k$. When the algorithm accelerates $||x^k - \bar{x}^k|| \le \varepsilon_k/2$. Also, $\varepsilon_{k+1} \le \varepsilon_k/2$ always. Hence, since each zero of f is isolated, one easily obtains the fact that the sequence x^k converges to a single zero x^* of f. We now use the fact that $f(\bar{x}^k) = f(x^*) + Df(x^*)(\bar{x}^k - x^*) + e'(\bar{x}^k)$, $||e'(\bar{x}^k)|| \le \frac{1}{2}\kappa\, ||\bar{x}^k - x^*||^2$ to obtain

$$||\bar{x}^k - x^* - \bar{D}_k^{-1} f(\bar{x}^k)|| = o(\varepsilon_k)$$

so that $||\bar{x}^k - x^*|| \sim \varepsilon_k$. Now the fact seen above that the ε_k's converge quadratically to zero establishes that $\{\bar{x}^k\}$ converges quadratically to x^* . For details see Saigal [11], Saigal and Todd [12]. We therefore have

Theorem 5.9: Under condition (R), the algorithm as described above produces a sequence $\{\bar{x}^k\}$ converging quadratically to a zero of f .

□

We remark that the technique used above for proving the (n+1)-step property via lemma 5.2 can be used in other simplicial algorithms also. For example, quadratic convergence of a variant of van der Laan and Talman's "conical" algorithm was proved in [15] using artificial vertices $\delta(u - (n+1)u^i + w)$ $i = 1, \ldots, n$ and $\delta(u + w)$. Using the present technique one can use the simpler vertices
$\delta(u - (n+1)u^i)$, $i = 1, \ldots, n$ and δu .

Finally we note that the analysis above holds also for

$$w = (2 + \frac{n}{n+1}, \frac{n-1}{n+1}, \ldots, \frac{1}{n+1})^T$$; that is the approximate zero is

moved to the centroid of a simplex two units further down the x_1-axis than that considered before. This modification is appropriate in computing solutions to antipodal fixed-point problems [18] where the function is not smooth at the origin but is smooth a small distance along the first coordinate axis. We now find that $n+3$ function evaluations suffice to obtain y^k from x^{k-1} for large k . The increase of two is due to the generation of three 1-simplices,
$[0, u^1]$, $[u^1, 2u^1]$, $[2u^1, 3u^1]$, rather than just one. However the proof above carries over with very minor modifications; we must consider the extra 1-simplices in the idealized situation, and the bounds on the terms in (19) are slightly larger. For large k we obtain an (n+3)-step property and quadratic convergence of the $\bar{x}^k$'s .

REFERENCES

1. Allgower, E.; Georg, K., Simplicial and Continuation Methods for Approximating Fixed Points, Institute for Applied Mathematics, University of Bonn, August 1978.

2. Eaves, B.C., On the Basic Theorem of Complementarity, Mathematical Programming 1, 68-75 (1971).

3. Eaves, B.C., A Short Course in Solving Equations with PL Homotopies, in Nonlinear Programming, Proceedings of the Ninth SIAM-AMS Symposium in Applied Mathematics, R.W. Cottle; C.E. Lemke (Eds.), SIAM, Philadelphia 1976.

4. Eaves, B.C., Computing Stationary Points, Mathematical Programming Study 7, 1-14 (1978).

5. Kojima, M., Strongly Stable Stationary Solutions in Nonlinear Programs, Mathematics Research Center Summary Report #1920, University of Wisconsin-Madison, February 1979.

6. van der Laan, G.; Talman, A.J.J., A Restart Algorithm without an Artificial Level for Computing Fixed Points on Unbounded Regions, Department of Actuarial Sciences and Econometrics, Free University, Amsterdam, August 1978.

7. van der Laan, G.; Talman, A.J.J., A Class of Simplicial Subdivisions for Restart Fixed Point Algorithms without an Extra Dimension, Technical Report 36, Department of Actuarial Sciences and Econometrics, Free University, Amsterdam, December 1978.

8. Merrill, O.H., Applications and Extensions of an Algorithm that Computes Fixed Points of Certain Upper Semi-Continuous Point to Set Mappings, Ph. D. Dissertation, Department of Industrial Engineering, University of Michigan 1972.

9. Ortega, J.M.; Rheinboldt, W., Iterative Solutions of Nonlinear Equations in Several Variables, Academic Press, New York 1970.

10. Reiser, P.M., A Modified Integer Labelling for Complementarity Algorithms, Manuscript, University of Zurich 1978.

11. Saigal, R., On the Convergence Rate of Algorithms for Solving Equations that are Based on Methods of Complementary Pivoting, Mathematics of Operations Research 2, 108-124 (1977).

12. Saigal, R.; Todd, M.J., Efficient Acceleration Techniques for Fixed-Point Algorithms, SIAM Journal on Numerical Analysis 15, 997-1007 (1978).

13. Todd, M.J., The Computation of Fixed Points and Applications, Springer, Berlin-Heidelberg-New York 1976.

14. Todd, M.J., Orientation in Complementary Pivot Algorithms, Mathematics of Operations Research 1, 54-66 (1976).

15. Todd, M.J., Fixed-Point Algorithms that Allow Restart without an Extra Dimension, Technical Report No. 379, School of Operations Research and Industrial Engineering, Cornell University, May 1978.

16. Todd, M.J., Traversing Large Pieces of Linearity in Algorithms that Solve Equations by Following Piecewise-Linear Paths, Technical Report No. 390, School of Operations Research and Industrial Engineering, Cornell University, September 1978.

17. Todd, M.J., Numerical Stability and Sparsity in Piecewise-Linear Algorithms, to appear in the Proceedings of a Symposium on Analysis and Computation of Fixed Points, Mathematics Research Center, University of Wisconsin-Madison, May 1979.

18. Todd, M.J.; Wright, A.H., A Variable-Dimension Simplicial Algorithm for Antipodal Fixed-Point Theorems, Technical Report No. 417, School of Operations Research and Industrial Engineering, Cornell Univeristy, April 1979.

Numerical Solution of Highly Nonlinear Problems
W. Forster (ed.)
© North-Holland Publishing Company, 1980

THE FLEX SIMPLICIAL ALGORITHM

C.B. Garcia and Willard I. Zangwill

University of Chicago,
Chicago, U.S.A.

Many algorithms have been suggested for obtaining fixed points, equilibria, or solutions to systems of equations that are based upon piecewise linear (PL) approximation. These algorithms use a fixed grid to triangulate the domain into simplices, and then move from simplex to simplex. The key idea behind their convergence proof and what prevents cycling is the principle of complementarity. In this paper we suggest a different PL algorithm. It also moves from simplex to simplex; however, no triangulation is necessary. The simplices can be almost anywhere, even on top of each other.

The work of the first author is supported in part by National Science Foundation Research Grant No. MCS 77-15509 and ARO Grant No. DAAG-29-78-G-0160. The work of the second author is supported in part by ARO Grant No. DAAG-29-78-G-0160.

Also, the vertices of the simplices can be selected adaptively or flexibly as the algorithm proceeds, since no fixed grid is required. Finally, the complementarity principle is not used; instead, the convergence proof is obtained by a different approach, path following.

1. BACKGROUND

It is indeed exciting that one can actually calculate fixed points, equilibria, or nonlinear equation solutions, and today a number of algorithms exist for this. Historically, Lemke [14] and Lemke and Howson [15] discovered the important principle of complementarity. Scarf [17] then brilliantly employed that principle to calculate fixed points, equilibria, and equation solutions. Since then, there has been continuing progress, and today a large number of algorithms have been developed for this type of problem. For example, Merrill [16] and, independently, Kuhn and MacKinnon [13] suggested the "sandwich" approach. Eaves [2] introduced the homotopic approach, and there have been numerous other variants and techniques proposed. See [1], [3], [9], [12], [18], for example.

All the algorithms in this genre share certain basic features. First, using a fixed grid, the space or domain is triangulated into simplices (or other polyhedra). Next, a piecewise linear (PL) approximation of the original function is made. Finally, the PL approximation is solved by moving from simplex to simplex within the triangulation. To assure no cycling can occur and that the algorithm does converge, the complementarity principle is invoked. That principle insures the algorithm cannot return to the same simplex, so it must converge.

Our objectives in this paper are threefold:

i) to describe a PL algorithm when the underlying path must be followed closely. Homotopy-type algorithms such as [2], [3], [13], [16] are no longer applicable in this case. For example, the path could be describing how an economic equilibrium changes over time in a dynamical system [8].

ii) to show that PL methods and differential equations methods are two parallel and inherently similar approaches for path-following.

iii) to show that PL methods can be freed of triangulations when the underlying path is followed closely.

The idea that the grid need not be selected and thus not fixed beforehand was suggested in [19]. That algorithm permitted the vertex points of the simplices to be chosen by considering the function values encountered. However, a triangulation is being dynamically generated as the algorithm proceeds, and the complementarity principle is utilized.

The algorithm in this paper does not use a grid, so it is flexible and adaptive and does not utilize the complementarity principle. Because no triangulation is needed, the simplices can overlap one another in a helter-skelter fashion. Instead of complementarity, a different method of proof is used. In brief, an underlying differentiable path exists for the original function. The PL path generated by the algorithm is shown to follow this path and thus converge.

We first establish that if our simplices are sufficiently small, all simplices generated by the PL algorithm are within a δ tubular neighbourhood of the underlying differentiable path. Then we establish that the PL path and the underlying path must be oriented similarly. The PL path generated by the algorithm therefore cannot get far from the underlying path, heads in a similar direction, so it must follow along.

2. STATEMENT OF THE PROBLEM

In order to calculate fixed points, solutions, and, in fact, to solve a variety of problems, we employ a homotopy function. Let R^n denote Euclidian n space. Suppose $D \subset R^n$ is an open set and $\overline{D}$ is its closure. A homotopy function is a mapping $H: \overline{D} \times T \to R^n$, where $\times$ denotes Cartesian product and:

$$T = \{t \mid 0 \le t \le 1\} .$$

We use the notation $y = (x , t) \in R^{n+1}$ where $x \in \overline{D}$ and $t \in T$. As an example of a homotopy function, consider:

$$H(y) = H(x, t) = (1 - t)\,G(x) + t\,F(x) \quad . \tag{2.1}$$

Of prime interest is the solution set to the equation

$$H(x, t) = 0 \quad . \tag{2.2}$$

Define

$$H^{-1} = \{y = (x, t) \mid H(x, t) = 0\ ,\ x \in \overline{D}\ ,\ t \in T\}$$

as that solution set. Also, for $t \in T$, let:

$$H^{-1}(t) = \{x \in \overline{D} \mid (x, t) \in H^{-1}\}$$

be the solution to (2.2) in x for a fixed t. For instance, if the homotopy is of the form (2.1), then $H^{-1}(0)$ is the solution set to

$$G(x) = 0 \quad , \tag{2.3}$$

and $H^{-1}(1)$ is the solution set to

$$F(x) = 0 \quad . \tag{2.4}$$

Under moderate conditions the set H^{-1} is quite nice. Let H' be the $n \times n+1$ Jacobian matrix of H. We say H is regular if $H'(y)$ is of full rank for $y \in H^{-1}$. If H is continuously differentiable and regular, then we have that H^{-1} consists solely of differentiable paths [7]. Also, with $\overline{D}$ compact, there are only a finite number of such paths.

Given only slightly more differentiability, namely, H twice differentiable, Sard's theorem assures us that H is regular with probability one [10]. Thus, under very moderate conditions, H^{-1} consists of continuously differentiable paths.

These paths are extremely useful for solving problems. Generally, G is chosen to be simply and trivially solved, so that a point $x^0 \in H^{-1}(0)$ is easily obtained. The point x^0 will be on some path in H^{-1}. The idea is to follow that path until, hopefully, $t = 1$. But then we have obtained a point $x^1 \in H^{-1}(1)$ that solves (2.4). This concept of starting from an easily obtained point $x^0 \in H^{-1}(0)$ and following the path until we find

$x^1 \in H^{-1}(1)$ underlies the homotopy approach.

Using this approach and varying G and H permit us to solve a variety of problems. For instance, let:

$$H(x, t) = (1 - t)(x - x^0) + t(F(x) - x) .$$

Starting at $x = x^0$ for $t = 0$, if we follow the path until $t = 1$ we obtain a fixed point x^1 where

$$F(x^1) = x^1 .$$

In problems such as determining fixed points or solving systems of equations, we are primarily interested in finding $x^1 \in H^{-1}(1)$. For other applications, we might be interested in the entire path from $t = 0$ to $t = 1$. Whence the homotopy algorithms such as [2], [3], [13], [16] are no longer applicable. In structural mechanics that path indicates how a system responds to the force or stress placed on it. Also, t could be a time parameter and the path would indicate how an economic equilibrium or other equilibrium changes dynamically over time. In such a situation, H is not a homotopy system, but rather a dynamical system [8].

In any case we observe the idea of following the path from $x^0 \in H^{-1}(0)$ until we reach $x^1 \in H^{-1}(1)$. Of course, not all paths from $t = 0$ reach $t = 1$, so we must be careful in structuring the problem [6]. Nevertheless, the algorithms emphasized in this paper are designed to follow a path in H^{-1} by making a PL approximation of it. Such algorithms are called PL algorithms.

Another technique to follow a path in H^{-1} is via a differential equation [5], [7], [11]. Although there are many means to accomplish this, one is of prime interest to us here because it will directly affect our PL algorithm. Let p be a parameter indicating distance travelled along a path H^{-1} from some starting point on the path. Also define H'_{-i} to be the $n \times n$ matrix formed by deleting column i from the Jacobian H'. That is, letting

$\frac{dH}{dy_i}$ be a column of the Jacobian

$$H'_{-i} = \left(\frac{dH}{dy_1}, \ldots, \frac{dH}{dy_{i-1}}, \frac{dH}{dy_{i+1}}, \ldots, \frac{dH}{dy_{n+1}} \right) .$$

Finally, specify the derivative

$$\dot{y} = \frac{dy}{dp} \tag{2.5}$$

where $\dot{y}_i = \dot{x}_i \quad i = 1, \ldots, n$ and $\dot{y}_{n+1} = \dot{t}$.

Now consider the differential equations, where det is determinant,

$$\dot{y}_i = (-1)^i \det (H'_{-i}) \qquad i = 1, \ldots, n+1 \ . \tag{2.6}$$

These differential equations are termed the Basic Differential Equations (BDE) . It was shown in [7] that given an initial point $y^0 = (x^0, t^0) \in H^{-1}$, a solution to the BDE $y(p)$ yields a path in H^{-1} as p changes. Furthermore, the vector $\dot{y}(p)$ is the tangent to such a path and points in the direction of increasing p .

These differential equations not only provide another means to follow the path but, as will be seen shortly, are a key to our convergence proof.

3. ALGORITHMIC PROCEDURES

The algorithm we are suggesting operates by taking a PL approximation to a path in H^{-1} . It is thus like other PL algorithms in that regard. However, its choice of simplices is different, as we will see.

Let

$$\sigma = \{v^i\}_{i=1}^{n+2}$$

denote a simplex σ in $\overline{D} \times T$ with $n+2$ vertices v^i , $i = 1, \ldots, n+2$. Given a homotopy function H , define a piecewise linear approximation of it, $G_\sigma : \sigma \to R^n$, on the simplex σ by

$$G\left[\sum_{i=1}^{n+2} \lambda_i v^i\right] = \sum_{i=1}^{n+2} \lambda_i H(v^i) \tag{3.1}$$

$$\sum_{i=1}^{n+2} \lambda_i = 1 \ , \qquad \lambda_i \geq 0 \quad \text{all } i \ ,$$

where note

$$y \equiv \sum_{i=1}^{n+2} \lambda_i v^i \in \sigma .$$

Thus, G_σ is H on the vertices of σ, and is linear on σ.

We want solutions to (2.2), and since G_σ approximates H, we are also interested in solutions to

$$G_\sigma(y) = 0 \quad . \qquad (3.2)$$

Define that solution set

$$G_\sigma^{-1} = \{y = (x, t) \mid G_\sigma(y) = 0 , \; y \in \sigma\} .$$

The solutions to (3.2) have been studied intensively [4]. Suppose $y \in G_\sigma^{-1}$. We say G_σ is regular if any $y \in G_\sigma^{-1}$ is either interior to σ or interior to a facet (an n-face) of σ. If G_σ is regular, then the solution set G_σ^{-1} is a straight line segment across σ. The line segment goes from the interior of one facet to the interior of another facet.

PL algorithms operate by starting with an initial solution to (3.2) for an initial simplex σ_0. Then, under regularity, they generate a PL path which solves (3.2) on each simplex. For example, let us suppose the path is generated from facet

$$\beta_{k-1} = \{v^i\}_{i=1}^{n+1} \quad \text{to} \quad \beta_k = \{v^i\}_{i=2}^{n+2}$$

across simplex $\sigma_k = \{v^i\}_{i=1}^{n+2}$. Here, v^1 is assumed to be the vertex dropped. Now a new vertex $\bar{v}^1$ is selected on the other side of the hyperplane through β_k from v^1.
The new simplex is then $\sigma_{k+1} = \{\bar{v}^1 , v^2 , \ldots , v^{n+2}\}$. The PL solution path to (3.2) then cuts across σ_{k+1} from β_k to another facet β_{k+1} of σ_{k+1}. This describes how a PL algorithm operates as the process continues in this manner from simplex to simplex. Clearly a PL path is thus generated. Moreover, the PL path is regular if it is on each simplex.

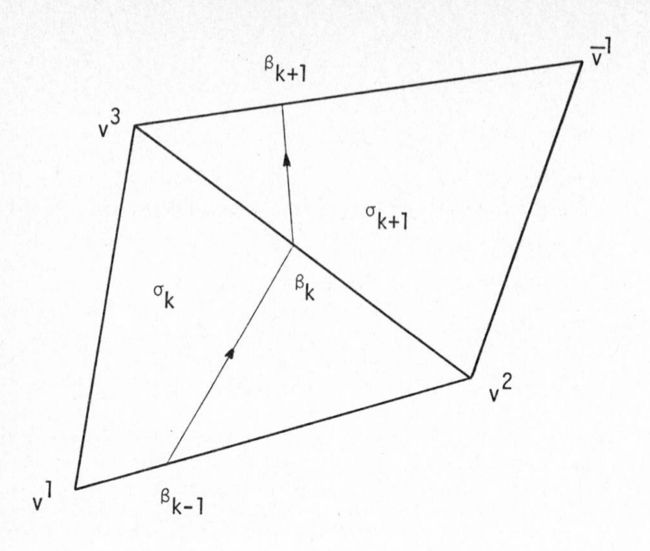

Figure 1 .

The difference between the flex simplicial algorithm of this paper and other PL algorithms is in how the new vertex $\bar{v}^1$ is selected. Previous PL algorithms select $\bar{v}^1$ according to a fixed predetermined grid. That is, once one arrives at facet β_k and has dropped v^1 , the new point $\bar{v}^1$ on the other side of β_k is already determined. In the flex simplicial algorithm, virtually any point on the other side of β_k can be selected for $\bar{v}^1$.

The only restriction we place on the selection of the new vertex $\bar{v}^1$ is that the new simplex σ_{k+1} maintain robustness. It is possible to have a simplex with the largest edge strictly positive yet with volume almost zero. A very flat skinny triangle could be an example. Such sickly shaped simplices can cause difficulties. Let $\ell(\sigma)$ be the length of the longest edge of the simplex σ , and $s(\sigma)$ the volume. Given an $\alpha > 0$, we require

$$\frac{s(\sigma)}{[\ell(\sigma)]^{n+1}} \geq \alpha \quad . \tag{3.3}$$

A simplex that satisfies this condition is called robust. Robustness is usually an easy condition to satisfy and, in fact, criteria other than (3.3) can be used for it [6].

Consider the G_σ defined in (3.1) and its Jacobian G'_σ. As will be discussed, if $\ell(\sigma)$ is sufficiently small, we can make G_σ arbitrarily close to H. Moreover, if σ is robust, we can make G'_σ arbitrarily close to H'.

To summarize, our algorithm operates the same as any other PL algorithm except in its choice of the new vertex $\overline{v}^1$ on the other side of the facet. That vertex can be selected arbitrarily as long as the new simplex formed maintains robustness.

Since we do not require a triangulation and the simplices generated can be on top of one another, the usual complementarity argument that prevents cycling cannot be applied. Indeed, without a triangulation, it would appear that the algorithm could cycle by generating similar points endlessly. In the next section we verify that cycling cannot occur.

4. PROOF OF CONVERGENCE

In this section, we establish that if the simplices generated are sufficiently small and robust, the proposed algorithm will converge. In the next section we elaborate on convergence for large simplices.

Given $D \subset R^n$ and $H : \overline{D} \times T \to R^n$ continuously differentiable, let us define a δ-tubular neighbourhood

$$N_\delta = \{y \in \overline{D} \times T \mid ||y - \overline{y}|| < \delta \ , \ \text{some } \overline{y} \in H^{-1}\} \tag{4.1}$$

where $||.||$ is the Euclidean norm. When δ is small, N_δ can be considered a collection of flexible tubes with each tube containing a path in H^{-1} running down its centre. The next theorem states that if a simplex is sufficiently small, any $y \in G_\sigma^{-1}$ must be in N_δ.

Theorem 4.1: Consider a continuously differentiable $H : \overline{D} \times T \to R^n$ for open, bounded $D \subset R^n$. Given $\delta > 0$, if a simplex σ is sufficiently small, and $y \in G_\sigma^{-1}$ then $y \in N_\delta$.

<u>Proof</u>: The set $\overline{D} \times T \sim N_\delta$ is compact. Also, if $y \in \overline{D} \times T \sim N_\delta$, then $||H(y)|| > 0$. Thus, for some $\varepsilon > 0$,

$$\min \{ ||H(y)|| \mid y \in \overline{D} \times T \sim N_\delta \} > \varepsilon > 0 .$$

Verbally, if y is at least a distance δ from all points $\overline{y}$ in H^{-1}, then $||H(y)|| > \varepsilon$.

It follows, that if

$$||H(y)|| \leq \varepsilon , \quad y \in \overline{D} \times T , \tag{4.2}$$

then

$$||y - \overline{y}|| < \delta$$

for some $\overline{y} \in H^{-1}$.

Now let $\sigma = \{v^i\}_1^{n+2}$ be any simplex in $\overline{D} \times T$. Via (3.1)

$$||H(y) - G_\sigma(y)|| \leq \sum_1^{n+2} \lambda_i \, ||H(y) - H(v^i)|| ,$$

for any $y = \sum_1^{n+2} \lambda_i v^i$, $\sum_1^{n+2} \lambda_i = 1$, $\lambda_i \geq 0$ all i. Since H is uniformly continuous in the compact set $\overline{D} \times T$,

$$||H(y) - G_\sigma(y)|| \leq \varepsilon \tag{4.3}$$

if $\ell(\sigma)$ is sufficiently small. It follows if $y \in G_\sigma^{-1}$, $G_\sigma(y) = 0$, so

$$||H(y)|| \leq \varepsilon .$$

But then, via (4.2)

$$||y - \overline{y}|| < \delta , \quad \text{for some } \overline{y} \in H^{-1} .$$

Hence, if $y \in G_\sigma^{-1}$, and σ is sufficiently small, $y \in N_\delta$. □

Suppose we generate a PL path where $G_\sigma(y) = 0$ for each simplex σ on that path. If the simplices are sufficiently small, then by Theorem 3.1, the PL path stays within a distance δ from an underlying path in H^{-1}. But staying close to an underlying path does not prevent it from cycling. To show the PL path does not cycle requires an orientation argument.

Orientation

By orientation of an underlying differentiable path, we mean the vector $\dot{y}$ which is tangent to the path and obtainable from the BDE. The PL path is not differentiable as a whole but is continuously differentiable (actually linear) on any simplex σ. Thus, at least on any simplex σ, the PL has orientation $\dot{y}^\sigma$ where, from the BDE,

$$\dot{y}_i^\sigma = (-1)^i \det(G_\sigma')_{-i} \qquad i = 1, \dots, n+1 . \tag{4.4}$$

Here $(G_\sigma')_{-i}$ is the $n \times n$ matrix remaining after deleting the i-th column of the Jacobian G_σ'. The solution $y^\sigma(p)$ to (4.4) will, of course, give the line segment $y^\sigma(p) \in G_\sigma^{-1}$ across σ as p changes.

The vector $\dot{y}^\sigma(p)$ is the tangent to $y^\sigma(p)$ as p increases. Yet this gives rise to a difficulty. It is not clear before examining (4.4) in which direction the path is traversed as p increases along a solution path. The path is a two-way street. In fact, just as solving (4.4) with $\dot{y}^\sigma$ describes a solution path, solving it with $\underline{\dot{y}}^\sigma = -\dot{y}^\sigma$ yields the same path but with the path traversed in the opposite direction.

Since it is known beforehand which direction for p will be selected, it is possible that the direction of p given on a simplex σ_k might be inconsistent with that obtained on the next simplex σ_{k+1}. Figure 2 illustrates that situation. There, the facet β_k is in the hyperplane $by = c$. On σ_k, p increases from y^1 to $\bar{y}$, while on σ_{k+1}, p increases from y^2 to $\bar{y}$. This p in Figure 2 behaves inconsistently.

We must prove that p is consistent. If p increases from y^1 to $\bar{y}$, it should increase from $\bar{y}$ to y^2. If it decreases from y^1 to $\bar{y}$, it should decrease from $\bar{y}$ to y^2. Since b is the normal to the hyperplane, these conditions become, letting k denote σ_k and k+1 denote σ_{k+1},

$$\text{If } b.\dot{y}^k > 0 \text{ then } b.\dot{y}^{k+1} > 0 \text{ , or} \qquad (4.5)$$
$$\text{If } b.\dot{y}^k < 0 \text{ then } b.\dot{y}^{k+1} < 0 \text{ .}$$

In fact, we shall prove the stronger result

$$b.\dot{y}^k = b.\dot{y}^{k+1} \quad . \qquad (4.6)$$

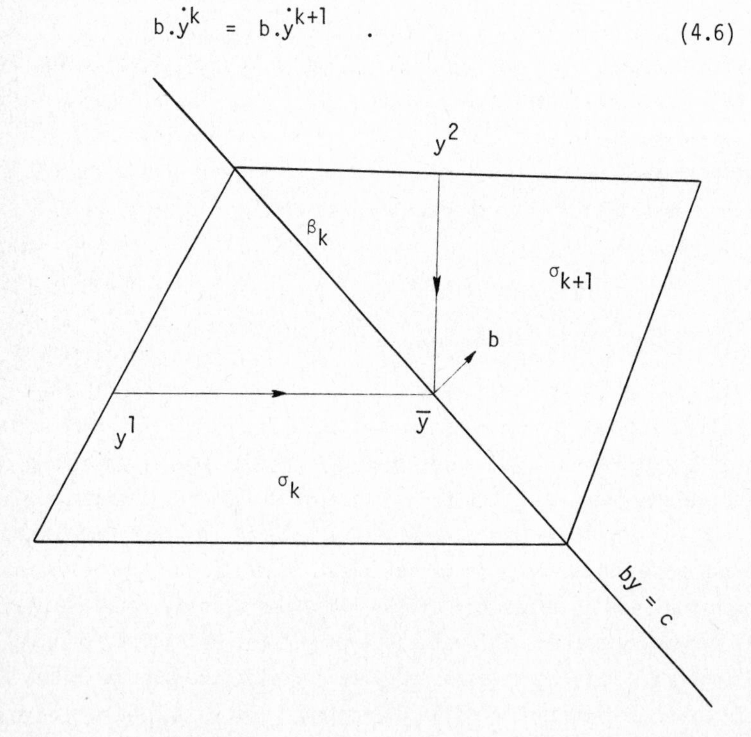

Figure 2 .

The arrows indicate $\dot{y}^k \equiv \dot{y}^{\sigma_k}$ and $\dot{y}^{k+1} \equiv \dot{y}^{\sigma_{k+1}}$, and the direction of increasing p on simplices σ_k and σ_{k+1}. Here $b.\dot{y}^k > 0$ and $b.\dot{y}^{k+1} < 0$. Thus, p is inconsistent.

To prove (4.6), we utilize some ideas in Eaves and Scarf [4]. First, by (4.4) note that for $j = k$ or $k+1$,

$$b.\dot{y}^j = \sum_{i=1}^{n+1} b_i \, (-1)^i \det(G_j')_{-i} \qquad (4.7)$$

$$= \det \begin{bmatrix} -b \\ \\ G_j' \end{bmatrix}$$

where $\begin{bmatrix} -b \\ \\ G_j' \end{bmatrix}$ is an $(n+1) \times (n+1)$ matrix, and we have simply expanded the determinant by its first row.

Recall the facet $\beta_k = \{v^i\}_2^{n+2}$ is in the hyperplane $by = c$, and being a facet, is n-dimensional. Thus the n vectors

$$z^i = v^i - v^2 \qquad i = 3, \ldots , n+2$$

are linearly independent. Also, since all the v^i are in the hyperplane $b.y = c$,

$$b.z^i = bv^i - bv^2 = c - c = 0 \qquad (4.8)$$

$$i = 3, \ldots , n+2 \; .$$

Letting $z = (z^i)$ be the $n+1 \times n$ matrix whose columns are the z^i , (4.8) becomes

$$bz = 0 \; . \qquad (4.9)$$

Moreover, the $(n+1) \times (n+1)$ matrix

$$(z, b)$$

must have full rank.

Notice, for $j = k$ or $k+1$,

$$\begin{bmatrix} -b \\ \\ G_j' \end{bmatrix} \begin{bmatrix} b, z \end{bmatrix} = \begin{bmatrix} -(b)^2 & -bz \\ & , \\ G_j'b & G_j'z \end{bmatrix} .$$

Yet using (4.9)

$$\det \begin{bmatrix} -b \\ \\ G_j' \end{bmatrix} = -\det (z, b)^{-1} (b)^2 \det (G_j' z) , \quad \text{for } j = k \text{ or } k+1 . \tag{4.10}$$

Thus, by (4.7) and (4.10), $b.\dot{y}^k = b.\dot{y}^{k+1}$ if and only if

$$\det (G_k' z) = \det (G_{k+1}' z) . \tag{4.11}$$

To establish (4.11), note that for $i = 3, \ldots , n+2$

$$G_k(v^i) = G_k(v^2) + G_k'(v^i - v^2) \tag{4.12}$$

and similarly for G_{k+1} . But $G_k(v^i) = G_{k+1}(v^i)$, for $i = 2, \ldots , n+2$ since by definition (3.1), G_k and G_{k+1} are equal on the facet β_k .

We thus see that (4.12) reduces to

$$G_k' z^i = G_{k+1}' z^i \qquad i = 3, \ldots , n+2 .$$

Or, in matrix form

$$G_k' z = G_{k+1}' z .$$

But this then verifies (4.11) and, hence, (4.6). We have just proved the following theorem.

Theorem 4.2: Suppose the algorithm generates a PL path which is regular on each simplex σ. Then on any simplex σ_k, $\dot{y}^k$ as expressed by (4.4) is the orientation of the path and p changes consistently from simplex to simplex.

□

We can thus rest assured that our orientation is well defined for the PL path. If we start from the initial simplex σ_0 and increase p, then increasing p will keep us moving along the PL path from simplex to simplex. Similarly, if we decrease p, decreasing p will move us along the path from simplex to simplex.

With the PL path $y^k(p)$ oriented consistently, we need only verify that its orientation is similar to an underlying path $y(p) \in H^{-1}$. That will verify that it cannot cycle.

If the simplices are robust and sufficiently small, say, $\ell(\sigma) \leq \overline{\ell}$ for some $\overline{\ell} > 0$ and all σ on the path, then G'_σ can be made arbitrarily close to H' on any σ (see [6]). Examining the BDE and (4.4), we observe that we can then ensure that for any y^k on the PL path and any y on the underlying path

$$\dot{y}\dot{y}^k > 0$$

if y and y^k are two nearby points. That is, the orientation of the PL path and the orientation of the underlying path are similar. In particular, the tangent to the PL path is always at an acute angle to the tangent to the differentiable path. But this provides our convergence proof. (See Figure 3 .)

Theorem 4.3: Given a continuously differentiable regular $H : \overline{D} \times T \to R^n$ with $D \subset R^n$ open bounded, suppose the algorithm generates a PL path that is regular. If the simplices are robust and sufficiently small, the PL path does not cycle but follows an underlying path $y(p) \in H^{-1}$.

Proof: Choose a δ-tubular neighbourhood of $y(p) \in H^{-1}$,

$$N_\delta(y(p)) = \{y \in \overline{D} \times T \mid \|y - y(\overline{p})\| < \delta, \text{ some } y(\overline{p}) \in y(p)\}.$$

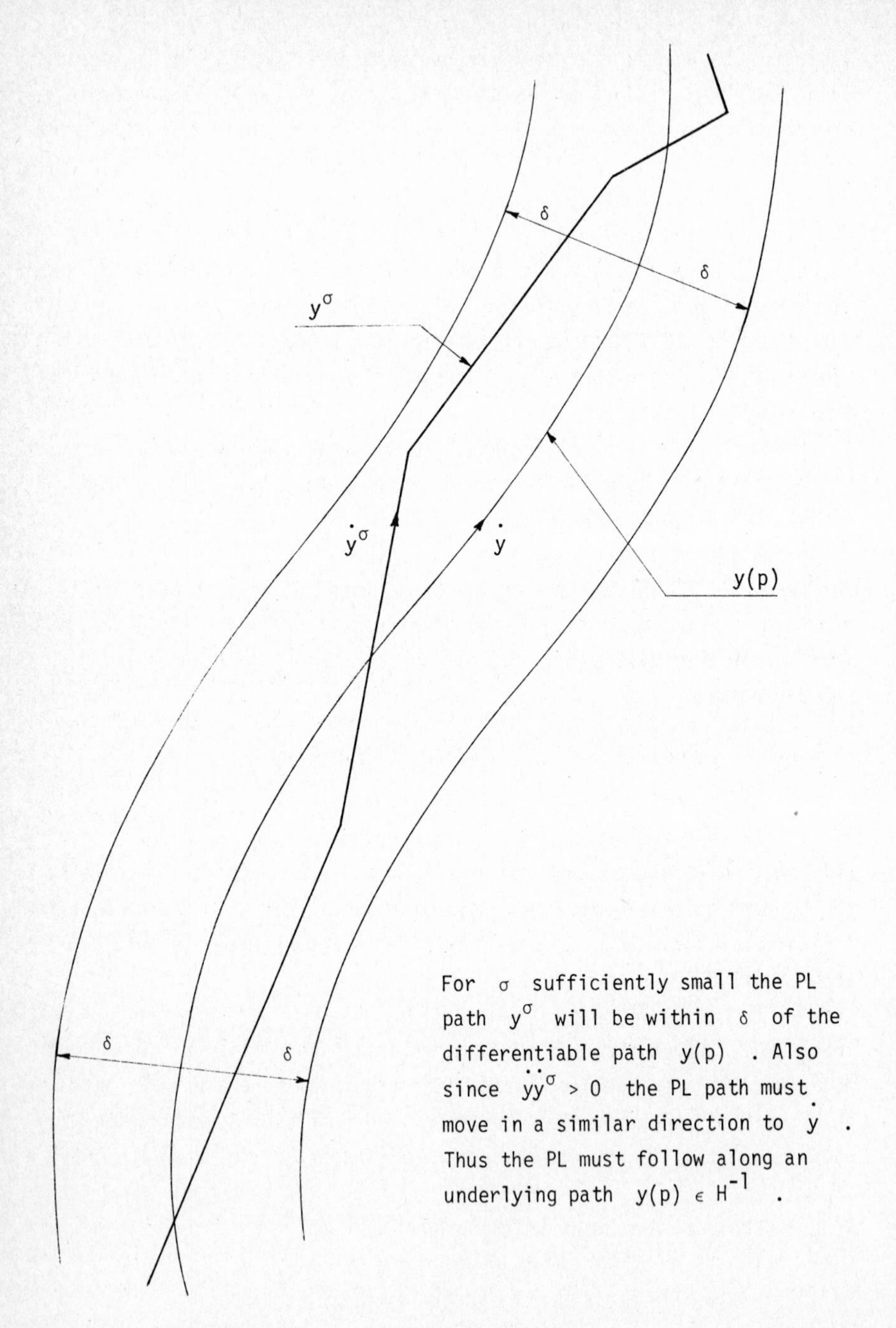

For σ sufficiently small the PL path y^σ will be within δ of the differentiable path $y(p)$. Also since $\dot{y}\dot{y}^\sigma > 0$ the PL path must move in a similar direction to $\dot{y}$. Thus the PL must follow along an underlying path $y(p) \in H^{-1}$.

Figure 3 .

Also choose an $\bar{\ell} > 0$ such that, if σ is a robust simplex in $N_\delta(y(p))$, and $\ell(\sigma) \le \bar{\ell}$, we have

$$\dot{y}\dot{y}^\sigma > 0 \tag{4.13}$$

where $y \in y(p)$ is any point sufficiently near σ, and $\dot{y}_i^\sigma$ satisfies (4.4). Moreover, select $\delta > 0$ small enough so that a δ-neighbourhood of any $y \in y(p)$ contains only a single piece of $y(p)$.

The PL path keeps within $N_\delta(y(p))$ for simplices robust and sufficiently small. It cannot cycle in this tubular neighbourhood, for then (4.13) would be violated. I.e., the two paths would have to go in roughly the opposite directions, which is impossible by (4.13). The PL path thus follows along the underlying path in H^{-1}.

□

For PL algorithms that utilize triangulations, the grid is constructed so that the simplices must be robust. Then, if the simplices are also sufficiently small, the corresponding PL path will also follow an underlying path in H^{-1}. For sufficiently small simplices, therefore, the convergence properties of triangulation PL algorithms and the flex simplicial algorithm presented here are the same.

5. LARGE SIMPLICES

For certain problems, for example in elasticity, economics, or catastrophe theory [8] it is essential to follow the entire path in H^{-1} accurately. Then small simplices are essential. For other problems, say, calculating fixed points or solutions, it is only required to obtain a point $x^1 \in H^{-1}(1)$. For such problems, large simplices are often used, at least in the beginning of an algorithm, in an effort to hopefully get close to $t = 1$ rapidly. Near $t = 1$, the simplices are typically made smaller to obtain greater accuracy.

The flex simplicial algorithm also works for large simplices. As before, we selected the new point $\bar{v}^1$ on the other side of β_k so that the new simplex σ_k is robust. We also require that, as the iteration k gets large, the simplices σ_k become arbitrarily small. In triangulation PL algorithms such as in the Eaves approach [2], conditions are imposed that force the simplices to get arbitrarily small as $k \to \infty$. Since the flex simplicial algorithm has no grid or triangulation, we require explicitly that simplices get arbitrarily small.

Again, the flex simplicial algorithm will converge in the same sense as a triangulation PL algorithm as long as H^{-1} has no loops. A loop is a path that is entirely in the interior of $\bar{D} \times T$ and if we follow it around from any point on it, we return to that point.

Theorem 5.1: Let a continuously differentiable and regular $H : \bar{D} \times T \to R^n$ where $D \subset R^n$ is open and bounded be given. Suppose the flex simplicial algorithm generates robust simplices and the path is regular. Let the size of the simplex σ_k get arbitrarily small as the iteration $k \to \infty$. Then, if H^{-1} has no loops, the flex simplicial algorithm converges.

Proof: As $k \to \infty$ the simplices get arbitrarily small. Via Theorem 4.1 , they must get within a δ-tubular neighbourhood of a path in H^{-1} . Also, the PL path must be oriented in the same direction as the path in H^{-1} . Since by assumption there are no loops, the PL path must either terminate at a boundary of $\bar{D} \times T$ or get into the δ-tubular neighbourhood of a path that is heading for a boundary of $\bar{D} \times T$.

□

6. OBSERVATIONS

1) In solving a particular problem, say to find a fixed point or a solution to a system of equations, the homotopy is often set up so that a path in H^{-1} must go to $t = 1$. Then, a PL algorithm, whether a triangulation one or a flexible one, will latch onto such a path and find an (approximate) point $x^1 \in H^{-1}(1)$. Artificial labellings or special boundary conditions are commonly used to ensure this behaviour.

2) While σ_k is still large, the flex simplicial algorithm might keep revisiting a region. Also, there are instances when it may be difficult to get σ_k smaller and maintain robustness. In those situations, imposing a "local" triangulation grid that it must follow might be helpful until the

algorithm departs from that region. Indeed, this use of a "local" grid when necessary may permit the flex simplicial algorithm to also handle loops in H^{-1} .

3) The regularity assumptions are quite mild and means to overcome them have been well discussed in the literature [4], [7]. Indeed, the fact that the vertices in the flex simplicial algorithm can be selected as the algorithm iterates may make it easier to ensure regularity of the PL path. For instance, if the path intersects a lower dimensional face of a simplex with one choice of a vertex, it might not with another. Conversely, deliberately selecting a vertex so that the path is not regular might aid in the study of bifurcations.

4) The flex simplicial algorithm will also work if H is merely continuous. Of course, then the underlying paths in H^{-1} need not exist. However, we may approximate H by functions that are continuously differentiable and regular in the usual manner [7].

5) There could be instances where the flex simplicial algorithm may perform better on polyhedra other than simplices. For example, if H is separable, i.e., there are no crossproduct terms in the definition of each $H_i(y)$, then rectangular solids, rather than simplices, could be the figures generated by the algorithms.

6) The flex simplicial algorithm permits the vertex points to be selected rather freely. Depending upon the specific form of the functions, it might be possible to select these points astutely and thus speed convergence.

REFERENCES

1. Allgower, E.; Georg, K., Simplicial and Continuation Methods for Approximating Fixed Points, Institute for Applied Mathematics, University of Bonn, August 1978.

2. Eaves, B.C., Homotopies for Computation of Fixed Points, Math. Programming 3, 1-22 (1972).

3. Eaves, B.C.; Saigal, R., Homotopies for Computation of Fixed Points on Unbounded Regions, Math, Programming 3, 225-237 (1972).

4. Eaves, B.C.; Scarf, H.E., The Solution of Systems of Piecewise Linear Equations, Math. Oper. Res. 1, 1-27 (1976).

5. Garcia, C.B.; Gould, F.J., A Theorem on Homotopy Paths, Math. Oper. Res. 3, 282-289 (1978).

6. Garcia, C.B.; Zangwill, W.I., Finding All Solutions to Polynomial Systems and Other Systems of Equations, Math. Programming 16, 159-176 (1979).

7. Garcia, C.B.; Zangwill, W.I., An Approach to Homotopy and Degree Theory, to appear in Math. Oper. Res.

8. Garcia, C.B.; Zangwill, W.I., Path Following for Catastrophe Theory, to appear in Proceedings of a Symposium on Analysis and Computation of Fixed Points, S. Robinson (Ed.), Academic Press, New York.

9. Gould, F.J.; Tolle, J.W., A Unified Approach to Complementarity in Optimization, Discrete Mathematics 7, 225-271 (1974).

10. Guillemin, V.; Pollack, A., Differential Topology, Prentice Hall, New York 1974.

11. Kellogg, R.B.; Li, T.Y.; Yorke, J., A Constructive Proof of the Brouwer Fixed Point Theorem and Computational Results, SIAM J. Numerical Analysis 13, 473-483 (1976).

12. Kojima, M., On the Homotopic Approach to Systems of Equations With Separable Mappings, Mathematical Programming Study 7, 170-184 (1978).

13. Kuhn, H.W.; MacKinnon, J.G., The Sandwich Method for Finding Fixed Points, Journal of Optimization Theory and Its Applications 17, 189-204 (1975).

14. Lemke, C.E., Bimatrix Equilibrium Points and Mathematical Programming, Management Science 11, 681-689 (1965).

15. Lemke, C.E.; Howson, J.T., Equilibrium Points of Bimatrix Games, SIAM 12, 413-423 (1964).

16. Merrill, O.H., Applications and Extensions of an Algorithm that Computes Fixed Points of Certain Upper Semi-Continuous Point-To-Set Mappings, Ph.D. Thesis, University of Michigan 1972.

17. Scarf, H.E., The Approximation of Fixed Points of Continuous Mappings, SIAM 15, 1328-1343 (1967).

18. Todd, M.J., Improving the Convergence of Fixed Point Algorithms, Mathematical Programming Study 7, 151-169 (1978).

19. Zangwill, W.I., An Eccentric Barycentric Fixed Point Algorithm, Math. Oper. Res. 2, 343-359 (1977).

Numerical Solution of Highly Nonlinear Problems
W. Forster (ed.)
© North-Holland Publishing Company, 1980

AN N-DIMENSIONAL BISECTION METHOD FOR SOLVING SYSTEMS OF N EQUATIONS IN N UNKNOWNS

Martin Stynes

Colaiste na hOllscoile,
Corcaigh, Ireland

Let P^n be an n-polyhedron in R^n with $F: P^n \to R^n$ continuous. Assume that the topological degree $d(F, P^n, \Theta^n)$ (Θ^n: origin in R^n) is defined and nonzero. An algorithm is described which generates a sequence $\{x_k\}$ of points in P^n with $\text{dist}(x_k, F^{-1}(\Theta^n)) \to 0$ as $k \to \infty$. This algorithm is a direct generalisation of the classical one-dimensional bisection method. An analysis of the type of points in $F^{-1}(\Theta^n)$ which can be found by the method is included.

1. INTRODUCTION

Let $[a , b] \subset R$ with $F: [a , b] \to R$ continuous. The classical bisection method [3, p. 220] for approximating a solution of the equation $F(x) = 0$ assumes that $F(a) F(b) < 0$ and proceeds by repeatedly bisecting subintervals of $[a , b]$ on which F has the same change of sign property. Compared with the secant method or Newton's method this method has a slow rate of convergence to a root of $F(x) = 0$, but has the advantage that such convergence is guaranteed.

In this paper we present a generalisation of the classical bisection method to the n-dimensional case. Let P^n be an n-polyhedron (definition in §2) in R^n, with $F: P^n \to R^n$ continuous. We assume that the topological degree $d(F, P^n, \Theta^n)$ (Θ^n: origin in R^n) is defined and nonzero (for a discussion of topological degree see [9]). When $n = 1$ this becomes the assumption above that $F(a)\,F(b) < 0$. We repeatedly bisect certain n-simplexes in P^n (those which lie near an approximate solution of $F(x) = \Theta^n$) and obtain a sequence $\{x_k\}$ of points in P^n for which

$$\mathrm{dist}(x_k, F^{-1}(\Theta^n)) = \inf_{y \in F^{-1}(\Theta^n)} ||x_k - y||_2 \to 0 \text{ as } k \to \infty .$$

The rate of convergence is in general expected to be inferior to say Newton's method, but this convergence is guaranteed.

The assumption that $d(F, P^n, \Theta^n) \neq 0$ cannot be discarded. For $n = 1$, to say that $d(F, P^1, \Theta^1) = 0$ means that $F(a)\,F(b) > 0$ and the bisection of subintervals of $[a, b]$ may or may not yield an approximate solution to $F(x) = 0$. In certain well-known situations we automatically have $d(F, P^n, \Theta^n) \neq 0$, e.g. in finding Brouwer fixed points of a continuous function F: convex $P^n \to \text{int } P^n$ (the interior of P^n), where we take $F = I - G$, I the identity map [9, pp. 34-35].

Several algorithms for computation of Brouwer and related fixed points exist (see e.g. [1], [5], [8], [10], [16]). In general their convergence proofs do not use degree theory (note however [11]) although certain results have been obtained (see e.g. [2], [12]) using this concept. Nevertheless extra simplifying assumptions such as differentiability have usually been introduced, whereas here we work with the minimum of hypotheses.

2. BACKGROUND MATERIAL

Definitions: The $m+1$ points $v_0, v_1, \ldots, v_m$ in R^n, $n \geq m \geq 1$, are said to be linearly independent if the vectors $v_1 - v_0, v_2 - v_0, \ldots, v_m - v_0$ are linearly independent in the usual sense. An m-simplex S^m is the closed convex hull of $m+1$ linearly independent points $v_0, v_1, \ldots, v_m$. These points are called its vertices. We write $S^m = (v_0, v_1, \ldots, v_m)$. The closed convex hull of any r $(1 \leq r \leq m+1)$ of these points is an $(r-1)$-simplex and is referred to as an $(r-1)$-dimensional face of the original m-simplex. A 1-dimensional face of a simplex is called an edge. An m-polyhedron P^m is the union of a finite number of m-simplexes such that the intersection of every two of these simplexes is either empty or a common face. We say that the m-simplexes form a triangulation of P^m.

For any m-simplex $S^m = (x_0 \, x_1 \ldots x_m)$ define its diameter to be

$$d(S^m) = \max_{0 \leq i, j \leq m} ||x_i - x_j||_2 .$$

Define the mesh of a triangulation of P^m to be the maximum diameter attained by its m-simplices.

Given any edge $(x_i \, x_j)$ of S^m, by "divide S^m using $(x_i \, x_j)$ " we mean "replace $S^m = (x_0 \ldots x_i \ldots x_m)$ by the two m-simplexes $(x_0 \ldots x_i \ldots x_{ij} \ldots x_m)$ and $(x_0 \ldots x_{ij} \ldots x_j \ldots x_m)$ where $x_{ij} = (x_i + x_j)/2$ ". By "bisect S^m " we mean "divide S^m using $(x_i \, x_j)$ where $||x_i - x_j||_2 = d(S^m)$ ".

For the rest of the paper, let P^n be a connected n-polyhedron in R^n with $F = (f_1, f_2, \ldots, f_n) : P^n \to R^n$ continuous, and suppose that $F(x) \neq \Theta^n$ for all $x \in b(P^n)$, the $(n-1)$-dimensional topological boundary of P^n. Then the topological degree $d(F, P^n, \Theta^n)$ of F at Θ^n is defined [9].

It is straightforward to show that $b(P^n)$ is an $(n-1)$-polyhedron. Any triangulation of P^n induces a triangulation of $b(P^n)$ in the obvious way. We say that we have a fine triangulation on P^n if for each S^{n-1} in the induced triangulation of $b(P^n)$ some f_i, $i \in \{1, 2, \ldots, n\}$, is nonzero on S^{n-1}.

In our algorithm we shall require that P^n be finely triangulated. This is in order that an approximate solution x_k to the equation

$F(x) = \Theta^n$ can be found at every stage of the algorithm. We defer until §6 the question of how one obtains a fine triangulation of P^n.

Proposition 2.1: Suppose we have a fine triangulation of P^n. Let $L_F : P^n \to R^n$ be the function obtained by defining $L_F(v_i) = F(v_i)$ for every vertex of every n-simplex in the triangulation of P^n and extending L_F affinely over these simplexes. Then $d(L_F , P^n , \Theta^n)$ is defined and $d(L_F , P^n , \Theta^n) = d(F , P^n , \Theta^n)$.

Proof: Let S^{n-1} be any (n-1)-simplex in $b(P^n)$. Since P^n is finely triangulated, some f_i, $i \in \{1 , 2 , .. , n\}$, say f_I, is nonzero on S^{n-1}. Writing $L_F = (\ell_1 , \dots , \ell_n)$ it follows from the definition of L_F that $\ell_I(y) f_I(y) > 0$ for all y in S^{n-1}. Hence $L_F \neq \Theta^n$ on S^{n-1}, so $L_F \neq \Theta^n$ on $b(P^n)$ and $d(L_F , P^n , \Theta^n)$ is defined [9].

Consider the homotopy $H: P^n \times [0 , 1] \to R^n$ defined by $H(x , t) = t F(x) + (1 - t) L_F(x)$. Write $H = (h_1 , \dots , h_n)$. Then $h_I \neq 0$ on S^{n-1}, hence $H(x , t) \neq \Theta^n$ for all $x \in b(P^n)$, $0 \le t \le 1$. From the homotopy invariance of degree [9, Theorem 2.1.2 (2)] it follows that $d(L_F , P^n , \Theta^n) = d(F , P^n , \Theta^n)$.

□

Definition: Let $c : [0 , \infty) \to R^n$ be defined by $c(t) = (t , t^2 , .. , t^n)$, $t > 0$. We say that an n-simplex S^n is ε-complete [5], [1, p. 39] if there exists $\varepsilon_0 > 0$ such that $c(t) \in S^n$ for $0 \le t \le \varepsilon_0$.

Note that if $\Theta^n \in \text{int } S^n$ then S^n is ε-complete.

Lemma 2.2: Let S^n be an n-simplex in R^n which is not ε-complete. Then there exists $\varepsilon_0 > 0$ such that $c(t) \notin S^n$ for $0 < t \le \varepsilon_0$.

Proof: Suppose the conclusion is false. Then there exist sequences $\{t'_n\}$, $\{t''_n\}$ of distinct points in $[0 , \infty)$ with $t'_n \to 0$ as $n \to \infty$, and $c(t'_n) \in S^n$, $c(t''_n) \notin S^n$ for all n. Hence $\{c(t) : t > 0\}$ must intersect $b(S^n)$ in an infinite number of points. In particular some (n-1)-dimensional face of S^n must contain an infinite number of points of $\{c(t) : t > 0\}$. But this is impossible since n+1 distinct points $c(t_i)$, $i = 0 , 1 , \dots , n$, are linearly independent (their coordinate matrix $[a_{ij}]$, $a_{ij} = (t_i)^j$ for $0 \le i , j \le n$, is Vandermonde and hence [3, p. 369] has nonzero determinant).

□

Lemma 2.3: Let $L: P^n \to R^n$ be affine on every n-simplex S_j^n in a triangulation of P^n and assume that $d(L , P^n , \Theta^n)$ is defined. Let $L x = A_j x + b_j$ on each S_j^n , where A_j is an $n \times n$ matrix and $b_j \in R^n$. Then

$$d(L , P^n , \Theta^n) = \sum_{j \in J} \text{sign det } A_j$$

where $J = \{j : L(S_j^n)$ is an ε-complete n-simplex$\}$.

Proof: By Lemma 2.2 and the definition of ε-complete we can choose a $t^* > 0$ such that for $0 < t \le t^*$ and every S_j^n in the triangulation, $c(t) \in L(S_j^n)$ iff $L(S_j^n)$ is ε-complete. In fact the proof of Lemma 2.2 shows that we can assume that $L(S_j^n)$ ε-complete implies that $c(t) \in \text{int } L(S_j^n)$ for $0 < t \le t^*$.

By [9, Theorem 2.1.3] we can choose $t' \in (0 , t^*]$ such that

$$d(L , P^n , \Theta^n) = d(L , P^n , c(t'))$$

$$= \sum_{j \in J} d(L , S_j^n , c(t'))$$

[9, Theorem 2.2.1]

$$= \sum_{j \in J} \text{sign det } A_j$$

by [9, Definition 1.1.3], since $j \in J$ implies A_j invertible using the linear independence of any $n+1$ distinct points $c(t_i)$, $i = 0 , .. , n$.

□

Theorem 2.4: Assume the hypotheses of Proposition 2.1 . Let $L_F x = A_i x + b_i$ on each n-simplex S_i^n in the fine triangulation of P^n , where A_i is an $n \times n$ matrix and $b_i \in R^n$. Then

$$d(F , P^n , \Theta^n) = \sum_{i \in I} \text{sign det } A_i ,$$

where $I = \{i : L_F(S_i^n)$ is an ε-complete n-simplex$\}$.

Proof: Combine Proposition 2.1 and Lemma 2.3 .

□

This theorem is in fact Theorem 7.1 of Allgower and Georg [1]. It was obtained here independently (with a rather different proof). It can be regarded as a generalisation of the main theorem of [4] and of [2, Lemma 3], and as an extension of the definition of topological degree for differentiable functions [9, Definition 1.1.3]. Here we use it for the following.

Corollary 2.5: Assume the hypotheses of Proposition 2.1 . Then there are $|d(F , P^n , \Theta^n)| + 2k$ n-simplexes S_i^n such that $L_F(S_i^n)$ is ε-complete, where k is a non-negative integer.

Proof: Immediate from Theorem 2.4 on recalling that $i \in I$ implies that $\det A_i \neq 0$.

□

3. THE ALGORITHM

Let $\{S_i^n : 1 \le i \le N\}$ be a fine triangulation of P^n , so

$$P^n = \bigcup_{i=1}^{N} S_i^n .$$

Assume that $d(F , P^n , \Theta^n) \neq 0$.

We shall associate with every n-simplex S_m^n in this algorithm a non-negative integer C_m called its division counter. For all original simplexes set $C_m = 0$. In the course of the algorithm, whenever an n-simplex is bisected or created by the bisection of another n-simplex its division counter is set equal to zero. Whenever an n-simplex is divided or created by the division of another n-simplex the new n-simplex has division counter set equal to one more than that of its immediate ancestor. If $C_m = n$, then S_m^n is immediately bisected. In §4 it is shown that this device forces convergence of the algorithm.

We now give a description of the algorithm.

1. Set $m = 0$ and $C_i = 0$ for $1 \le i \le N$. Form a list of those n-simplexes S^n for which $L_F(S^n)$ is ε-complete. Choose an n-simplex S_i^n from this list.

2. Increase N to $N+1$. Bisect S_i^n into two n-simplexes S_i^n and S_N^n. Set $C_i = C_N = 0$.
3. For every n-simplex S_j^n , $j \neq i$, $j \neq N$, having as an edge the bisected edge of S_i^n , increase N to $N+1$ and using this edge divide S_j^n into two n-simplexes S_j^n and S_N^n . If this division is a bisection set $C_N = 0$, otherwise set $C_N = C_j + 1$. Set $C_j = C_N$.
4. Inspect all n-simplexes formed in steps 2 and 3. Hence form new lists of (i) n-simplexes S^n for which $L_F(S^n)$ is ε-complete (note that L_F is now defined with respect to the new triangulation of P^n) and (ii) n-simplexes S_k^n for which $C_k \geq n$.
5. If list (ii) of step 4 is non-empty, choose i such that $C_i \geq n$ and $d(S_i^n) = \max \{d(S_j^n) : C_j \geq n\}$, then go to step 2. Otherwise go to step 6 .
6. Choose i such that $L_F(S_i^n)$ is ε-complete. Find $x \in S_i^n$ such that $L_F\, x = \Theta^n$. Increase m to $m+1$. Set $x_m = x$. Go to step 2 .

<u>Remarks</u>: (i) In [1] a test is given for ε-completeness. By Proposition 2.1 one could more simply run the algorithm by searching for S^n such that $\Theta^n \in L_F(S^n)$. Recall of course that if $\Theta^n \in \operatorname{int} L_F(S^n)$, then $L_F(S^n)$ is ε-complete, and if $\Theta^n \notin L_F(S^n)$, then $L_F(S^n)$ is not ε-complete.

(ii) Step 5 can only return the algorithm to step 2 a finite number of times before proceeding to step 6 . This is so because of the proof of Theorem 4.2 and because any edge-bisection can increase the division counter of only those n-simplexes of greater diameter. We could replace n in step 5 by any positive integer; however computational experience in the two-dimensional case indicates this is in general optimal (unless F is differentiable and is known to have non-zero Jacobian at the root being found, in which case n should be replaced by a larger number).

(iii) By Corollary 2.5 and the assumption that $d(F , P^n , \Theta^n) \neq 0$ there is always at least one ε-complete n-simplex. One could attempt to run the algorithm even if $d(F , P^n , {}^n) = 0$, but then the existence of ε-complete n-simplexes is not guaranteed. See however the remark following Theorem 5.2 below.

(iv) In fact in step 6 one can choose x_m to be any point in S_i^n. However analytical arguments [13, Theorem 3.1] indicate that the specified x_m is optimal.

(v) Note that instead of defining the function L_F on P^n, we could attach an integer label to every vertex of every n-simplex in the triangulation of P^n as is done in many other algorithms (e.g. [6]). The the search for ε-complete n-simplexes becomes a search for n-simplexes whose associated labels are $\{0, 1, \ldots, n\}$. This may be worthwhile because of its relative simplicity, although a lessening of the rate of convergence to a root of F is possible. There is an appropriate analogue to Theorem 2.4 above [11, Theorem 1] and convergence of the algorithm is proven using Theorem 4.2 below and [6, Theorem 4.1] (with $f: C^n \to C^n$ replaced by $F: P^n \to R^n$). Moreover in this case precisely the same roots of F can be approximated as in Theorem 5.2 below, replacing $f: S^n \to S^n$ in [11, Theorem 4] by $F: P^n \to R^n$. This approach as well may be regarded as a direct generalisation of the classical bisection method.

4. CONVERGENCE OF THE ALGORITHM

Let $\varepsilon > 0$ be given. Let $D_\varepsilon = \{x \in P^n : \text{dist}(x, F^{-1}(\Theta^n)) \geq \varepsilon\}$. Assume that ε is so small that D_ε is non-empty. Let

$$m_\varepsilon = \min_{x \in D_\varepsilon} ||F(x)||_\infty .$$

Since D_ε is compact $m_\varepsilon > 0$. Fix an n-simplex S^n in the triangulation of P^n. Let $\omega_i(.)$ be the modulus of continuity of $f_i|_{P^n}$, $i = 1, 2, \ldots, n$.

<u>Lemma 4.1</u>: If $\Theta^n \in L_F(S^n)$ and $\max_{1 \leq i \leq n} \omega_i(d(S^n)) < m_\varepsilon$, then $S^n \subseteq P^n \backslash D_\varepsilon$.

<u>Proof</u>: Suppose not, i.e. suppose that there exists $q \in S^n \cap D_\varepsilon$. Then

$||F(q)||_\infty \geq m$, i.e. for some $i \in \{1, 2, \ldots, n\}$ we have $|f_i(q)| \geq m_\varepsilon$. But then for any $x \in S^n$

$$|f_i(x) - f_i(q)| \leq \omega_i(d(S^n)) < m_\varepsilon$$

so $f_i(x) = 0$. Thus f_i has constant sign on S^n . It follows that $\Theta^n \notin L_F(S^n)$, a contradiction.

□

This Lemma shows that in the algorithm "small" n-simplexes can only occur near roots of F .

Theorem 4.2: Let $\{x_m : m = 1, 2, 3, \ldots\}$ be a sequence of points generated by step 6 of the algorithm. Then a subsequence converges to a solution of $F(x) = \Theta^n$.

Proof: Write S^n_m for the n-simplex associated with x_m , $m = 1, 2, \ldots$, in step 6 of the algorithm. We show that

$$\inf_{m\geq 1} d(S^n_m) = 0 \quad ,$$

so that a sequence $\{S^n_{m_i} : i = 1, 2, \ldots\}$ can be chosen with $d(S^n_{m_i}) \to 0$ as $i \to \infty$. Then any convergent sequence $\{y_i\}$ with

$$y_i \in S^n_{m_i} \qquad \text{for } i = 1, 2, \ldots$$

must converge to a root of F by Lemma 4.1 . Such convergent sequences exist by compactness of P^n .

Suppose

$$\inf_{m\geq 1} d(S^n_m) = \delta > 0 .$$

Then as we see on examining the algorithm, any bisected edge has length at least δ . Consequently if an n-simplex T^n is divided into T^n_1 and T^n_2 , all edges in T^n_1 and T^n_2 which were not edges of T^n have their lengths bounded by $d_1 = \max\{d(T^n)/2 , [d^2(T^n) - \delta^2/4]^{1/2}\}$ [14, Lemma 5.2]. But since any bisected edge has length at least δ , $d(T^n) \geq \delta$. Hence $d_1(T^n) = [d^2(T^n) - \delta^2/4]^{1/2}$.

Let M be a fixed but arbitrary positive integer. Then we can choose a sequence of n-simplexes $\{T_r^n : r = 0, 1, 2, \ldots, M^n(n+1)\}$ such that T_r^n is obtained from T_{r-1}^n by one division or bisection for $r = 1, 2, \ldots, M^n(n+1)$. Let $T_0^n = (a_0, a_1, \ldots, a_n)$. Among the a_j, $j = 0, 1, \ldots, n$, suppose without loss of generality that only $a_0, a_1, \ldots, a_k$, $1 \le k \le n$, are endpoints of edges exceeding $d_1(T_0^n)$ in length. There is at least one bisection in every n+1 divisions, caused by step 5 of the algorithm. If more than one of $a_0, \ldots, a_k$ is a vertex, a bisection eliminates one of them. Thus at most one of $a_0, \ldots, a_k$ is a vertex of $T_{k(n+1)}^n$. By choice of k and definition of d_1, $d(T_{k(n+1)}^n) \le d_1(T_0^n)$. Hence $d(T_{n(n+1)}^n) \le d_1(T_0^n)$, since $\{d(T_r^n)\}$, $r = 1, 2, \ldots$ is clearly a monotonically decreasing sequence. Applying this inequality M times gives

$$\begin{aligned} d^2(T_{M(n+1)}^n) &\le d_1^2(T_{(M-1)(n+1)}^n) \\ &= d^2(T_{(M-1)(n+1)}^n) - \delta^2/4 \\ &\le d_1^2(T_{(M-2)(n+1)}^n) - \delta^2/4 \\ &\quad \cdots\cdots\cdots\cdot \\ &\le d^2(T_0^n) - M\,\delta^2/4 \quad . \end{aligned}$$

But M is arbitrary, and this inequality cannot hold if M is large. This contradicts the supposition that $\inf_{m \ge 1} d(S_m^n) > 0$. By the opening remarks we are done.

□

We note that if F is continuously differentiable with its derivatives DF satisfying a Lipschits condition, then from [13] one can make more specific statements about convergence.

<u>Theorem 4.3</u>: [13, Theorem 3.1] .
In the notation of Theorem 4.2 , $||F(x_m)||_\infty \le K\, d^2(S_m^n)/2$, where $K > 0$ is the Lipschitz constant.

Theorem 4.4: [13, Theorem 3.2] .
Suppose that the subsequence $\{y_i\}$ of $\{x_m\}$ converges to y , and $DF(y)$ is nonsingular. Then there exists $L \geq 1$ and $\Theta > 0$ such that for $i \geq L$, $||y_i - y||_\infty \leq \Theta\, d^2(S_i^n)$, where $L_F(S_i^n)$ is the ε-complete n-simplex of step 6 of the algorithm in which y_i lies.

Our computational experience (§7) indicates that in many cases there exists I such that for $i \geq I$, the S_i^n of Theorem 4.3 contains a root of F , and from this stage onwards the algorithm converges by repeated bisection of simplexes, i.e., $S_{i+1}^n \subset S_i^n$ for $i \geq I$. Of relevance here is the main result of [7]:

Theorem 4.5: [7, Theorem 3.1] .
If in the sequence S_0^n , S_1^n , ... of n-simplexes S_r^n is obtained by a bisection of S_{r-1}^n for $r = 1, 2, \ldots$, then

$$d(S_r^n) \leq (3/2)^{[r/n]} d(S_0^n)$$

for $r = 1, 2, \ldots$, where $[z]$ denotes the integer part of z .

5. ROOTS OF F FOUND BY THE ALGORITHM

In this section we show that certain solutions of $F(x) = \Theta^n$ can always be found by using a modified version of the algorithm. Even in the one-dimensional case the classical bisection method may not find all roots of F . An isolated solution at which F does not change sign would in theory only be discovered if it itself were sampled. On the other hand an isolated solution at which F does change sign will certainly be found if the triangulation of $[a, b] = P^1$ has sufficiently small mesh. As the theorems of this section illustrate, these two possiblities are in a sense typical.

Theorem 5.1: Let C be a connected component of $F^{-1}(\Theta^n)$. Let $U \subset P^n$ be an open neighbourhood of C whose closure $\bar{U}$ contains no other points in $F^{-1}(\Theta^n)$, so that $d(F , \bar{U} , \Theta^n)$ is defined. Then if the mesh μ of the triangulation of P^n is sufficiently small, there are $|d(F , \bar{U} , \Theta^n)| + 2k$ n-simplexes S^n in U such that $L_F(S^n)$ is ε-complete, where k is a non-negative integer.

Proof: Let

$$\varepsilon = \text{dist}(C , b(U)) = \inf_{x \in C} (\text{dist}(x , b(U)) .$$

Then $\varepsilon > 0$ by compactness of C . Suppose $\mu \leq \varepsilon/3$. Let Q^n be the n-polyhedron formed by taking the union of all n-simplexes in the triangulation of P^n which intersect the set $\{x \in P^n : \text{dist}(x , C) \leq \varepsilon/3\}$. Then $C \subset Q^n \subset U$, and $C \cap b(Q^n)$ is empty. In fact if $x \in b(Q^n)$, $\varepsilon/3 < \text{dist}(x , C) \leq 2\varepsilon/3$.

Let $m = \min \{||F(x)||_\infty : \varepsilon/3 \leq \text{dist}(x , C) \leq 2\varepsilon/3$. Then $m > 0$. If μ is sufficiently small, then $\omega_i(\mu) < m$, where $\omega_i(.)$ is the modulus of continuity of $f_i|_{\bar{U}}$, $i = 1 , 2 , \ldots , n$. Hence as in the proof of Lemma 4.1 it follows that some f_i is nonzero on each (n-1)-simplex S^{n-1} in $b(Q^n)$. We thus have a fine triangulation of Q^n . By Proposition 2.1 applied to Q^n instead of P^n

$$d(L_F , Q^n , \Theta^n) = d(F , Q^n , \Theta^n)$$

$$= d(F , \bar{U} , \Theta^n)$$

by [9, Theorem 2.2.1 (2)] since $F^{-1}(\Theta^n) \cap (\bar{U} \backslash Q^n)$ is empty. Applying Corollary 2.5 to Q^n instead of P^n gives the required result.

□

Theorem 5.2: Using the notation of Theorem 5.1 , if $d(F , \bar{U} , \Theta^n) \neq 0$ then for some $y \in C$ a sequence y_m : $m = 1 , 2 , \ldots$ of points with $y_m \to y$ as $m \to \infty$ can be generated by modifying the algorithm of §3 .

Proof: We give a version of the "impartial bisection method" of [15, §4] which ensures that the mesh of the triangulation of P^n approaches zero as the algorithm progresses (the notation is that of §3 above).

0. Set $m = 0$.
1. Set $r = 0$, $d = \max_{1 \le i \le N} d(S_i^n)$,

   increase m to $m+1$. Form a list of those n-simplexes S_ν^n , $\nu = 1 , 2 , \dots , N'$, for which $L_F(S_\nu^n)$ is ε-complete. Choose the unique point $x_{m,\nu}$ in each of the S_ν^n satisfying $L_F \, x_{m,\nu} = \Theta^n$.
2. Set $i = 1$, $k = N$, increase r to $r+1$. If then $r = n+1$ go to step 1 ; if $r < n+1$ go to step 3 .
3. Set $d_i = d(S_i^n)$. If $d_i \le 2d/\sqrt{5}$, increase i to $i+1$ and begin step 3 again; if $d_i < 2d/\sqrt{5}$ go to step 4.
4. Increase N to $N+1$. Bisect S_i^n into S_i^n and S_N^n . For every n-simplex S_j^n , $j \ne i$, $j \ne N$, having as an edge the bisected edge of the former S_i^n , increase N to $N+1$ and using this edge divide S_j^n into two n-simplexes S_j^n and S_N^n .
5. Inspect all n-simplexes formed in step 4. Hence form a new list of those n-simplexes S^n for which $L_F(S^n)$ is ε-complete.
6. If $i < k$, increase i to $i+1$ and go to step 3 ; if $i = k$ go to step 2 .

From [15, §4] it follows that if μ_0 is the mesh of the original triangulation of P^n and μ_m is the mesh of the triangulation when step 6 returns us to step 2 for the mth time, then $\mu_m \le (2/\sqrt{5})^m \mu_0$, $m = 1 , 2 , \dots$. Consequently $\mu_m \to 0$ as $m \to \infty$.

Let $A_m = \{x_{m,r} : r = 1 , 2 , \dots , N'\}$, $m = 1 , 2 , \dots$. If $z_m \in A_m$ for $m = 1 , 2 , \dots$ and $z_m \to z$, then $z \in F^{-1}(\Theta^n)$ by Lemma 4.1 and the fact that $\mu_m \to 0$. Given $\varepsilon > 0$, for all sufficiently large m there exists $y' \in A_m$ such that $\mathrm{dist}(y'_m , C) \le \varepsilon$ by Theorem 5.1 . Hence for some $y \in C$ there exists $\{y_m : m = 1 , 2 , \dots\}$ with $y_m \in A_m$ for every m and $y_m \to y$ as $m \to \infty$.

□

We observe that even if $d(F , P^n , \Theta^n) = 0$ the hypotheses of Theorem 5.2 may still be satisfied for some connected component C of $F^{-1}(\Theta^n)$. In this case the modified algorithm will yield approximations to a point of C .

Definition 5.3: Let C be a connected component of $F^{-1}(\Theta^n)$. We say that C is a locally extreme component of $F^{-1}(\Theta^n)$ if for some open set U with $C \subset U \subset P^n$,

$$\Theta^n = \sum_{i=0} \lambda_i F(x_i)$$

for $x_i \in U$ and $\lambda_i \in [0 , 1]$, $i = 0 , 1 , 2 , \ldots , n$, with

$$\sum_{i=0} \lambda_i = 1 ,$$

imply $x_i \in C$ for some i .

For example if $C = \{z\}$ and there exists $v \in R^n$ such that the inner product $\langle F(.) , v\rangle$ has a local minimum or maximum at z , then C is a local extreme component of $F^{-1}(\Theta^n)$. This is the case if any $f_i(.)$, $1 \le i \le n$, has an isolated local minimum or maximum at z . In the one-dimensional case a point which is a locally extreme component of $F^{-1}(0)$ is an isolated root of F at which F does not change sign.

Theorem 5.4: Let C be a local extreme component of $F^{-1}(\Theta^n)$. Then if S^n is any n-simplex of any triangulation of P^n with $d(S^n)$ and

$$\min_{x \in S^n , y \in C} ||x - y||_\infty$$

both sufficiently small, $\Theta^n \in L_F(S)$ iff some vertex of S^n lies in C .

Proof: If $d(S^n)$ and

$$\min_{x \in S^n , y \in C} ||x - y||_\infty$$

are both sufficiently small, then $S^n \subset U$, where U is the open set of Definition 5.3 . Suppose $\Theta^n \in L_F(S^n)$. Then

$$\Theta^n = \sum_{i=0} \lambda_i F(x_i)$$

where $0 \le \lambda_i \le 1$ for all i, $\sum_{i=0} \lambda_i = 1$, and the x_i are the vertices of S^n.
By Definition 5.3 some x_i lies in C as required.

The converse implication ($F(x_i) = \Theta^n$ implies $\Theta^n \in L_F(S^n)$) holds trivially.

□

Thus the probability of closely approximating any point of a locally extreme component C of $F^{-1}(\Theta^n)$ via our algorithm is zero, if C has zero-dimensional Lebesgue measure (as is usually the case).

Suppose C is a locally extreme component of $F^{-1}(\Theta^n)$, and let U be the open set of Definition 5.3 . Then we may assume that $\bar{U} \cap F^{-1}(\Theta^n) = C$, so $d(F, \bar{U}, \Theta^n)$ is defined. In fact $d(F, \bar{U}, \Theta^n) = 0$ as can be seen by combining Theorems 5.1 and 5.4 . However the converse is false for $n > 1$, i.e. $d(F, \bar{U}, \Theta^n)$ for U an open neighbourhood of a component C of $F^{-1}(\Theta^n)$ with $\bar{U} \cap F^{-1}(\Theta^n) = C$ does not imply that C is a local extreme component of $F^{-1}(\Theta^n)$. For example take P^2 to include in its interior the point Θ^2 and $F: P^2 \to R^2$ to be $F(x, y) = (x^2 - y, x^2 + y)$. Then $F^{-1}(\Theta^2) = \Theta^2$. It can be seen [9, Theorem 2.1.1 and 2.1.3] that $d(F, \bar{U}, \Theta^2) = 0$ for any open neighbourhood $U \subseteq P^2$ of Θ^2, since for arbitrary small $\varepsilon > 0$ the set $F^{-1}(-\varepsilon, 0)$ is empty. However $C = \{\Theta^2\}$ is not a locally extreme component of $F^{-1}(\Theta^2)$. For if $U \subset P^2$ is any neighbourhood of Θ^2, then for $\delta > 0$ sufficiently small the points $(0, \delta)$, $(0, -\delta)$ and $(\delta, 0)$ lie in U, and it is easy to check that

$$(1/2)\, F(0, \delta) + (1/2)\, F(0, -\delta) + (0)\, F(\delta, 0) = \Theta^2 .$$

Note however that if $n = 1$ locally extreme components and components of degree zero are one and the same.

6. FINE TRIANGULATIONS OF P^n

If $d(F, P^n, \Theta^n)$ is defined, i.e., if $F(x) = \Theta^n$ implies $x \notin b(P^n)$, then a fine triangulation of P^n can be obtained in a finite number of steps. This is done by implementing the algorithm of [15, §4] (which is quite similar to the modified algorithm of Theorem 5.2). This algorithm of [15] repeatedly bisects the (n-1)-simplexes of $b(P^n)$; here it must be modified by making these bisections induce divisions of the n-simplexes of P^n in the obvious way.

If we know that the f_i , $i = 1, 2, \ldots, n$, are Lipschitz continuous, and know moreover their Lipschitz constants, we can use the algorithm of [14]. This algorithm also must be modified as in the previous paragraph. Its advantage is that it automatically terminates when a fine triangulation of P^n has been reached, unlike the algorithm of [15].

In computing Brouwer fixed points one often has a fine triangulation *a priori*. Suppose that

$$P^n = \prod_{i=1}^{n} [a_i , b_i]$$

is an n-dimensional rectangular box with $G: P^n \to \text{int } P^n$ continuous, and that we want a fixed point of G . Define F by $F(x) = x - G(x)$, $x \in P^n$, so that fixed points of G are roots of F . Then $d(F, P^n, \theta^n)$ is defined and in fact equals 1 [9, Theorem 3.1.1] . Also for $j = 1, 2, \ldots, n$, $f_j = 0$ on the (n-1)-dimensional faces

$$Q_j^- = a_j \times \prod_{\substack{i=1 \\ i \neq j}}^{n} [a_i , b_i] \qquad \text{and}$$

$$Q_j^+ = b_j \times \prod_{\substack{i=1 \\ i \neq j}}^{n} [a_i , b_i]$$

of P^n . Hence any triangulation of P^n is in fact a fine triangulation.

Similarly for the non-linear complementarity problem, where P^n is as above, $H = (h_1, \ldots, h_n) : P^n \to R^n$ is continuous and satisfies $h_j > 0$ on $Q_j^- \cup Q_j^+$ for all j , and we want $x \geq 0$ in R^n such that $H(x) \geq 0$ and $\langle x , H(x)\rangle = 0$ (here " $\geq$ " means the usual coordinate partial ordering of R^n). For $y \in R^n$ define y^+ to be $(\max(y_1, 0), \ldots, \max(y_n, 0))$. Define $G: P^n \to R^n$ by $G(x) = (x - H(x))^+$. Then $x^* \in R^n$ solves our problem iff x^* is a fixed point of G . Define $F = (f_1, \ldots, f_n) : P^n \to R^n$ by $F(x) = x - G(x)$, so we require a root of F . Then for

$j = 1, 2, \ldots, n$, $f_j \neq 0$ on $Q_j^- \cup Q_j^+$ so any triangulation of P^n is in fact a fine triangulation.

7. COMPUTATIONAL EXPERIENCE

The algorithm was tested on an IBM 370 using a variety of two-dimensional problems from papers concerning other methods of solving systems of equations. Initial triangulations usually had mesh 4, and the program refined them if during an iteration (= passage through step 6 of the algorithm) no root of L_F was found. Computation terminated when a 2-simplex of diameter at most 10^{-4} was found containing a root of L_F. For roots of F with non-zero Jacobian, after 10 - 15 iterations a 2-simplex was usually found in which the root of F actually lay, and from then on the algorithm proceeded by repeatedly bisecting this 2-simplex. This generally took 30 - 45 iterations altogether. Increasing n in step 5 had little effect on the number of iterations, but decreased the work by returning the algorithm to step 2 less often. For roots of F without the benefit of a non-zero Jacobian the number of iterations always exceeded 100 and increasing n in step 5 increased the number of iterations required.

REFERENCES

1. Allgower, E.L.; Georg, K., Simplicial and Continuation Methods for Approximating Fixed Points (preprint).

2. Charnes, A.; Garcia, C.B.; Lemke, C.E., Constructive Proofs of Theorems Relating to $F(x) = y$ with Applications (report).

3. Dahlquist, G.; Bjorck, A., Numerical Methods , Prentice-Hall, Englewood Cliffs, New Jersey 1974.

4. Eaves, B.C., An Odd Theorem, Proc. A.M.S. 26, 509 - 513 (1970).

5. Eaves, B.C., Homotopies for Computation of Fixed Points, Math. Programming 3, 1 - 22 (1972).

6. Jeppson, M.M., A Search for the Fixed Points of a Continuous Mapping, in Mathematical Topics in Economic Theory and Computation, R.H. Day; S.M. Robinson (Eds.), SIAM, Philadelphia 1972.

7. Kearfott, R.B., A Proof of Convergence and an Error Bound for the Method of Bisection in R^n , Math. Comp. 32, 1147 - 1153 (1978).

8. Kellog, R.B.; Li, T.Y.; Yorke, J., A Constructive Proof of the Brouwer Fixed Point Theorem and Computational Results, SIAM J. Numer. Anal. 13, 473 - 483 (1976).

9. Lloyd, N.G., Degree Theory, Cambridge University Press, Cambridge 1978.

10. Merrill, O.H., Applications and Extensions on an Algorithm that Computes Fixed Points of Certain Upper Semi-Continuous Point to Set Mappings, Ph.D. Dissertation, Dept. of Industrial Engineering, U. of Michigan 1972.

11. Prufer, M., Sperner Simplices and the Topological Fixed Point Index, Sonderforschungsbereich 72, Universität Bonn, preprint no. 134 (1977).

12. Saigal, R., On Piecewise Linear Approximations to Smooth Mappings, Northwestern Univ. 1978 (preprint).

13. Saigal, R., On the Convergence Rate of Algorithms for Solving Equations that are Based on Complementarity Pivoting, Math. of Operations Research 2, 108-124 (1977).

14. Stynes, M., An Algorithm for Numerical Calculation of Topological Degree, Applicable Anal. (to appear).

15. Stynes, M., On the Construction of Sufficient Refinements for Computation of Topological Degree, Numer. Math. (submitted).

16. Todd, M.J., The Computation of Fixed Points and Applications, Springer-Verlag, New York 1976.

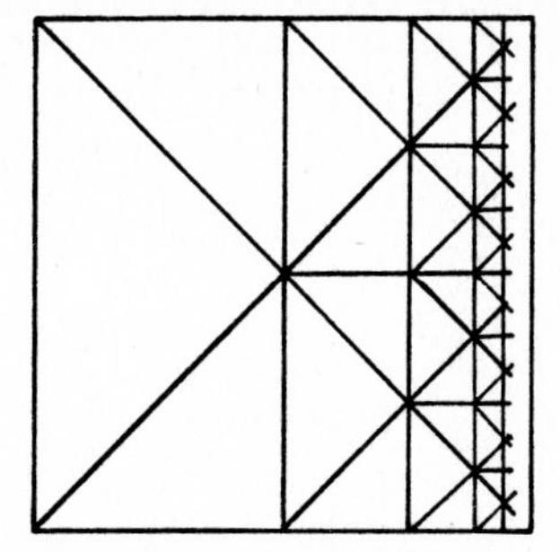

The papers in this section deal with various computational aspects.

Numerical Solution of Highly Nonlinear Problems
W. Forster (ed.)
© North-Holland Publishing Company, 1980

AN EFFICIENT WAY OF PROGRAMMING EAVES' FIXED POINT ALGORITHM

R. Jansen and A.S. Louter

Erasmus University Rotterdam,
Rotterdam, The Netherlands

1. ABSTRACT

The recently developed fixed point algorithms have provided the possibility of applying advanced numerical computation procedures in a broad field of economics. In particular, the pioneer work of Scarf [4], [5] has shown its importance in the computation of economic equilibria. Moreover, Eaves [3] has succeeded in constructing an algorithm based more or less on the same principles as used by Scarf but providing in general a more efficient way of computing fixed points. Therefore his algorithm may be preferred especially in its application to rather extensive problems.

This paper deals with the formulation of Eaves' method such that with a minimum of storage and computation time his algorithm may be prepared for the computer. The algorithm as described in [3] can substantially be improved by reformulating the replacement step and by application of an efficient inverse routine developed by Bartels [1] on the pivot step.

A very concise introduction to fixed point algorithms is given in section 2 , where the basic ideas behind this method are given, omitting for the sake of legibility the mathematical justifications. For those who are interested in the latter aspect we refer to the original publications.

In section 3 Eaves' method is considered with an explanation of the replacement step in section 3.1 and its reformulation in section 3.2 . Sections 3.3 and 3.4 are dealing with the pivot step and the labelling respectively, while in section 4 a complete FORTRAN computer program of the algorithm is given.

2. INTRODUCTION TO FIXED POINT ALGORITHMS

Since the algorithms of Scarf and Eaves are developed to approximate fixed points we first summarize two important fixed point theorems, which are often used to prove the existence of a general equilibrium in economic models.

Theorem 1 (Brouwer): If S is a nonempty, compact, convex subset of R^n and if f is a continuous function from S to S , then f has a fixed point, i.e. there is an $x^* \in S$ such that $f(x^*) = x^*$.

□

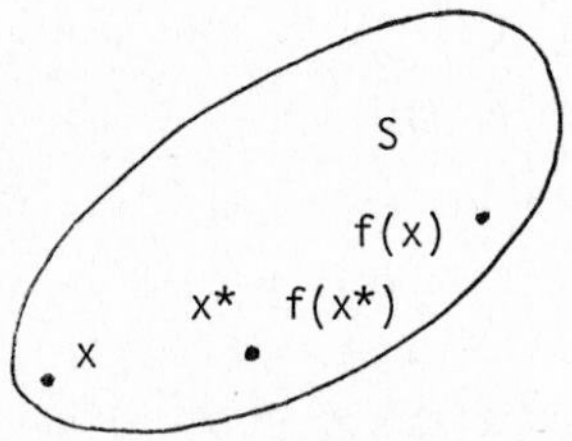

Figure 1A : Brouwer fixed point.

Theorem 2 (Kakutani): If S is a nonempty, compact, convex subset of R^n and if Φ is an upper semicontinuous correspondence from S to S such that for all $x \in S$ the set $\Phi(x)$ is convex (nonempty), then Φ has a fixed point, i.e. there is an $x^* \in S$ such that $x^* \in \Phi(x^*)$.

□

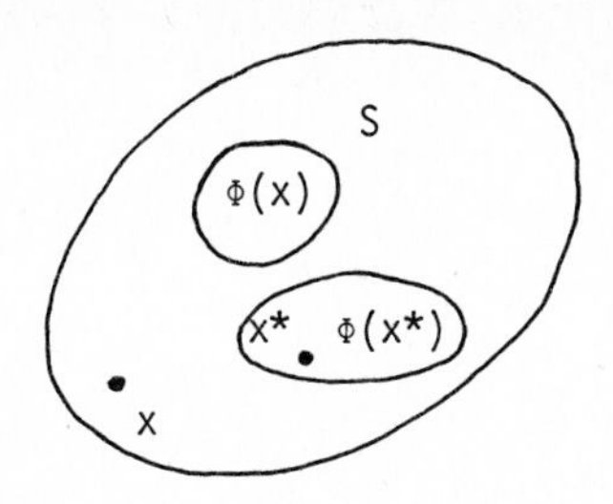

Figure 1B : Kakutani fixed point.

The algorithms of Scarf and Eaves are both dealing with unit simplices, defined as the collection of vectors $x = \{x_1, \ldots, x_n\}$ with $x_i \geq 0$ for all i, and

$$\sum_{i=1}^{n} x_i = 1 \quad .$$

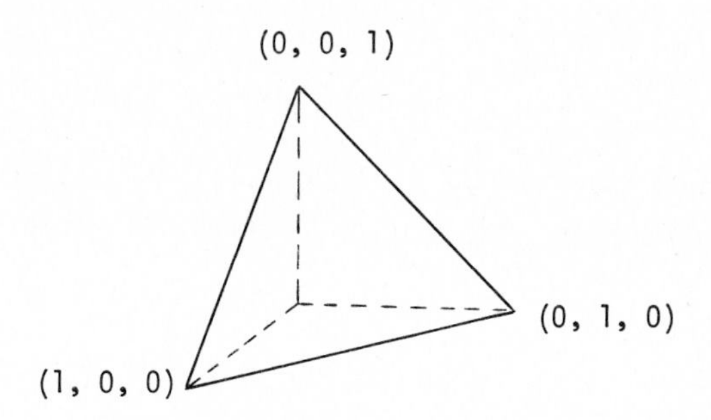

Figure 1C : Unit simplex.

On the simplex we define a collection of points, which play an important role in the search procedure for finding an approximate fixed point.

<u>Definition</u> 1 (Regular grid): A regular grid defined on the simplex S is the collection of points x in S of the form $x = (1/N)(m_1, \ldots, m_n)$ with m_i non-negative integers summing to N. □

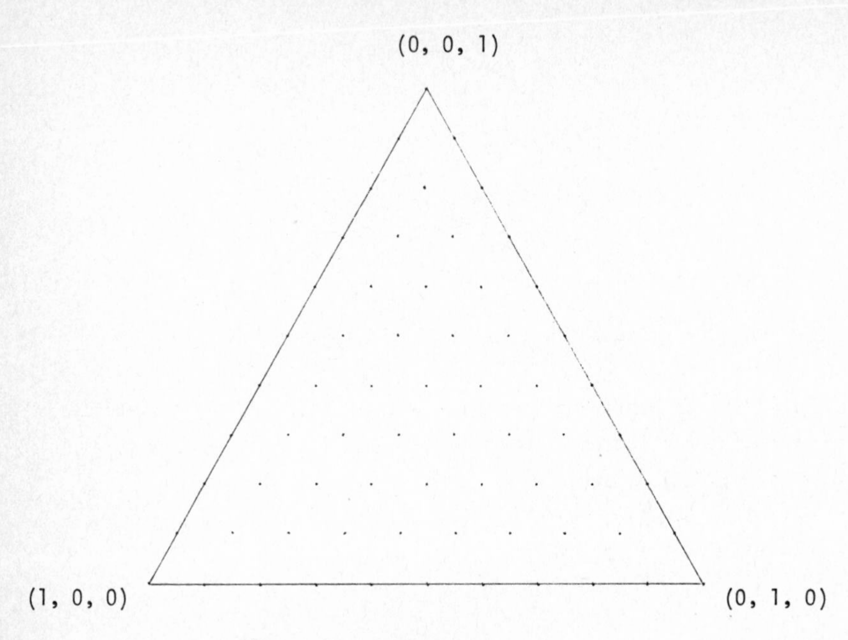

Figure 2A : Simplex with $n = 3$ and $N = 10$.

We connect these points as in Fig. 2B and this triangulation divides S into regular subsimplices which conform to the definition of primitive sets introduced by Scarf. Each subsimplex includes n-vectors of the regular grid. Scarf's idea for finding a fixed point is to walk through the simplex, starting at a vertex of the simplex and passing from one primitive set to the next primitive set as illustrated in Fig. 2B .

The algorithm of Scarf is completely described by the following procedures.

a) a replacement step on primitive sets,

b) a proper labelling of the vectors in each primitive set.

ad a) : The replacement step describes how to find a new primitive set when we eliminate one vector from the original primitive set. In case of a regular grid this procedure turns out to be very simple.

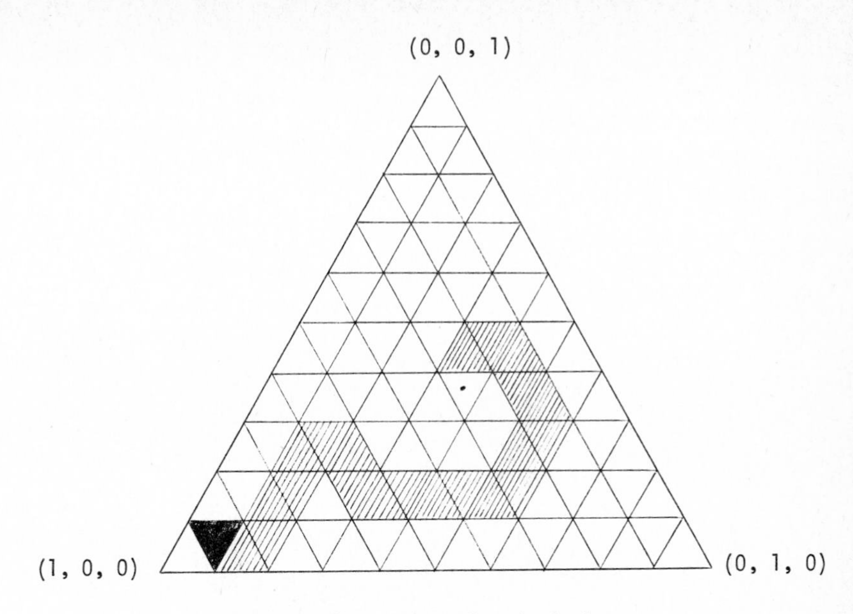

Figure 2B : Triangulated simplex with path indicated.

Definition 2 (Replacement step): If we put the vertices of the primitive set in a matrix and order them in a lexicographically increasing [1] way and suppose that the j-th column is eliminated, then the new vector to be introduced has to be calculated as follows:

$$x'^{1} = x^{n} + x^{2} - x^{1} \quad ; \quad j = 1 \quad ,$$

$$x'^{j} = x^{j-1} + x^{j+1} - x^{j} \quad ; \quad j = 2, \ldots, n-1 \quad ,$$

$$x'^{n} = x^{n-1} + x^{1} - x^{n} \quad ; \quad j = n \quad . \qquad \square$$

---

[1] The vector $a = (a_1, a_2, \ldots, a_n)$ is lexicographically larger than $b = (b_1, b_2, \ldots, b_n)$ if the first nonzero element in the sequence $a_1 - b_1, a_2 - b_2, \ldots, a_n - b_n$ is positive.

Example 1: We consider the following primitive set

$$\frac{1}{10}\begin{bmatrix} 3 & 3 & 4 \\ 5 & 6 & 5 \\ 2 & 1 & 1 \end{bmatrix} . \qquad \text{(PS)}$$

We form a new primitive set by replacing the second vector and we obtain

$$\frac{1}{10}\begin{bmatrix} 3 & 4 & 4 \\ 5 & 4 & 5 \\ 2 & 2 & 1 \end{bmatrix} .$$

Replacing the last vector in (PS) leads to

$$\frac{1}{10}\begin{bmatrix} 3 & 3 & 2 \\ 5 & 6 & 6 \\ 2 & 1 & 2 \end{bmatrix} \quad \text{or rearranged} \quad \begin{bmatrix} 2 & 3 & 3 \\ 6 & 5 & 6 \\ 2 & 2 & 1 \end{bmatrix} .$$

ad b) : A label of a primitive set vector is a characteristic of each vector in the primitive set, which makes it possible to select the vector to be eliminated, and assures the convergence of the algorithm to the neighbourhood of a fixed point.

We will elucidate this matter by means of an example of labelling described by Scarf.
Labelling: Suppose that we have a regular grid on our simplex with a certain grid size. We associate a label with each vector x in the grid by the following rule:

$$\text{Compute } f(x) - x = \begin{bmatrix} f_1(x) - x_1 \\ f_2(x) - x_2 \\ \cdot \quad \cdot \\ \cdot \quad \cdot \\ f_n(x) - x_n \end{bmatrix} ;$$

if $f_j(x) - x_j$ is the first nonnegative element of $f(x) - x$, x is given the label j .

This labelling is chosen so that the fixed point is reached as soon as a primitive set is "completely labelled", which means that the vectors in that primitive set have all different labels.

Example 2: We shall try to find the fixed point for the following simple problem

$$f\begin{bmatrix} x_1 \\ x_2 \end{bmatrix} = \begin{bmatrix} x_2 \\ x_1 \end{bmatrix} \quad ;$$

$$x_i \geq 0 \quad , \quad i = 1, 2 \quad , \quad \sum_i x_i = 1 \quad .$$

We define a regular grid on our simplex with a grid size of ten, and start the algorithm with a primitive set consisting of a vertex of the simplex and a point in the grid which is nearest to this vertex. So the initial primitive set is

$$\frac{1}{10}\begin{bmatrix} 9 & 10 \\ 1 & 0 \end{bmatrix} \quad .$$

The labels of the vectors

$$\begin{bmatrix} 9 \\ 1 \end{bmatrix} \quad \text{and} \quad \begin{bmatrix} 10 \\ 0 \end{bmatrix}$$

are computed by the described procedure, yielding for both vectors the label 2 .

In the initial position of the algorithm the first vector in the primitive set to be replaced is the side vector. So by applying the replacement step we find the next primitive set

$$\begin{bmatrix} 8 & 9 \\ 2 & 1 \end{bmatrix} \quad .$$

Since the new vector

$$\begin{bmatrix} 8 \\ 2 \end{bmatrix}$$

has also label 2 we replace the second vector and obtain

$$\begin{bmatrix} 7 & 8 \\ 3 & 2 \end{bmatrix} .$$

The subsequent steps are:

$$\begin{bmatrix} 6 & 7 \\ 4 & 3 \end{bmatrix}$$

and finally

$$\begin{bmatrix} 5 & 6 \\ 5 & 4 \end{bmatrix} .$$

The labelling is now complete and the fixed point will be in the neighbourhood of this primitive set, which is correct since $x_1 = 5 = x_2$ is the fixed point. Scarf pointed out that it would be more efficient in general to associate <u>vector</u> labels with the primitive set vectors instead of integer labels. In a crucial theorem he stated that the fixed point is approximated by a primitive set of which the corresponding associated label vectors form a feasible basis [1] for a system of linear equalities $Ay = b$, with A the matrix of associated (label) vectors and b some nonnegative vector. For the associated vectors he chose

---

[1] The columns $j_1, \ldots, j_n$ of the matrix A form a feasible basis if the equations $Ay = b$ have a unique, nonnegative solution with $y_j = 0$ unless $j = j_1, \ldots, j_n$.

$f(x^i) - x^i + \iota$, with x^i the i-th primitive set vector not on the boundary of the simplex, and $\iota = (1, 1, . . . , 1)'$ and b is defined to be ι .

For the primitive set vectors representing points on the boundary of the simplex unit vectors are associated by the following rule: Each primitive set vector with the first zero element in the i-th place will be associated with the i-th unit vector. The procedure to be followed for finding the fixed point can be described as follows.

The aim of the algorithm is to find a primitive set of which the corresponding label vectors form a feasible basis for the equations $Ay = \iota$. In order to reach this correspondence we start the algorithm with a feasible basis for $Ay = \iota$ formed by $n - 1$ label vectors corresponding with $n - 1$ vectors in the initial primitive set and one additional label vector generally not corresponding with the remaining primitive set vector. During the algorithm we follow a path through the simplex running from one primitive set to another such that the associated matrices (consisting throughout of $n - 1$ corresponding label vectors and one additional vector) form a feasible basis for $Ay = \iota$. As soon as we have found a primitive set completely corresponding with the associated feasible basis the algorithm is terminated, since then a fixed point is reached.

In order to elucidate the procedure with vector labels we follow the algorithm in a number of subsequent steps. The initial primitive set is formed near a vertex of the simplex (say near the vertex

$$\begin{bmatrix} 1 \\ 0 \\ . \\ . \\ 0 \end{bmatrix})$$

and consists of $n - 1$ side vectors and one vector interior to the simplex (in this case the vector in the grid with the largest first coordinate). In the 3-dimensional case with a grid size of 10 we have:

$$\begin{bmatrix} 8 & 9 & 9 \\ 1 & 0 & 1 \\ 1 & 1 & 0 \end{bmatrix} .$$

The initial associated matrix is formed completely by unit vectors, since this matrix forms a feasible basis for $Ay = \iota$. In doing so the associated matrix corresponds with the primitive set except for the vector interior to the simplex (see the labelling rule for side vectors).

The initial position of the algorithm may then be described as follows:

Primitive set consisting of vectors $x^j , x^2 , \ldots , x^n$;
Associated matrix consisting of n unit vectors $a^1 , \ldots , a^n$.

In the primitive set x^j is the interior vector and $x^2 , \ldots , x^n$ the $n - 1$ side vectors. The associated label vectors of the primitive set vectors are recognized by the same indices in this notation.

In order to reach complete correspondence between the primitive set and the associated matrix we introduce the label vector of x^j in the associated matrix by eliminating one of the unit vectors in this matrix such that the new associated matrix again forms a feasible basis for $Ay = \iota$. This operation is carried out by means of a pivot step (familiar to the pivot step known from linear programming, discussed in detail in section 3.3).

If the vector to be eliminated turns out to be the first unit vector the algorithm is terminated, since correspondence is reached. If another vector is eliminated (say the i-th one) we are in the next stage:

Primitive set $x^j , x^2 , \ldots , x^n$;
Associated matrix $a^j , a^2 , \ldots , a^{i-1} , a^{i+1} , \ldots , a^n$.

We now replace the i-th primitive set vector by applying the replacement step from Definition 2 . In doing so another vector is introduced and if this is not the first side vector (= a vector with zero first coordinate) we insert its corresponding label vector in the feasible basis by means of a pivot step, and so on.

The algorithms of Scarf and Eaves are both based on the described procedure. Scarf, however, deals with a specific grid size and tries to reach the fixed point by starting in general near a vertex of the simplex. Since the accuracy of the approximation of the fixed point depends on the chosen grid size this algorithm should be combined with some numerical optimization method in order to refine the last primitive set at the required accuracy level. Moreover, the number of iterations (replacement and pivot steps) might be substantial since the starting point may lie relatively far beyond the definite fixed point.

Eaves therefore developed an algorithm using more or less the same procedures but without the disadvantage just described, at the price that his method is much more complicated than Scarf's method.

3. EAVES' METHOD

The principle of Eaves' algorithm is based on a subsequent extension of the grid size during the algorithm.

The crucial idea behind this method is that knowing the final primitive set with respect to a certain grid size it would cost less effort to find another primitive set which is the best approximation to a fixed point in a grid size finer than the preceding one. Therefore the original simplex is multiplied by 2^k , $k = 0, 1, 2, \ldots$, obtaining an infinite sequence of n-simplices [1] S_k . In Figure 3 this is illustrated in case $n = 3$.

On each simplex a regular grid is defined in the usual way. In S_k the grid consists of all vectors that can be written as $(m_1, \ldots, m_n)$ with $m_1, \ldots, m_n$ nonnegative integers summing to 2^k . In doing so a triangulation of each n-simplex S_k in subsimplices is obtained as shown in Figure 4 for $n = 3$ and $k = 2$.

---

[1] An n-simplex is defined to be any simplex having n vertices.

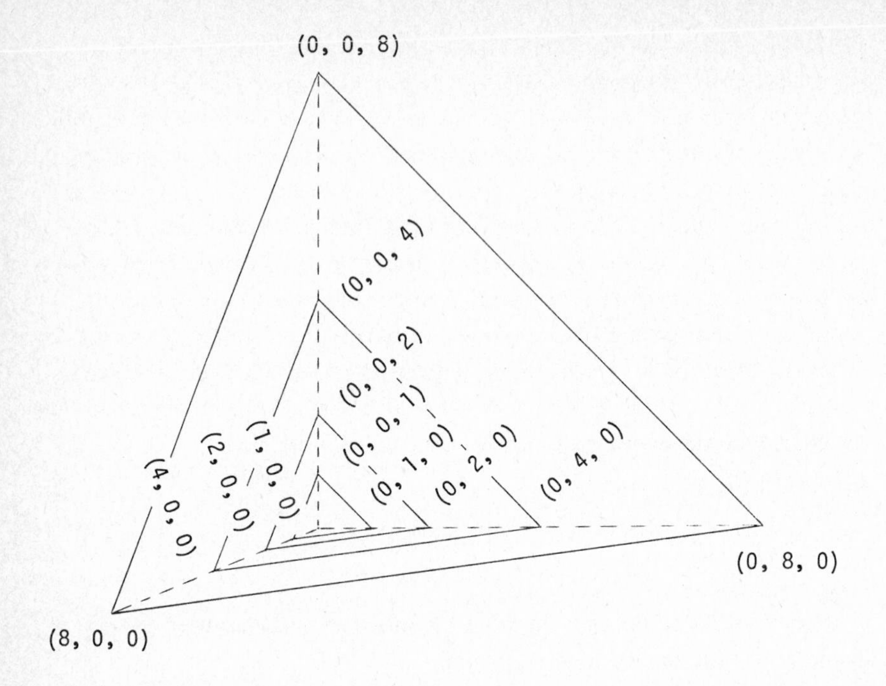

Figure 3 .

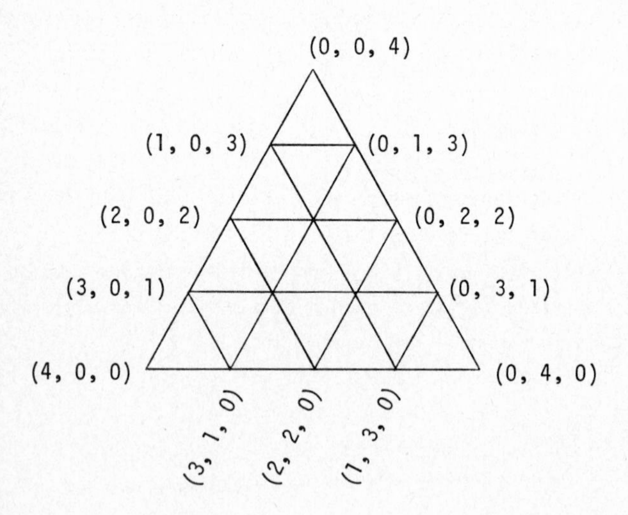

Figure 4 .

The algorithm is aimed at finding a path through the positive orthant of the n-dimensional space (denoted by D), running from primitive sets in S_k to primitive sets in S_{k+1} for arbitrary k . We therefore have to extend the triangulation such that the spaces between two adjacent simplices are partitioned too. This extension is performed by dividing the space between S_k and S_{k+1} into (n+1)-subsimplices with vertices corresponding to the vertices of the subsimplices (primitive sets) in S_k and S_{k+1} .

Eaves has defined such a triangulation on

$$D' = \{x \in D : \sum x_i \leq 2\}$$

in the following way:

If τ is an (n+1)-subsimplex in D' consisting of $n + 1$ vertices $v^i = (v_1^i, \ldots, v_n^i)$, $i = 1, \ldots, n+1$, which are ordered in a lexicographically decreasing way ($v^1 > v^2 > \ldots > v^{n+1}$) with v^i in S_0 or S_1 for all i and let v^i be generated according to the following scheme:

$$v^{i+1} = v^i + q(\gamma_i) \quad , \quad i = 1, \ldots, n \quad ,$$

with $\gamma_1, \ldots, \gamma_n$ some permutation of $1, \ldots, n$ and $q(j)$ the j-th column of the $n \times n$ matrix

$$\begin{bmatrix} -1 & 0 & . & . & . & . & 0 & 0 \\ +1 & -1 & . & . & . & . & 0 & 0 \\ 0 & +1 & . & & & . & 0 & 0 \\ 0 & 0 & . & & & . & 0 & 0 \\ . & . & . & & & . & . & . \\ . & . & . & & & . & . & . \\ 0 & 0 & . & & & . & -1 & 0 \\ 0 & 0 & . & & & . & +1 & -1 \end{bmatrix}$$

such that the v generated is nonnegative or zero, then the collection M of all such τ in D' form a triangulation of D' and each τ is completely described by v^1 and γ and we may characterize τ therefore by $\tau(v^1, \gamma)$.

Furthermore, we notice that any n-simplex σ in S_k may be characterized by $\sigma(u^1, \beta)$, where σ is generated from u^1 in S_k according to:

$$u^{i+1} = u^i + p(\beta_i) \quad , \quad i = 1, \ldots, n-1 \quad ,$$

with $\beta_1, \ldots, \beta_{n-1}$ some permutation of $1, \ldots, n-1$ such that the u generated is nonnegative or zero and $p(j)$ the j-th column of the $n \times (n-1)$ matrix

$$\begin{bmatrix} -1 & . & . & . & . & . & 0 \\ +1 & & & & & & 0 \\ 0 & & & & & & 0 \\ . & & & & & & . \\ . & & & & & & . \\ . & & & & & & . \\ 0 & & & & & & -1 \\ 0 & . & . & . & . & . & +1 \end{bmatrix} .$$

The complete triangulation on D' is pictured in Figure 5 for the 3-dimensional case.

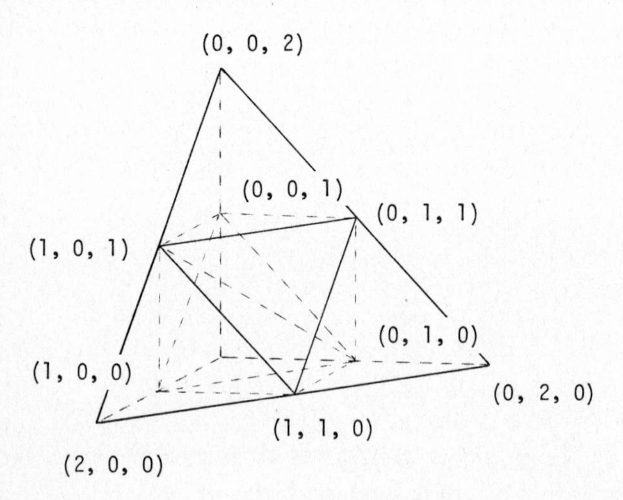

Figure 5 .

It appears that in this example 7 possible τ-simplices may be distinguished, which are summarized below with their corresponding γ-vectors.

Example 3:

$$\tau_1 = \begin{bmatrix} 2 & 1 & 1 & 1 \\ 0 & 1 & 0 & 0 \\ 0 & 0 & 1 & 0 \end{bmatrix}, \quad \tau_2 = \begin{bmatrix} 1 & 0 & 0 & 0 \\ 1 & 2 & 1 & 1 \\ 0 & 0 & 1 & 0 \end{bmatrix},$$

$$\gamma = (1, 2, 3) \qquad \gamma = (1, 2, 3)$$

$$\tau_3 = \begin{bmatrix} 1 & 1 & 0 & 0 \\ 1 & 0 & 1 & 1 \\ 0 & 1 & 1 & 0 \end{bmatrix}, \quad \tau_4 = \begin{bmatrix} 1 & 1 & 1 & 0 \\ 1 & 0 & 0 & 1 \\ 0 & 1 & 0 & 0 \end{bmatrix},$$

$$\gamma = (2, 1, 3) \qquad \gamma = (2, 3, 1)$$

$$\tau_5 = \begin{bmatrix} 1 & 0 & 0 & 0 \\ 0 & 1 & 0 & 0 \\ 1 & 1 & 2 & 1 \end{bmatrix}, \quad \tau_6 = \begin{bmatrix} 1 & 0 & 0 & 0 \\ 0 & 1 & 1 & 0 \\ 1 & 1 & 0 & 1 \end{bmatrix},$$

$$\gamma = (1, 2, 3) \qquad \gamma = (1, 3, 2)$$

$$\tau_7 = \begin{bmatrix} 1 & 1 & 0 & 0 \\ 0 & 0 & 1 & 0 \\ 1 & 0 & 0 & 1 \end{bmatrix}$$

$$\gamma = (3, 1, 2)$$

We are now prepared to extend the complete triangulation on D' to D. Namely, if σ is any n-simplex in S_k, $k = 0, 1, 2, \ldots$ and τ is any (n+1)-simplex in M, then all (n+1)-simplices in D are of the form $T_\sigma(\tau)$, since if we have a $\sigma(u^1, \beta)$ and a $\tau(v^1, \gamma)$, we may find an (n+1)-simplex $T_\sigma(\tau)$ in D by:

$$(3.1) \qquad T_\sigma(\tau) = (T_\sigma(v^1), \ldots, T_\sigma(v^{n+1})) =$$

$$= \left(\sum_{i=1}^{n} u^i v_i^1, \ldots, \sum_{i=1}^{n} u^i v_i^{n+1} \right).$$

<u>Example</u> 4: For example if

$$\sigma = \begin{bmatrix} 4 & 3 & 3 \\ 2 & 3 & 2 \\ 2 & 2 & 3 \end{bmatrix}$$

and
$$\tau = \begin{bmatrix} 2 & 1 & 1 & 1 \\ 0 & 1 & 0 & 0 \\ 0 & 0 & 1 & 0 \end{bmatrix},$$

we obtain
$$T_\sigma(\tau) = \begin{bmatrix} 8 & 7 & 7 & 4 \\ 4 & 5 & 4 & 2 \\ 4 & 4 & 5 & 2 \end{bmatrix}.$$

3.1. REPLACEMENT STEP ON D

Eaves pointed out that a replacement step on the (n+1)-simplices in D may be carried out by a replacement step in D', since there exists a correspondence $T_\sigma(\tau)$ between all τ in D' and the (n+1)-simplices in D.

Eaves describes this replacement step in the following way:
Suppose we find ourselves in a certain stage of the algorithm and that we are in the position to move from $T_\sigma(\tau)$ to $T_{\sigma'}(\tau')$ and suppose that the vector $T_\sigma(v^i)$ should be replaced.

We then generate a vector $v = (v_1, \ldots, v_n)$ by means of the replacement step stated in Definition 2 on $\tau = \{v^1, \ldots, v^{n+1}\}$, replacing v^i.

There are four cases that might occur:

1) $v \in D'$,

2) $\sum v_i > 2$,

3) $v_j < 0$, for some $j = 1, \ldots, n$, and

4) $v = 0$.

We illustrate the possible occurence of these four cases by the following example.

Example 5:

a) Suppose that in τ_3 of Example 3 v_3 is replaced, we then obtain by applying the replacement operation on v_3:

$$v = v_2 + v_4 - v_3 = \begin{bmatrix} 1 \\ 0 \\ 0 \end{bmatrix}, \text{ which refers to case 1)}.$$

b) If we replace v_4 in τ_1 we obtain:

$$v = v_3 + v_1 - v_4 = \begin{bmatrix} 2 \\ 0 \\ 1 \end{bmatrix}, \text{ which leads us to case 2)}.$$

c) If we replace v_3 in τ_1, we have case 3) since

$$v = \begin{bmatrix} 1 \\ 1 \\ -1 \end{bmatrix}.$$

d) If v_1 in τ_7 is replaced, v becomes the zero vector.

Since $\sigma(u^1, \beta)$ and $\tau(v^1, \gamma)$ are known at any stage of the algorithm and since u^1, β and v^1, γ define σ and τ completely, Eaves described the procedure of the replacement step in the four cases in terms of changes in u^1, β and v^1, γ instead of considering the complete σ and τ .

Case 1) ($v \in D'$):

(u^1, β) are unchanged and the new (v^1, γ) are computed according to Table 1 .

index of the replaced vector	v^1 becomes	γ becomes
$i = 1$	$v^1 + q(\gamma_1)$	$(\gamma_2, \ldots, \gamma_n, \gamma_1)$
$2 \leq i \leq n$	v^1	$(\gamma_1, \ldots, \gamma_i, \gamma_{i-1}, \ldots, \gamma_n)$
$i = n + 1$	$v^1 - q(\gamma_n)$	$(\gamma_n, \gamma_1, \ldots, \gamma_{n-1})$

Table 1 .

Case 2) ($\sum v_i > 2$):

u^1 becomes $\Theta_1 u^1 + \sum_{i=1}^{n-1} \Theta_{i+1} q(\beta_i)$ where

$$\Theta_i = \sum_{j=i}^{n} v_j^1 \quad , \quad i = 1, \ldots, n \; ,$$

β becomes $(\beta_{\gamma_1}, \ldots, \beta_{\gamma_{n-1}})$, v^1 becomes $(1, 0, \ldots, 0, 1)$ and γ becomes $(n, 1, \ldots, n-1)$.

Case 3) ($v_j < 0$):
First we note that j is unique. The new (u^1, β) is computed according to Table 2 where only u^j is replaced. The new (v^1, γ) is computed according to Table 3 (the (i, j) combinations not listed cannot occur).

index of negative entry	u^1 becomes	β becomes
$j = 1$	$u^1 + q(\beta_1)$	$(\beta_2, \ldots, \beta_{n-1}, \beta_1)$
$1 < j < n$	u^1	$(\beta_1, \ldots, \beta_j, \beta_{j-1}, \ldots, \beta_{n-1})$
$j = n$	$u^1 - q(\beta_{n-1})$	$(\beta_{n-1}, \beta_1, \ldots, \beta_{n-2})$

Table 2 .

index of negative entry j and index of replaced vector i	v becomes	γ becomes
$j = 1$ $i = 1$	$(v_2^2, v_3^2, \ldots$ $\ldots, v_n^2, 0)$	$(\gamma_2 - 1, \ldots$ $\ldots, \gamma_{k-1} - 1, n - 1,$ $\gamma_k, \gamma_{k+1} - 1, \ldots$ $\ldots, \gamma_n - 1)$ where $\gamma_k = n$
$1 < j < n$ $1 \leq i \leq n+1$	v^1	γ
$j = n$ $2 \leq i \leq n$	$(1, v_1^1 - 1, v_2^1,$ $\ldots, v_{n-1}^1)$	$(1, \gamma_1 + 1, \ldots$ $\ldots, \gamma_{i-2} + 1, n,$ $\gamma_{i+1} + 1, \ldots$ $\ldots, \gamma_n + 1)$

Table 3 .

Case 4) ($v = 0$):
There is a unique set $\eta \subseteq \{1, \ldots, n-1\}$ such that

$$y = (1/2)(u^1 - \sum_{i \in \eta} q(i)) \quad \text{has integer components.}$$

The flow chart generates η and y ; " <= " means "becomes" .
Observe that components of

$$(1/2)u^1 \quad \text{are } 0 \text{ or } (1/2) \text{ modulo } 1 .$$

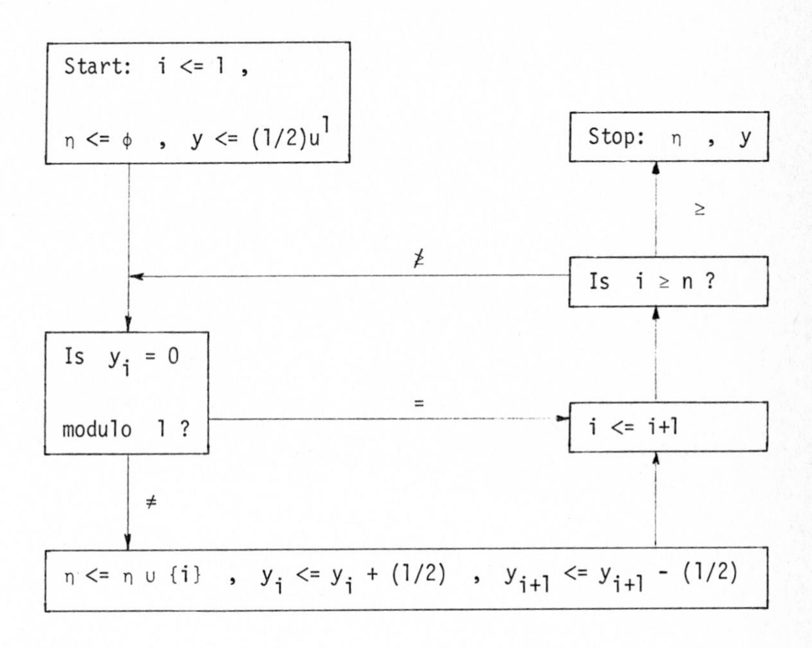

Let $\xi = \{1, \ldots, n-1\} \sim \eta$ and let $\eta = (\eta_1, \ldots, \eta_k)$ and $\xi = (\xi_1, \ldots, \xi_{n-k-1})$ inherit order from $\beta = (\beta_1, \ldots, \beta_{n-1})$.

After computing the new (v^1, γ) according to Table 4 , the new (u^1, β) becomes $(y, (\eta, \xi))$.

	v^1 becomes	γ becomes
$k = 0$	$(2, 0, \ldots, 0)$	$(1, \ldots, n)$
$k > 0$	$(z_0, \ldots, z_n)$ where $z_\ell = \begin{cases} 1 & \ell = 0, \ldots, k \\ 0 & \text{else} \end{cases}$	$(m_1, \ldots, m_{n-1}, n)$ where $m_\ell = \begin{cases} h & \beta_\ell = \eta_h \\ k+h & \beta_\ell = \xi_h \end{cases}$

Table 4 .

We will try to elucidate the various operations in the replacement step by the following example for the 3-dimensional space.

Example 6:
Suppose that at some stage of the algorithm σ , τ and $T_\sigma(\tau)$ are:

$$\sigma = \begin{bmatrix} 1 & 1 & 0 \\ 1 & 0 & 1 \\ 0 & 1 & 1 \end{bmatrix} \qquad \tau = \begin{bmatrix} 2 & 1 & 1 & 1 \\ 0 & 1 & 0 & 0 \\ 0 & 0 & 1 & 0 \end{bmatrix} \qquad T_\sigma(\tau) = \begin{bmatrix} 2 & 2 & 1 & 1 \\ 2 & 1 & 2 & 1 \\ 0 & 1 & 1 & 0 \end{bmatrix}$$

$\beta = (2, 1)$ $\qquad \gamma = (1, 2, 3)$

and suppose we want to replace $T_\sigma(v^4)$.

First we compute v :

$$v = v^3 + v^1 - v^4 = (2,\ 0,\ 1) \quad \text{, which corresponds with case 2).}$$

We folow the scheme valid for this case:

$$\Theta_1 = 2 \rightarrow u^1 \Rightarrow 2u^1 = \begin{bmatrix} 2 \\ 2 \\ 0 \end{bmatrix} \quad ; \text{ and } \beta \Rightarrow (2,\ 1) \ ;$$

$$v^1 = \begin{bmatrix} 1 \\ 0 \\ 1 \end{bmatrix} \quad ; \text{ and } \gamma \Rightarrow (3,\ 1,\ 2) \ .$$

So the new σ , τ and $T_\sigma(\tau)$ become :

$$\sigma = \begin{bmatrix} 2 & 2 & 1 \\ 2 & 1 & 2 \\ 0 & 1 & 1 \end{bmatrix} \qquad \tau = \begin{bmatrix} 1 & 1 & 0 & 0 \\ 0 & 0 & 1 & 0 \\ 1 & 0 & 0 & 1 \end{bmatrix} \qquad T_\sigma(\tau) = \begin{bmatrix} 3 & 2 & 2 & 1 \\ 4 & 2 & 1 & 2 \\ 1 & 0 & 1 & 1 \end{bmatrix}$$

$$\beta = (2,\ 1) \qquad \gamma = (3,\ 1,\ 2)$$

We now want to replace $T_\sigma(v^3)$. Computing v yields:

$v = (1,\ -1,\ 1)$; corresponding with case 3) ; $j = 2$, $i = 3$.

From Table 2 it is seen that u^1 remains unchanged while β becomes $(1,\ 2)$. Table 3 shows that v^1 and γ do not change. So the following situation has been achieved.

$$\sigma = \begin{bmatrix} 2 & 1 & 1 \\ 2 & 3 & 2 \\ 0 & 0 & 1 \end{bmatrix} \qquad \tau = \begin{bmatrix} 1 & 1 & 0 & 0 \\ 0 & 0 & 1 & 0 \\ 1 & 0 & 0 & 1 \end{bmatrix} \qquad T_\sigma(\tau) = \begin{bmatrix} 3 & 2 & 1 & 1 \\ 4 & 2 & 3 & 2 \\ 1 & 0 & 0 & 1 \end{bmatrix}$$

Suppose $T_\sigma(v^1)$ has to be replaced. In that case $v = (0, 0, 0)$, which leads us to case 4) .

Following the flow chart we find:
$i = 1$, $\eta = \phi$, $y = (1, 1, 0)$; so $y_1 = 0$ modulo 1 which holds for all y_j , $j = 2, 3$ and therefore $\eta = \phi$ and $\xi = (1, 2)$. Computing v^1 and γ according to Table 4 we have

$$v^1 \Rightarrow \begin{bmatrix} 2 \\ 0 \\ 0 \end{bmatrix} \qquad \text{and} \quad \gamma \Rightarrow (1, 2, 3) \quad .$$

Finally $u^1 \Rightarrow \begin{bmatrix} 2 \\ 1 \\ 0 \end{bmatrix}$ and $\beta \Rightarrow (1, 2)$.

Summarizing the new situation:

$$\sigma = \begin{bmatrix} 1 & 0 & 0 \\ 1 & 2 & 1 \\ 0 & 0 & 1 \end{bmatrix} \qquad \tau = \begin{bmatrix} 2 & 1 & 1 & 1 \\ 0 & 1 & 0 & 0 \\ 0 & 0 & 1 & 0 \end{bmatrix} \qquad T_\sigma(\tau) = \begin{bmatrix} 2 & 1 & 1 & 1 \\ 2 & 3 & 2 & 1 \\ 0 & 0 & 1 & 0 \end{bmatrix}$$

$\beta = (1, 2)$ $\qquad$ $\gamma = (1, 2, 3)$

The quite compact description of the replacement operation might unjustly suggest that the derivation of a new subsimplex requires extensive computational effort, since it seems to be necessary to store u^1 , β , v^1 and γ at each step of the algorithm and moreover, that at each step $T_\sigma(\tau)$ has to be computed according to the definition stated in equation (3.1) .
Since it can be seen that a replacement step on $T_\sigma(\tau)$ influences only one vector in $T_\sigma(\tau)$, one might wonder if it is not possible to formulate the replacement step in such a way that the costly operation on σ and τ can be avoided. For this reason we have searched for a reformulation of the replacement step, so that the calculations can be seen more directly on the subsimplices $T_\sigma(\tau)$ we are interested in.

3.2. REFORMULATION OF THE REPLACEMENT STEP

The reformulated replacement step is pictured in the flow chart in Figure 6 . The symbols used in this flow chart have the following meaning:

JMOUT = the index of the vector in $T_\sigma(\tau)$ to be replaced.

JIN = the index of the vector to be introduced in $T_{\sigma'}(\tau')$.

T^i = the i-th vector of $T_\sigma(\tau)$.

TIN = the vector to be introduced in $T_\sigma(\tau')$ after replacing T^{JMOUT} in $T_\sigma(\tau)$.

NRLH = the index of the last vector on the highest level.

The other symbols correspond with the notation used in the previous text. The equivalence of the reformulated replacement step with the replacement step discussed in section 3.1. is proved in Appendix A .

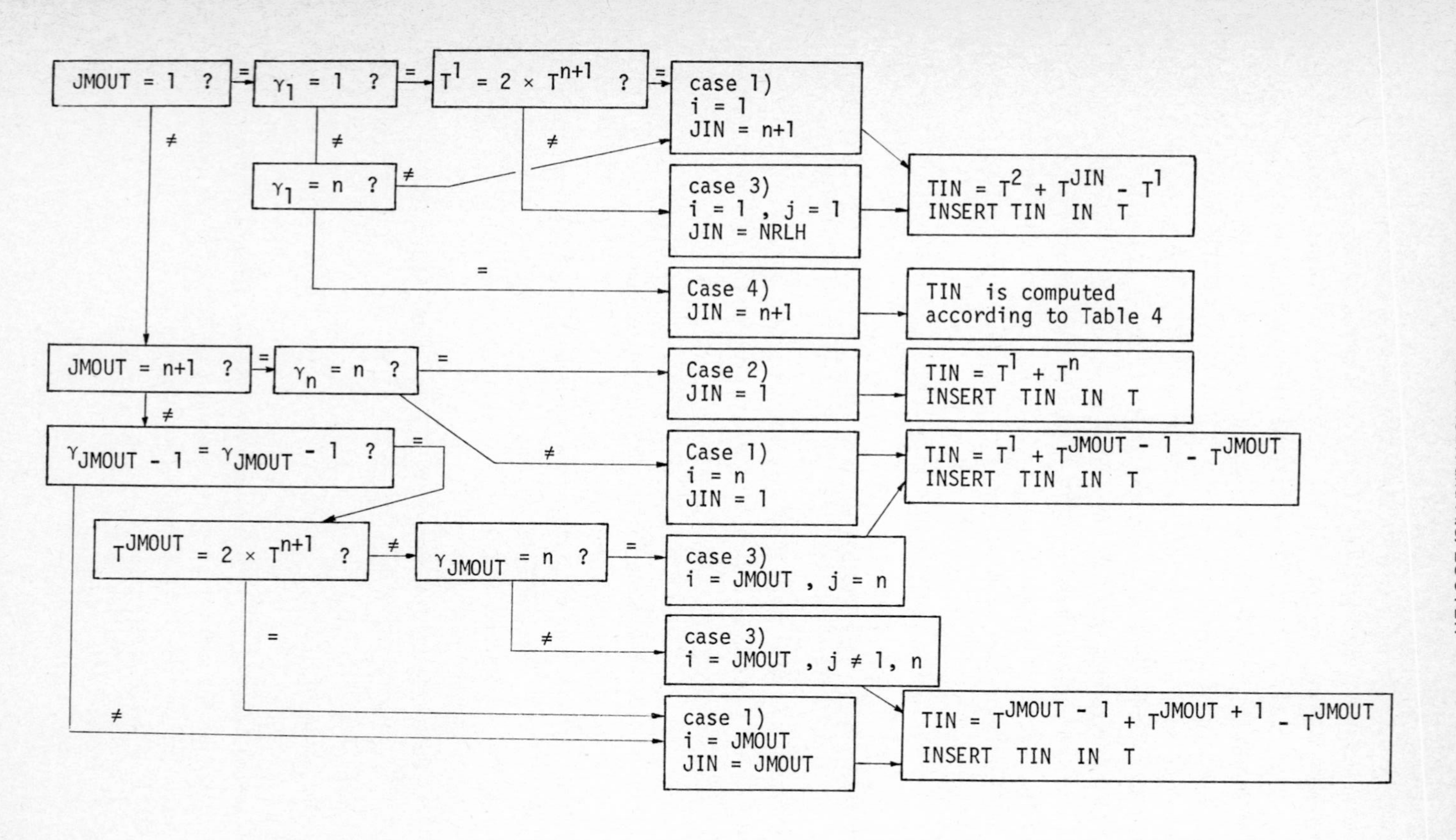

Figure 6 : Flow chart of the replacement step

From the flow chart in Figure 6 it can be seen that with the knowledge of γ and the structure of $T_\sigma(\tau)$ we are able to calculate the vector to be introduced in $T_{\sigma'}(\tau')$ by applying a number of simple tests. In order to adapt γ at each step of the algorithm we follow Eaves' tables from section 3.1.

The cases 1, 2, 3 and 4 to which these tables correspond are explicitly given in the flow chart. When case 4) appears we have to apply Table 4 from which we can see that we have to know u^1 and β. Fortunately, however, this problem can be solved immediately since the last n vectors of $T_\sigma(\tau)$ in this case are the vectors that form σ.

Summarizing we may conclude that since only storage and adaption is required with respect to γ the reformulated replacement step requires less information compared with the description of Eaves' replacement step. From the computational point of view important progress has been made by avoiding the extensive calculations of $T_{\sigma'}(\tau')$ given by equation (3.1).

3.3. PIVOT STEP

As already stated in section 2 the pivot step is used for the case of vector labelling and provides a procedure for finding the vector to be replaced in the primitive set. This procedure will be elucidated by the following example.

Suppose we are at a certain stage of the algorithm with a primitive set consisting of n+1 vectors in which the j-th vector has just been introduced by means of the replacement step. So the associated matrix A consists of vectors corresponding with the primitive set vectors except for the j-th primitive set vector.

With the notation of section 2 we have:

primitive set	$x^i, x^2, \ldots, x^{n+1}$,
associated matrix	$a^1, a^2, \ldots, a^{n+1}$.

The matrix A is such that it satisfies $Ay = \iota$, with y strictly positive.

Since the j-th primitive set vector has just been introduced, the associated vector $(a^*)^j$ will be introduced in the A-matrix such that the new A-matrix (say A^*) forms a feasible basis for $A^*p = \iota$.

Now we consider the problem of selecting the vector in A to be removed. Therefore, we write $(a^*)^j$ as a linear combination of the originally associated vectors. We have

$$(3.3.1) \qquad a^1 z_1 + a^2 z_2 + \ldots + a^k z_k + \ldots + a^{n+1} z_{n+1} = (a^*)^j \quad .$$

Moreover, from $Ay = \iota$ we have

$$(3.3.2) \qquad a^1 y_1 + a^2 y_2 + \ldots + a^{n+1} y_{n+1} = \iota \quad .$$

Reformulating (3.3.1) yields

$$(3.3.3) \qquad - a^1 \frac{z_1}{z_k} - a^2 \frac{z_2}{z_k} - \ldots + \frac{(a^*)^j}{z_k} - \ldots - a^{n+1} \frac{z_{n+1}}{z_k} = a^k \quad ; \quad z_k \neq 0 \, ,$$

and substituting a^k into (3.3.2) gives

$$(3.3.4) \qquad a^1 y_1 \left(1 - \frac{z_1}{y_1}\frac{y_k}{z_k}\right) + a^2 y_2 \left(1 - \frac{z_2}{y_2}\frac{y_k}{z_k}\right) + \ldots + (a^*)^j \frac{y_k}{z_k} + \ldots$$

$$\ldots + a^{n+1} y_{n+1} \left(1 - \frac{z_{n+1}}{y_{n+1}}\frac{y_k}{z_k}\right) = \iota \quad ,$$

which must be equivalent to $A'p = \iota$ if a^k is the vector to be removed from A in favour of $(a^*)^j$. Since all p_i should be strictly positive, the first thing to be noticed is

$$(3.3.5) \qquad p_k = \frac{y_k}{z_k} > 0 \quad \text{, which implies}$$

$$(3.3.6) \qquad z_k > 0 \quad \text{, since} \quad y_k > 0 \quad ,$$

and

$$(3.3.7) \qquad p_i = y_i \left(1 - \frac{z_i}{y_i}\frac{y_k}{z_k}\right) > 0 \quad , \quad i \neq k \quad ,$$

or equivalently

$$(3.3.8) \qquad \frac{z_i}{y_i} < \frac{z_k}{y_k} \quad , \quad i \neq k \quad .$$

(3.3.6) and (3.3.8) may be satisfied by selecting k such that $z_k > 0$ and

$$\frac{z_k}{y_k} = \max_m \left(\frac{z_m}{y_m} \right) \quad , \quad m = 1, \ldots, n+1 \quad .$$

As a result of linear programming such a k can in general always be found. If it is not unique a criterion must be formulated such that the pivot step is carried out consistently during the algorithm.

Since there is no need for reordering the A-matrix after each pivot step this may be accomplished easily by selecting the first k for which

$$\frac{z_k}{y_k} = \max_m \left(\frac{z_m}{y_m} \right) \quad .$$

It can be seen that for finding the pivot column we have to invert A in order to compute $y = A^{-1}\iota$ and $z = A^{-1}(a^*)^j$. Dohmen and Schoeber [2] suggested in their paper that by applying a result of Bartels [1] a procedure may be carried out avoiding the computation of the complete A^{-1} at each pivot step.

Bartels pointed out that if A^{-1} is known and the k-th column of A is replaced by $(a^*)^k$ $(A^*)^{-1}$ may be computed by Gauss-Jordan elimination as $(A^*)^{-1} = D A^{-1}$ with

$$D = I - \frac{1}{z_k}(z - e^{(k)})\, e^{(k)^T} \quad ,$$

in which $e^{(k)}$ is the k-th unit vector.

In order to avoid too large round-off errors it is recommended to invert the associated matrix after a number of iterations completely.

3.4. LABELLING

In section 2 some examples of labellings were given for the Scarf method. With respect to Eaves' method a labelling can also be formulated. What we need is a so-called proper labelling defined by Eaves as following.

Definition 3: A proper labelling of vectors in the original unit simplex S must be such that:

a) the set of vertices of S is completely labelled, which means in our context that if we are dealing with vector labels, the associated vectors of the vertices of S form a feasible basis for $Ay = \iota$,
b) no facet of S contains a completely labelled set,
c) if a sequence of complete sets tends to $x \in S$, then x is a fixed point of f .

In our computer program in section 4 we use the following labels (= associated vectors) which in all problems we examined turned out to be sufficient [1])

$$a^j = f(x^j) - x^j + \iota \qquad \text{if } x^j > 0 \;,$$

and if x^j_{k+1} is the first nonzero element $k+1 \leq n$ in x^j after the first zero element in x^j then

$$a^j = \text{the k-th unit vector or}$$
$$a^j = \text{the n-th unit vector if } x^j_n \text{ appears to be still zero starting from the first zero element in } x^j \;.$$

This labelling has the advantage that we may start the algorithm with I_n as associated matrix and that during the algorithm associated vectors corresponding with side-vectors (slack vectors) can be introduced in the associated matrix avoiding the costly pivot operation. This can be illustrated by the description of the computer program in the next section.

---

[1]) A formal proof for the sufficiency of this labelling will not be given here.

4. A FORTRAN COMPUTER PROGRAM FOR EAVES' METHOD

LIST OF THE MAIN CHARACTERS

PRIMITIVE SET

ITMAT (I, J)	=	Matrix $T_\sigma(\tau)$ in Eaves or T in flow chart of Figure 6 ,
NP 1	=	Number of rows in ITMAT ,
NP 2	=	Number of columns in ITMAT ,
JMOUT	=	Column index of the vector in ITMAT to be replaced ,
ITM(I)	=	The vector to be introduced in ITMAT after a replacement step (= TIN in Figure 6),
JIN	=	Column index of the vector ITM in ITMAT,
ITNUM(I)	=	Vector of numbers joined to the primitive set vectors in order to visualize the correspondence with the label vectors in the associated matrix.

ASSOCIATED MATRIX OF LABEL VECTORS

AINV (I, J)	=	The inverse of the associated matrix of label vectors,
AIN(I)	=	The label vector to be introduced in the associated matrix after a replacement step,
JAOUT	=	The index of the pivot column in the associated matrix,
PHULP(I)	=	AINV.AIN ,
W(I)	=	Vector with unit elements (in $Ay = \iota$, see section 2),
QHULP	=	Element of AINV.W ,
CRITER	=	Small real number introduced to correct round-off errors,
IANUM(I)	=	Vector of numbers joined to the label vectors in order to visualize the correspondence with ITNUM(I) ,
IAACT(I)	=	Vector of numbers joined to the unit vectors in the associated matrix .

OTHER IMPORTANT CHARACTERS

IGAM(I)	=	γ according to Eaves and Figure 6 ,
IBETA(I)	=	β according to Eaves,
N	=	Order of IBETA ,
KSLEK	=	k standing for the k-th slack vector,
ITRE	=	Number of iterations,
INVE	=	Number of inversions on the associated matrix,
KDIM	=	Reserved storage capacity with respect to the number of rows in ITMAT .

The examples were executed on a DEC-10 computer system.

```
C**********************************************************************
C
C  MAIN PROGRAM WHICH USES EAVES' PROCEDURE FOR FINDING
C  THE FIXED POINT OF F(X)
C
C  IN THE EXAMPLE F(X) IS DEFINED AS:
C       F(X(I)) = X(I+1)  FOR  I=1,..,N-1
C       F(X(N)) = X(1)
C
C**********************************************************************
C
      IMPLICIT REAL*8 (A-H,O-Z)
      DIMENSION  AINV(40,40),AIN(40),IANUM(40),IAACT(40)
      DIMENSION  ITMAT(40,41),ITNUM(41),ITM(40)
      DIMENSION  IGAM(40),IBETA(39),PHULP(40),W(40)
      DATA  KDIM,NP1,CRITER,W/40,20,1.D-7,40*1.D0/
      CALL MINIT(ITMAT,KDIM,NP1,NP2,ITNUM,IGAM,AINV,IANUM,
     $      IAACT,ITRE,INVE,JMOUT,ITMNU)
      N = NP1-1
      MGRID = 2**17
      NGRID = 2
      WRITE(3,5000)
      WRITE(3,5001)  NP1
      WRITE(3,5002)
 350  MGRIT = 0
      DO 400  I = 1,NP1
      MGRIT = MGRIT+ITMAT(I,1)
 400  CONTINUE
      IF(MGRIT-NGRID)  600,450,600
 450  DO 455  I = 1,NP1
      ITM(I) = ITMAT(I,1)
 455  CONTINUE
      DMAX = DABS((ITM(1)-ITM(NP1))/DFLOAT(MGRIT))
      DO 460  I = 2,NP1
      FXX = DABS((ITM(I)-ITM(I-1))/DFLOAT(MGRIT))
      IF(FXX.GT.DMAX)  DMAX=FXX
 460  CONTINUE
      WRITE(3,4004)  MGRIT,ITRE,INVE,DMAX
      NGRID = NGRID*2
      IF(MGRIT-MGRID)  600,500,500
 500  WRITE(3,4001)  (ITMAT(J,1),J=1,NP1)
      CALL EXIT
 600  ITRE = ITRE+1
      CALL EAVES(ITMAT,KDIM,N,NP1,NP2,ITNUM,IGAM,
     *        IBETA,JMOUT,ITM,ITMNU)
      CALL LABELS(ITM,NP1,AIN,NACT)
      IF(NACT.NE.0)  GOTO 800
      CALL LABELF(ITM,NP1,AIN)
 800  CALL PIVOT(AINV,KDIM,AIN,W,PHULP,CRITER,NP1,NP2,IAACT,NACT,
     $      INVE,IANUM,ITNUM,JMOUT,ITMNU)
      GOTO  350
4001 FORMAT (///' FIRST COLUMN OF FINAL MATRIX ''ITMAT'''/
     $   (1H0,10I7))
4004 FORMAT (1H ,I7,9X,I8,12X,I8,D26.7)
5000 FORMAT (1H1,'EXAMPLE OF THE USE OF EAVES'' PROCEDURE'/
     $  1H0,'FUNCTION  F(X(I)) = X(I+1)  FOR  I=1,..,N-1'/
     $  1H ,10X,'F(X(N)) = X(1)'/)
5001 FORMAT (1H0,'N =',I3)
5002 FORMAT (//' GRIDSIZE',5X,'# OF ITERATIONS'
     $  ,5X,'# OF INVERSIONS',7X,'MAX(ABS(F(X)-X))')
      END
```

```
      SUBROUTINE  MINIT(ITMAT,KDIM,NP1,NP2,ITNUM,IGAM,
     $   AINV,IANUM,IAACT,ITRE,INVE,JMOUT,ITMNU)
      IMPLICIT REAL*8 (A-H,O-Z)
      DIMENSION  ITMAT(KDIM,1),ITNUM(1),IGAM(1)
      DIMENSION  AINV(KDIM,1),IANUM(1),IAACT(1)
C
C     ***************************
C     * INITIALIZE ALL VARIABLES *
C     ***************************
C
      NP2 = NP1+1
      ITRE = -1
      INVE = 0
      DO 10  I = 1,NP2
      ITNUM(I) = I
 10   CONTINUE
      DO 20  J = 1,NP1
      IAACT(J) = J
      IANUM(J) = J+1
      IGAM(J) = J
 20   CONTINUE
      IANUM(NP1) = 1
      DO 40  I = 1,NP1
      DO 30  J = 1,NP2
      ITMAT(I,J) = 0
 30   CONTINUE
      ITMAT(I,I) = 1
 40   CONTINUE
      JMOUT = NP2
      ITMNU = NP2
      DO 50  I = 1,NP1
      DO 45  J = 1,NP1
      AINV(I,J) = 0.D0
 45   CONTINUE
      AINV(I,I) = 1.D0
 50   CONTINUE
      RETURN
      END
```

```
      SUBROUTINE  EAVES(ITMAT,KDIM,N,NP1,NP2,ITNUM,IGAM,
     *           IBETA,JMOUT,ITM,ITMNU)
      DIMENSION  ITMAT(KDIM,1)
      DIMENSION  ITM(1),ITNUM(1),IGAM(1),IBETA(1)
C
C     ***********************************
C     * REPLACEMENT STEP IN PRIMITIVE SET *
C     ***********************************
C
      IF(JMOUT.NE.1)  GOTO 400
      IG = IGAM(1)
      IF(IG.NE.1)  GOTO 200
      DO 112  I = 1,NP1
      IF(ITMAT(I,1).NE.2*ITMAT(I,NP2))  GOTO 115
 112  CONTINUE
      GOTO  210
C
C     JMOUT=1
C     CASE(3), I=1,J=1
C
 115  IDUM = 1
      I = 1
 120  II = I+IDUM
      IGAM(I) = IGAM(II)-1
      IF(IGAM(I).NE.N)  GOTO 140
      IDUM = 0
      I = I+1
      IGAM(I) = NP1
      JIN = I
 140  I = I+1
      IF(I.GT.NP1)  GOTO 230
      GOTO 120
 200  IF(IG.EQ.NP1)  GOTO 300
C
C     JMOUT=1
C     CASE(1), I=1
C
 210  DO 220  I = 2,NP1
      IGAM(I-1) = IGAM(I)
 220  CONTINUE
      IGAM(NP1) = IG
      JIN = NP2
 230  DO 240  I = 1,NP1
      ITM(I) = ITMAT(I,2)+ITMAT(I,JIN)-ITMAT(I,1)
 240  CONTINUE
      GOTO  800
```

```
C
C      JMOUT=1
C      CASE(4)
C
 300   DO 320  J = 1,N
       DO 305  I = 1,NP1
       IF(ITMAT(I,J+1).GT.ITMAT(I,J+2))  GOTO 310
 305   CONTINUE
 310   IBETA(J) = I
 320   CONTINUE
       DO 325  I = 1,NP1
       ITM(I) = ITMAT(I,2)
 325   CONTINUE
       KK = 0
       DO 350  I = 1,N
       IF(ITM(I)/2*2.EQ.ITM(I))  GOTO 335
       ITM(I) = ITM(I)+1
       ITM(I+1) = ITM(I+1)-1
       KK = KK+1
       GOTO  350
 335   DO 340  J = 1,N
       IF(I.EQ.IBETA(J))  GOTO 345
 340   CONTINUE
 345   IBETA(J) = -IBETA(J)
 350   CONTINUE
       IH = 0
       IHK = KK
       DO 370  I = 1,N
       IF(IBETA(I).GT.0)  GOTO 365
       IHK = IHK+1
       IGAM(I) = IHK
       GOTO  370
 365   IH = IH+1
       IGAM(I) = IH
 370   CONTINUE
       IGAM(NP1) = NP1
       DO 375  I = 1,NP1
       ITM(I) = ITMAT(I,2)-ITM(I)/2
 375   CONTINUE
       JIN = NP2
       GOTO  800
```

```
 400  IF(JMOUT.LT.NP2)  GOTO 600
      IG = IGAM(NP1)
      IF(IG.EQ.NP1)  GOTO 500
C
C     JMOUT=NP2
C     CASE(1), I=NP1
C
      DO 430  I = 2,NP1
      K = NP1-I+1
      IGAM(K+1) = IGAM(K)
 430  CONTINUE
      IGAM(1) = IG
 440  JIN = 1
      DO 450  I = 1,NP1
      ITM(I) = ITMAT(I,1)+ITMAT(I,JMOUT-1)-ITMAT(I,JMOUT)
 450  CONTINUE
      GOTO  850
C
C     JMOUT=NP2
C     CASE(2)
C
 500  DO 510  I = 1,NP1
      IGAM(I) = I-1
 510  CONTINUE
      IGAM(1) = NP1
      DO 520  I = 1,NP1
      ITM(I) = ITMAT(I,1)+ITMAT(I,NP1)
 520  CONTINUE
      JIN = 1
      GOTO  850
 600  IF(IGAM(JMOUT-1).NE.IGAM(JMOUT)-1)  GOTO 642
      DO 620  I = 1,NP1
      IF(ITMAT(I,JMOUT).NE.2*ITMAT(I,NP2))  GOTO 700
 620  CONTINUE
C
C     1 < JMOUT < NP2
C     CASE(1), I=JMOUT
C
 642  IDUM = IGAM(JMOUT-1)
      IGAM(JMOUT-1) = IGAM(JMOUT)
      IGAM(JMOUT) = IDUM
 660  DO 665  I = 1,NP1
      ITM(I) = ITMAT(I,JMOUT-1)+ITMAT(I,JMOUT+1)-ITMAT(I,JMOUT)
 665  CONTINUE
      JIN = JMOUT
      GOTO  900
```

```
 700  IF(IGAM(JMOUT).NE.NP1)  GOTO 660
C
C     1 < JMOUT < NP2
C     CASE(3), I=JMOUT,J=NP1
C
      DO 730  I = 1,NP1
      IGAM(I) = IGAM(I)+1
 730  CONTINUE
      DO 740  I = 2,JMOUT
      K = JMOUT-I+1
      IGAM(K+1) = IGAM(K)
 740  CONTINUE
      IGAM(1) = 1
      GOTO  440

 800  DO 820  J = 2,JIN
      ITNUM(J-1) = ITNUM(J)
      DO 820  I = 1,NP1
      ITMAT(I,J-1) = ITMAT(I,J)
 820  CONTINUE
      GOTO  900
 850  DO 870  J = 2,JMOUT
      JJ = JMOUT-J+1
      ITNUM(JJ+1) = ITNUM(JJ)
      DO 870  I = 1,NP1
      ITMAT(I,JJ+1) = ITMAT(I,JJ)
 870  CONTINUE

 900  DO 910  I = 1,NP1
      ITMAT(I,JIN) = ITM(I)
 910  CONTINUE
      ITNUM(JIN) = ITMNU
      RETURN
      END
```

```
      SUBROUTINE  LABELS(ITM,NP1,AIN,NACT)
      IMPLICIT  REAL*8  (A-H,O-Z)
      DIMENSION  ITM(1),AIN(1)
C
C     **********************************
C     * ASSOCIATE LABEL TO SLACK-VECTOR *
C     **********************************
C
      DO 10  I = 1,NP1
      II = I
      IF(ITM(II).EQ.0)  GOTO 15
 10   CONTINUE
      NACT = 0
      RETURN
 15   DO 20  K = 1,NP1
      IK = II+K
      IF(IK.GT.NP1)  IK=IK-NP1
      IF(ITM(IK).GT.0)  GOTO 25
 20   CONTINUE
 25   KSLEK = IK-1
      IF(KSLEK.EQ.0)  KSLEK=NP1
      DO 30  I = 1,NP1
      AIN(I) = 0.D0
 30   CONTINUE
      AIN(KSLEK) = 1.D0
      NACT = KSLEK
      RETURN
      END
```

```
      SUBROUTINE  LABELF(ITM,NP1,AIN)
      IMPLICIT  REAL*8  (A-H,O-Z)
      DIMENSION  ITM(1),AIN(1)
C
C     *************************************
C     * ASSOCIATE LABEL TO INTERIOR VECTOR *
C     *************************************
C
      NGRID = 0
      DO 10  I = 1,NP1
      NGRID = NGRID+ITM(I)
 10   CONTINUE
      GRID = NGRID
      DO 30  I = 2,NP1
      AIN(I-1) = (ITM(I)-ITM(I-1))/GRID+1.D0
 30   CONTINUE
      AIN(NP1) = (ITM(1)-ITM(NP1))/GRID+1.D0
      RETURN
      END
```

```
      SUBROUTINE  PIVOT(AINV,KDIM,AIN,W,PHULP,CRITER,NP1,NP2,IAACT,NACT,
     $     INVE,IANUM,ITNUM,JMOUT,ITMNU)
      IMPLICIT REAL*8 (A-H,O-Z)
      DIMENSION  AINV(KDIM,KDIM),AIN(1),W(1),PHULP(1)
      DIMENSION  IAACT(1),IANUM(1),ITNUM(1)
C
C     *************
C     * PIVOT STEP *
C     *************
C
      IF(NACT.EQ.0)  GOTO 215
      DO 205  I = 1,NP1
      JAOUT = I
      IF(IAACT(I).EQ.NACT)  GOTO 235
  205 CONTINUE
  215 XMAX = 0.D0
      DO 218  I = 1,NP1
      PHULP(I) = 0.D0
      DO 218  J = 1,NP1
      PHULP(I) = PHULP(I)+AINV(I,J)*AIN(J)
  218 CONTINUE
      DO 230  I = 1,NP1
      IF(PHULP(I).LE.CRITER)  GOTO 230
      QHULP = 0.D0
      DO 222  J = 1,NP1
      QHULP = QHULP+AINV(I,J)*W(J)
  222 CONTINUE
      IF(QHULP.GT.CRITER)  GOTO 226
      JAOUT = I
      GOTO  232
  226 QUOT = PHULP(I)/QHULP
      IF(QUOT.LE.XMAX)  GOTO 230
      XMAX = QUOT
      JAOUT = I
  230 CONTINUE
  232 INVE = INVE+1
C
C     BARTELS' PROCEDURE
C
      DO 234  IC = 1,NP1
      CONST = AINV(JAOUT,IC)/PHULP(JAOUT)
      DO 233  JR = 1,NP1
      AINV(JR,IC) = AINV(JR,IC)-PHULP(JR)*CONST
  233 CONTINUE
      AINV(JAOUT,IC) = CONST
  234 CONTINUE

  235 IAOUD = IANUM(JAOUT)
      IANUM(JAOUT) = ITMNU
      IAACT(JAOUT) = NACT
      DO 270  I = 1,NP2
      IF(ITNUM(I).EQ.IAOUD)  GOTO 300
  270 CONTINUE
  300 JMOUT = I
      ITMNU = ITNUM(JMOUT)
      RETURN
      END
```

```
EXAMPLE OF THE USE OF EAVES' PROCEDURE

FUNCTION  F(X(I)) = X(I+1)   FOR  I=1,..,N-1
          F(X(N)) = X(1)


N = 10

GRIDSIZE      # OF ITERATIONS       # OF INVERSIONS          MAX(ABS(F(X)-X))
       2              0                    0                  0.5000000D+00
       4             10                    0                  0.5000000D+00
       8             38                    0                  0.2500000D+00
      16            117                    0                  0.1250000D+00
      32            188                   47                  0.6250000D-01
      64            206                   65                  0.3125000D-01
     128            224                   83                  0.1562500D-01
     256            242                  101                  0.7812500D-02
     512            260                  119                  0.3906250D-02
    1024            278                  137                  0.1953125D-02
    2048            296                  155                  0.9765625D-03
    4096            314                  173                  0.4882813D-03
    8192            332                  191                  0.2441406D-03
   16384            350                  209                  0.1220703D-03
   32768            368                  227                  0.6103516D-04
   65536            386                  245                  0.3051758D-04
  131072            404                  263                  0.1525879D-04


FIRST COLUMN OF FINAL MATRIX 'ITMAT'

  13107  13108  13106  13108  13108  13106  13108  13106  13108  13107

END OF EXECUTION
CPU TIME: 1.94  ELAPSED TIME: 1:2.78
EXIT
```

```
EXAMPLE OF THE USE OF EAVES' PROCEDURE

FUNCTION  F(X(I)) = X(I+1)   FOR  I=1,..,N-1
          F(X(N)) = X(1)


N = 20
```

GRIDSIZE	# OF ITERATIONS	# OF INVERSIONS	MAX(ABS(F(X)-X))
2	0	0	0.5000000D+00
4	20	0	0.5000000D+00
8	78	0	0.2500000D+00
16	287	0	0.1250000D+00
32	846	0	0.6250000D-01
64	1558	423	0.3125000D-01
128	1620	485	0.1562500D-01
256	1682	547	0.7812500D-02
512	1744	609	0.3906250D-02
1024	1806	671	0.1953125D-02
2048	1868	733	0.9765625D-03
4096	1930	795	0.4882813D-03
8192	1992	857	0.2441406D-03
16384	2054	919	0.1220703D-03
32768	2116	981	0.6103516D-04
65536	2178	1043	0.3051758D-04
131072	2240	1105	0.1525879D-04

```
FIRST COLUMN OF FINAL MATRIX 'ITMAT'

   6553   6554   6554   6554   6552   6554   6554   6554   6554   6554

   6552   6554   6554   6554   6554   6552   6554   6554   6554   6553

END OF EXECUTION
CPU TIME: 25.76 ELAPSED TIME: 1:51.16
EXIT
```

APPENDIX A:

PROOF OF THE EQUIVALENCE OF THE REFORMULATED AND EAVES' ORIGINAL REPLACEMENT STEP

In our proof we will follow the flow chart of Figure 6 .

1) JMOUT = 1, $\gamma_1 = 1$ and $T^1 = 2 \times T^{n+1}$.

$T^1 = 2 \times T^{n+1}$ implies $T_\sigma(v^1) = 2 \times T_\sigma(v^{n+1})$ and since T_σ is a linear transformation on σ having full rank we may conclude that

$$v^1 = 2\, v^{n+1} = 2(v^1 + \sum_{i=1}^{n} q(\gamma i)) =$$

$$= 2(v^1 + \begin{bmatrix} -1 \\ 0 \\ . \\ . \\ 0 \end{bmatrix})$$

and therefore $v^1 = \begin{bmatrix} 2 \\ 0 \\ . \\ . \\ 0 \end{bmatrix}$ and $v^{n+1} = \begin{bmatrix} 1 \\ 0 \\ . \\ . \\ 0 \end{bmatrix}$. Moreover, $v^2 = \begin{bmatrix} 1 \\ 1 \\ 0 \\ . \\ 0 \end{bmatrix}$.

Computing v we find

$$v = v^{n+1} + v^2 - v^1 = \begin{bmatrix} 1 \\ 0 \\ . \\ . \\ 0 \end{bmatrix} + \begin{bmatrix} 1 \\ 1 \\ 0 \\ . \\ 0 \end{bmatrix} - \begin{bmatrix} 2 \\ 0 \\ \\ \\ 0 \end{bmatrix} = \begin{bmatrix} 0 \\ 1 \\ 0 \\ . \\ 0 \end{bmatrix} ,$$

yielding case 1), i = 1 .

Case 1) shows that $\sigma(u^1 , \beta)$ does not change and so

[1)]

$$T_{\sigma'}(v) = T_\sigma(v) = T_\sigma(v^{n+1}) + T_\sigma(v^2) - T_\sigma(v^1)$$

or in our notation

$$TIN = T^{n+1} + T^2 - T^1 \quad .$$

Furthermore, Table 1 shows that the new τ , denoted by τ' , is built up in the following way

$$\tau' = \{v^2 , v^3 , . . . , v^{n+1} , v\}$$

and so TIN must be introduced in T in the (n+1)-st place, (JIN = n+1) .

2) JMOUT = 1, $\gamma_1 = 1$ and $T^1 \neq 2 \times T^{n+1}$.

$T^1 \neq 2 \times T^{n+1}$ implies

$v^1 \neq \begin{bmatrix} 2 \\ 0 \\ . \\ . \\ 0 \end{bmatrix}$ and since $\gamma_1 = 1$, the first element of v^1 , denoted by v_1^1 must be 1 and $v_1^2 = 0$.

From 1 we saw that $v^1 - v^{n+1} = \begin{bmatrix} 1 \\ 0 \\ . \\ . \\ 0 \end{bmatrix}$ and so $v_1^{n+1} = 0$.

[1)] All expressions with a prime will refer throughout this appendix to situations after the replacement step has been executed.

Computing

$$v_1 = v_1^{n+1} + v_1^2 - v_1^1 = -1$$

we arrive at case 3), with $i = 1$ and $j = 1$.
Table 2 shows that in this case

$$\sigma' = \{u^2 , . . . , u^n , u^n + q(\beta_1)\}$$

and using the changes in v^1 and γ as stated in Table 3 , we find

$$T_{\sigma'}(v^{1'}) \overset{\text{def}}{=} \sum_{i=1}^{n} u^{i'} v_i^{1'} = \sum_{i=1}^{n-1} u^{i+1} v_{i+1}^2 \text{ , since } v_n^{1'} = 0 ,$$

$$= \sum_{i=1}^{n} u^i v_i^2 \text{ , since } \gamma_1 = 1 \text{ and so } v_1^2 = 0 ,$$

$$= T_\sigma(v^2) .$$

$$T_{\sigma'}(v^{2'}) \overset{\text{def}}{=} \sum_{i=1}^{n} u^{i'} v_i^{2'} = \sum_{i=1}^{n-1} u^{i+1} v_{i+1}^3 \text{ , since } v_n^{2'} = 0$$

$$\text{and } \gamma_1' = \gamma_2 - 1 ,$$

$$= \sum_{i=1}^{n} u^i v_i^3 = T_\sigma(v^3) \text{ , since } v_1^3 = 0 .$$

This procedure may be continued until $T_{\sigma'}(v^{k'})$, $\gamma_k = n$.
From Table 3 we learn that

$$T_{\sigma'}(v^{k'}) \overset{\text{def}}{=} \sum_{i=1}^{n} u^i v_i^{k'} = \sum_{i=1}^{n-1} u^{i+1} v_{i+1}^k - u^n v_n^k +$$

$$+ u^n (v_n^k - 1) + u^n + q(\beta_1) = T_\sigma(v^k) + q(\beta_1) .$$

It is immediately clear that this is the vector to be introduced in T, since $T_{\sigma'}(v^{i'}) = T_\sigma(v^i)$, $i = k+1$, $n+1$, so JIN = k , and since $\gamma_k = n$, k = NRLH .

It is straightforward to prove the calculation rule

$$\text{TIN} = T^{\text{NRLH}} + T^2 - T^1$$

$$= T_\sigma(v^k) + \sum_i u^i v_i^2 - \sum_i u^i v_i^1$$

, and since $\gamma_1 = 1$ we may write

$$= T_\sigma(v^k) + \sum_i u^i v_i^1 - u^1 + u^2 - \sum_i u^i v_i^1 = T_\sigma(v^k) + q(\beta_1)$$

$$= T_{\sigma'}(v^{k'}) \quad .$$

3) JMOUT = 1, $\gamma_1 = n$.

From $v^1 - v^{n+1} = \begin{bmatrix} 1 \\ 0 \\ . \\ . \\ 0 \end{bmatrix}$ and $\gamma_1 = n$ it can be seen that v_1^1 and v_n^1 cannot be zero.

Therefore $v^1 = \begin{bmatrix} 1 \\ 0 \\ . \\ . \\ 1 \end{bmatrix}$, $v^2 = \begin{bmatrix} 1 \\ 0 \\ . \\ . \\ 0 \end{bmatrix}$ and $v^{n+1} = \begin{bmatrix} 0 \\ . \\ . \\ 0 \\ 1 \end{bmatrix}$.

Computing $v = v^2 + v^{n+1} - v^1 = \begin{bmatrix} 0 \\ . \\ . \\ . \\ 0 \end{bmatrix}$, we arrive at case 4) .

4) JMOUT = 1, $\gamma_1 \neq 1, n$.

From $v = v^2 + v^{n+1} - v^1$ or $v = v^2 - \begin{bmatrix} 1 \\ 0 \\ \cdot \\ \cdot \\ 0 \end{bmatrix}$ it can be seen that case 1)

occurs only if a) $v_1^2 \neq 0$, and b) $0 < \sum_i v_i \leq 2$.

ad a). Since $\gamma_1 \neq 1$ it follows that $v_1^2 = v_1^1 \neq 0$.

ad b). Since $\gamma_1 \neq n$, $\sum_i v_i^2 = \sum v_i^1 = 2$, and so $0 < \sum_i v_i \leq 2$.

The proof that TIN = $T^{n+1} + T^2 - T^1$ and JIN = n + 1 is completely analogue to 1 .

5) JMOUT = n + 1, $\gamma_n = n$.

From $\gamma_n = n$ it follows that $\sum_i v_i^n = 2$. So if we compute

$$v = v^1 + v^n - v^{n+1} = \begin{bmatrix} 1 \\ 0 \\ \cdot \\ \cdot \\ 0 \end{bmatrix} + v^n$$

, we arrive at case 2), since $\sum_i v_i = 3$.

We will prove now that the new vector to be introduced in T may be computed as TIN = $T^1 + T^n$ and that JIN = 1 .

The scheme of case 2) tells us that

$$T^{1'} = u_1' + u_n' \quad , \text{ since } \quad \gamma' = (1, 0, \ldots, 0, 1) \quad .$$

This may be rewritten as

$$T^{1'} = 2\, u_1' + \sum_{i=1}^{n-1} q(\beta_i)$$

$$= 2\, \{u^1\, \Theta_1 + \sum_{i=1}^{n-1} \Theta_{i+1}\, q(\beta_i)\} + \sum_i q(\beta_i)$$

$$= 2\, \{ \sum_{i=2}^{n} \Theta_i (u^i - u^{i-1}) + u^1\, \Theta_1 \} + \sum_i q(\beta_i)$$

$$= 2\, \{ \sum_{i=1}^{n-1} u^i (\Theta_i - \Theta_{i+1}) + u^n\, \Theta_n \} + \sum_i q(\beta_i)$$

$$= 2 \sum_{i=1}^{n} u^i\, v_i^1 + \sum_i q(\beta_i) = 2\, \sigma\, v^1 + \sum_i q(\beta_i)$$

$$= 2\, \sigma\, v^1 + \sum_{i=1}^{n-1} q(\gamma_i) \quad , \quad \text{since} \quad \gamma_n = n \quad ,$$

$$= 2\, \sigma\, v^1 + \sigma \sum_{i=1}^{n-1} q(\gamma_i)$$

$$= \sigma (v^1 + v^n) = T^1 + T^n \quad .$$

q.e.d.

6) JMOUT = $n + 1$, $\gamma_n \neq n$.

Since $\gamma_n \neq n$, $\sum_i v_i^n = 1$, computing

$$v = \begin{bmatrix} 1 \\ 0 \\ . \\ . \\ 0 \end{bmatrix} + v^n$$ it is immediately clear that $0 < \sum_i v_i \leq 2$ and $v_i \geq 0$,

for all i , which corresponds to case 1), $i = n+1$.

The proof that $TIN = T^1 + T^n - T^{n+1}$ and $JIN = 1$ is analogue to 1 .

7) $JMOUT \neq 1,\ n+1$, $\gamma_{JMOUT-1} = \gamma_{JMOUT} - 1$, $T^{JMOUT} = 2 \times T^{n+1}$.

In 1 we already stated that $T^{JMOUT} = 2 \times T^{n+1}$ implies

$v^{JMOUT} = 2 \times v^{n+1}$ and moreover that v^{JMOUT} can only have one entry differing from zero equal to 2 and since $JMOUT \neq 1$ this entry cannot be the first element of

$$v^{JMOUT} .$$

Suppose the k-th element of v^{JMOUT} is 2 . It can be seen immediately

that $v_k^{JMOUT-1} = 1$ and $v_{k-1}^{JMOUT-1} = 1$ and

$$v_k^{JMOUT+1} = 1 \quad \text{and} \quad v_{k+1}^{JMOUT+1} = 1 .$$

From $v = v^{JMOUT-1} + v^{JMOUT+1} - v^{JMOUT}$ it follows then that only

$v_{k-1} = v_{k+1} = 1$, while all other elements of v are zero.

So we have case 1) , $i = JMOUT$.

The new vector TIN may be computed as

$TIN = T^{JMOUT-1} + T^{JMOUT+1} - T^{JMOUT}$ and $JIN = JMOUT$, which is

easily proved according to 1 .

8) $JMOUT \neq 1,\ n+1$, $T^{JMOUT} \neq 2 \times T^{n+1}$, $\gamma_{JMOUT-1} = \gamma_{JMOUT} - 1$

Suppose $\gamma_{JMOUT} = k$ and $\gamma_{JMOUT-1} = k-1$.

Since $v_k^{JMOUT} \neq 2$ and $\gamma_{JMOUT} = k$, v_k^{JMOUT} must equal 1 and $v_k^{JMOUT+1} = 0$ and moreover $v_k^{JMOUT-1} = 0$, since

$\gamma_{JMOUT-1} = k - 1$. So $v_k = v_k^{JMOUT-1} + v_k^{JMOUT+1} - v_k^{JMOUT} = -1$,

corrsponding to case 3) .

a) $\gamma_{JMOUT} \neq n$.

Since $\gamma_{JMOUT} \neq 1, n$ it can be seen that $v_1 \neq -1$ and $v_n \neq -1$.

Therefore we are dealing with $1 < j < n$ in Table 2 and

$1 < j < n$, $1 \leq i \leq n+1$ in Table 3 .

From Table 2 is can be seen that σ is unchanged, except for

u^k , $u^{k'} = u^{k-1} + u^{k+1} - u^k$. So in general this may affect $T_\sigma(\tau)$

for all v^i , having a nonzero element in the k-th place. But from our

previous discussion it appears that v^{JMOUT} can only have a nonzero

element in the k-th place, implying that in $T_\sigma(\tau)$ only $T_\sigma(v^{JMOUT})$

can change; so JIN = JMOUT .

We compute TIN according to

$$
\begin{aligned}
TIN &= T^{JMOUT-1} + T^{JMOUT+1} - T^{JMOUT} \\
&\stackrel{def}{=} \sigma\, v^{JMOUT-1} + \sigma\, v^{JMOUT+1} - \sigma\, v^{JMOUT} \\
&= \sigma(v^{JMOUT} - q(k - 1) + \sigma(v^{JMOUT} + q(k)) - \sigma\, v^{JMOUT} \\
&= \sigma\, v^{JMOUT} + \sigma(q(k) - q(k - 1))
\end{aligned}
$$

$$= \sigma\, v^{JMOUT} + u^{k-1} - 2\, u^{k} + u^{k+1}$$

$$= \sigma\, v^{JMOUT} + u^{k'} - u^{k} = \sigma'\, v^{JMOUT} \quad , \text{ since}$$

$$v_k^{JMOUT} = 1 \ ,$$

$$= T_{\sigma'}(v^{JMOUT}) = T_{\sigma'}(v^{JMOUT'}) \quad , \text{ since } \tau(v^1 , \gamma)$$

does not change according to Table 3 .

b) $\gamma_{JMOUT} = n$.

In this case we are dealing with $j = n$ in Table 2 and

$j = n$, $1 \le i \le n$ in Table 3 .

If we are able to prove that

$$TIN = T^1 + T^{JMOUT-1} + T^{JMOUT} = T_{\sigma'}(v^{1'})$$

then it is clear that $T_{\sigma'}(v^{1'})$ is the new vector to be introduced, since $\{T^1 , \ldots , T^{JMOUT}\}$ consisting of vectors in the same level must form a linearly independent set. So in that case $JIN = 1$.

We will compute $T_{\sigma'}(v^{1'})$ in the following way

$$T_{\sigma'}(v^{1'}) \overset{def}{=} \sigma'\, v^{1'} \quad .$$

Using the definition of $v^{1'}$ from Table 3 , we may write

$$T_{\sigma'}(v^{1'}) = \sigma'\, v^{1'} = u^{1'} - u^{2'} + \sum_{i=2}^{n} u^{i'}\, v_{i-1}^{1}$$

and from Table 2 , this may be reformulated in terms of the original

$\sigma(u^1 , \beta)$ by

$$= u^1 - q(\beta_{n-1}) - u^1 + \sum_{i=1}^{n-1} u^i v_i^1$$

$$= u^{n-1} - u^n + \sum_{i=1}^{n} u^i v_i^1 \quad , \text{ since } v_n^1 \text{ must be zero,}$$

due to the fact that $\gamma_{JMOUT} = n$ and $\gamma_{JMOUT-1} = n-1$,

$$= - \sigma q(n-1) + T^1$$

$$= T^1 + \sigma(v^{JMOUT} - q(n-1)) - \sigma v^{JMOUT}$$

$$= T^1 + T^{JMOUT-1} - T^{JMOUT} \quad ,$$

q.e.d.

9) JMOUT $\neq$ 1, n + 1 , $\gamma_{JMOUT-1} \neq \gamma_{JMOUT} - 1$.

$\gamma_{JMOUT-1} \neq \gamma_{JMOUT} - 1$ implies that if $\gamma_{JMOUT} = k$,

$v_k^{JMOUT} = 1$, $v_k^{JMOUT+1} = 0$ and $v_k^{JMOUT-1} = 1$

and supposing further that $\gamma_{JMOUT-1} = \ell \neq k-1$, we are in the following position

$$v_\ell^{JMOUT-1} = 1 \quad , \quad v_\ell^{JMOUT} = 0 \quad ,$$

$$v_{\ell+1}^{JMOUT-1} = 0 \quad , \quad v_{\ell+1}^{JMOUT} = 1 \quad , \quad v_{\ell+1}^{JMOUT+1} = 1 \quad ,$$

$$v_k^{JMOUT-1} = 1 \quad , \quad v_k^{JMOUT} = 1 \quad , \quad v_k^{JMOUT+1} = 0 \quad .$$

Computing v it then follows that $v_i \geq 0$ for all i and $\sum_i v_i \leq 2$, corresponding to case 1) , i = JMOUT , which in its turn comes up for the same implementations as in 7 .

REFERENCES

1. Bartels, R.H., A Stabilization of the Simplex Method, Num. Math. 16, 414-434 (1971).

2. Dohmen, J.; Schoeber, J., Approximated Fixed Points, Research Memorandum EIT/55, Dept. of Econometrics, Tilburg University, Tilburg (1975).

3. Eaves, B.C., Homotopies for Computation of Fixed Points, Math. Programming 3, 1-22 (1972).

4. Scarf, H., The Approximation of Fixed Points of a Continuous Mapping, SIAM Journal of Applied Mathematics 15, 1328-1343 (1967).

5. Scarf, H., The Computation of Economic Equilibria, Yale University Press, New Haven 1973.

Numerical Solution of Highly Nonlinear Problems
W. Forster (ed.)
© North-Holland Publishing Company, 1980

FIXED POINT ALGORITHMS AND ARRAY PROCESSORS

Walter Forster

University of Southampton,
Southampton, England

1. INTRODUCTION

Fixed point algorithms allow the solution of e.g. systems of nonlinear equations in an n-dimensional Euclidian space. With increasing dimension n the algorithms become progressively slower. It is therefore important to try to improve the performance of fixed point algorithms. In [1] it is shown that the rate of convergence depends on the labelling and on the triangulation. To increase the speed of convergence to a completely labelled simplex (i.e. to an approximate solution) one can utilize e.g. special properties of the function [2], [3], or one can utilize special triangulations [4], etc. One has to remember that the number of simplices in a unit cube grows with $n!$, where n is the dimension of the cube. If we can find triangulations which minimize the number of simplices in a unit cube, then in general we will require a smaller number of steps to reach a completely labelled simplex. Mara [5] conjectures that there are triangulations of the n-dimensional unit cube with $(n! + 2^{n-1})/2$ simplices of dimension n. No constructive method is known to produce such triangulations for arbitrary dimension n.

This paper is concerned with utilizing an advanced computer structure to achieve an improved performance of fixed point algorithms. It will be shown on examples that it is possible to achieve considerable improvements.

For integer labelling e.g. it is possible to make the parallel machine equivalent of the number of operations required per pivot step independent of the dimension n of the problem. On ordinary computers the number of operations per pivot step increases with the dimension n of the problem and that means the computation time required becomes progressively longer. On an array processor it is possible to increase the speed by a factor of n without e.g. requiring the function to have special properties (i.e. one can avoid testing functions for special properties).

2. COMPUTER DESIGNS

Let us first give a brief description of conventional computers and array processors.

A conventional computer can be described as follows.

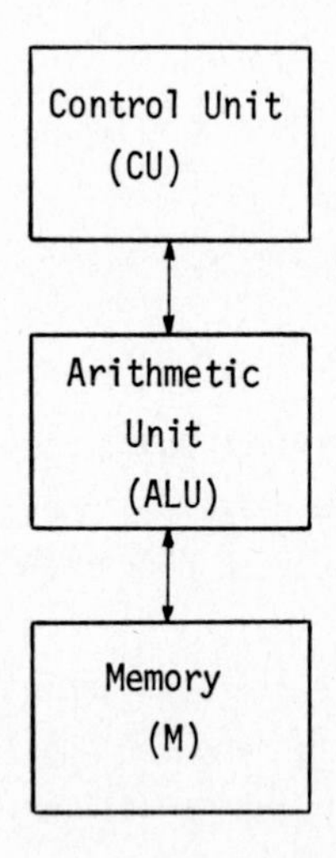

It consists of a control unit (CU), an arithmetic and logic unit (ALU), and a memory (M). In the ALU only one arithmetic or logical operation can be performed at a time. Input and output devices are not important in this context.

An array processor can be described as follows.

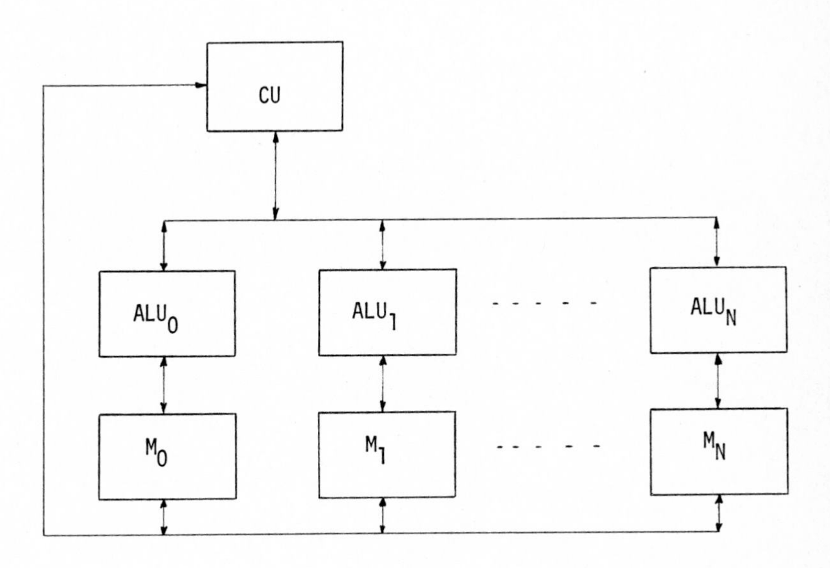

We have one control unit (CU), we have more than one e.g. $N+1$ arithmetic and logic units (ALU_0 to ALU_N), and each ALU has its own memory (M_0 to M_N). We call an ALU plus its M a processing element (PE). Each ALU can perform one arithmetic or logical operation at a time, but because of the structure of this computer one can perform e.g. $N+1$ multiplications at the same time. It is usually possible with this type of computer design to route data from each PE to adjacent neighbours. Furthermore, each processing element can receive instructions broadcast by the CU . For further details see e.g. [6].

In the next section we will show that this kind of computer design is very suitable for fixed point algorithms.

3. OPERATIONS COUNT FOR ONE PIVOT STEP AND TRIANGULATION K_2

We assume an n-dimensional simplex is embedded in the usual way in an R^{n+1}, e.g.

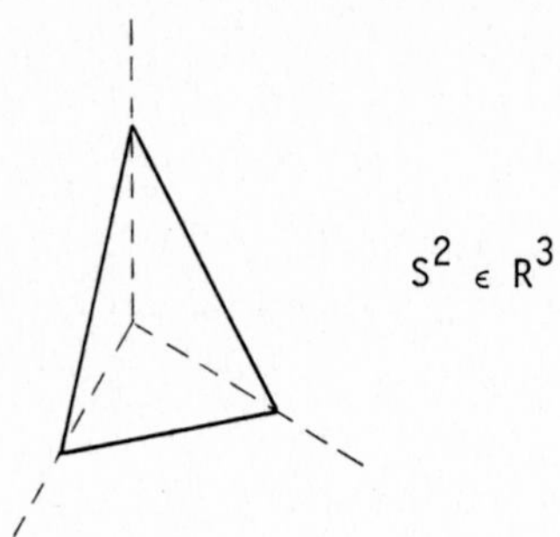

We have a triangulation called K_2, e.g. for S^2

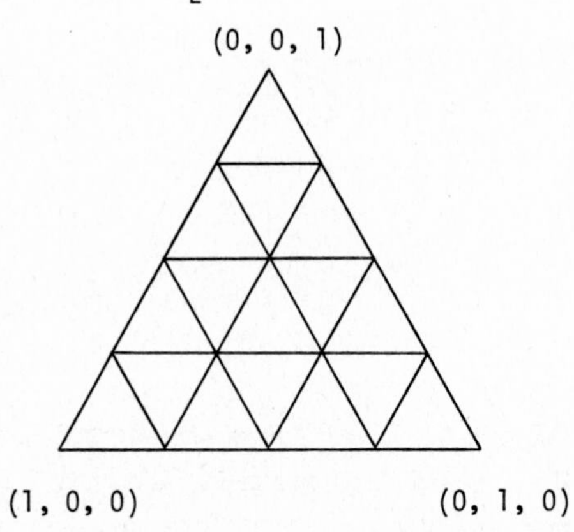

We have the following pivot rules (for details see e.g. [7]). We go from a simplex $\sigma = k(y^0, \pi)$ to a simplex $\tau = k(z^0, \rho)$ via the following pivot rules

	z^0	ρ	new vertex
$i = 0$	$y^0 + \frac{1}{m} q^{\pi(1)}$	$(\pi(2), \pi(3), \ldots, \pi(n), \pi(1))$	z^n
$0 < i < n$	y^0	$(\pi(1), \ldots, \pi(i+1), \pi(i), \ldots, \pi(n))$	z^i
$i = n$	$y^0 - \frac{1}{m} q^{\pi(n)}$	$(\pi(n), \pi(1), \pi(2), \ldots, \pi(n-1))$	z^0

with y^0 a vertex of σ ,

π a permutation of the numbers $\{0, 1, \ldots, n\}$,

z^0 a vertex of τ ,

ρ a permutation of the numbers $\{0, 1, \ldots, n\}$,

q^j the jth row of

$$\begin{bmatrix} -1 & 1 & 0 & . & . & . & 0 & 0 \\ 0 & -1 & 1 & 0 & . & . & 0 & 0 \\ & & & & & & & \\ . & . & . & . & . & . & . & . \\ & & & & & & & \\ 0 & 0 & & & & & -1 & 1 \end{bmatrix} ,$$

i.e. a vector in R^{n+1} .

The labelling function is

$$i = \min_j \{j \mid f_j \le y_j > 0\} .$$

We assume we use the artificial layer algorithm [8], [7].

We first determine the number of operations required for one pivot step on a conventional computer.

	number of steps
search for double label	n
renaming of vertices	n
computation of new vertex z^i	$n+1$
function evaluations	$(n+1)\#F$
determination of label of z^i	$\le 2(n+1)$
total number of operations for one pivot step and triangulation K_2 on a conventional computer	$\le 5n + 3 + (n+1)\#F$

The operation count shows that the number of operations necessary for one pivot step is growing linearly with the dimension n of the problem. This implies that the computation time will grow similarly with n.

If we consider an array processor with $N+1$ processing elements (PE), then we arrive at the following parallel machine equivalent (i.e. instead of performing $N+1$ operations one after the other, we can perform $N+1$ operations at the same time. The given pivoting rules and integer labelling allow parallel execution). We assume we have $(N+1) \geq (n+1)$ PE's.

	parallel machine equivalent
search for double label	1
renaming of vertices	1
computation of new vertex z^i	1
function evaluations	#F
determination of label of z^i	2
total number of operations (parallel machine equivalent) for one pivot step and triangulation K_2 on a parallel machine	5 + #F

Let us give some details.

Search for double label: The new label is in the CU. From the CU it is broadcast to n PE's and compared with the label stored in the memory. Only the PE where the two labels are equal replies to the CU. The old coordinats of the vertex with this label are then erased from all PE's.

Renaming of vertices: We have assumed that data can be communicated to adjacent neighbours.

Name of old vertices:

y^0

y^1

.

y^n

Name of new vertices:

z^0
z^1
.
z^n

If y^0 is dropped, we rename as follows:

y^0			dropped
y^1	$\rightarrow$	z^0	
y^2	$\rightarrow$	z^1	
.		.	
.		.	
y^{i-1}	$\rightarrow$	z^{i-2}	
y^i	$\rightarrow$	z^{i-1}	
y^{i+1}	$\rightarrow$	z^i	
.		.	
.		.	
y^{n-1}	$\rightarrow$	z^{n-2}	
y^n	$\rightarrow$	z^{n-1}	
		z^n	new vertex .

If y^i $0 < i < n$ is dropped, we rename as follows:

y^0	$\rightarrow$	z^0	
y^1	$\rightarrow$	z^1	
.		.	
.		.	
y^{i-1}	$\rightarrow$	z^{i-1}	
y^i			dropped
		z^i	new vertex
y^{i+1}	$\rightarrow$	z^{i+1}	
.		.	
.		.	
y^{n-1}	$\rightarrow$	z^{n-1}	
y^n	$\rightarrow$	z^n	.

If y^n is dropped, we rename as follows:

$$\begin{array}{lcll} & & z^0 & \text{new vertex} \\ y^0 & \rightarrow & z^1 & \\ y^1 & \rightarrow & z^2 & \\ \cdot & & \cdot & \\ \cdot & & \cdot & \\ y^i & \rightarrow & z^{i+1} & \\ \cdot & & \cdot & \\ \cdot & & \cdot & \\ y^{n-2} & \rightarrow & z^{n-1} & \\ y^{n-1} & \rightarrow & z^n & \\ y^n & & & \text{dropped} \end{array} .$$

Determination of label of z^i : To determine the label of z^i we first compare $f_j(y) \le y_j$ and then check $y_j > 0$. Each PE $0 \le p \le n$, where the conditions of the labelling function are satisfied, transmits $n - p$ single pulses to the CU . We assume these transmissions are synchronized, so the CU receives only one string of pulses corresponding to the longest string of $n - p$ pulses.

Example: Let us assume we have a 10-dimensional problem, i.e. $n = 10$.
E.g. on PE_3 the conditions $f_3(y) \le y_3$ and $y_3 > 0$ are satisfied. Then PE_3 sends a string of 7 pulses to the CU.
E.G. On PE_6 the conditions $f_6(y) \le y_6$ and $y_6 > 0$ are satisfied. Then PE_6 sends a string of 4 pulses to the CU.
We now assume the pulses are synchronized. The CU registers 7 pulses. This is then converted in the CU into the new label. Here the new label is 3 .

We can e.g. store the coordinates of the vertices of the simplex under consideration in the following manner in the memory.

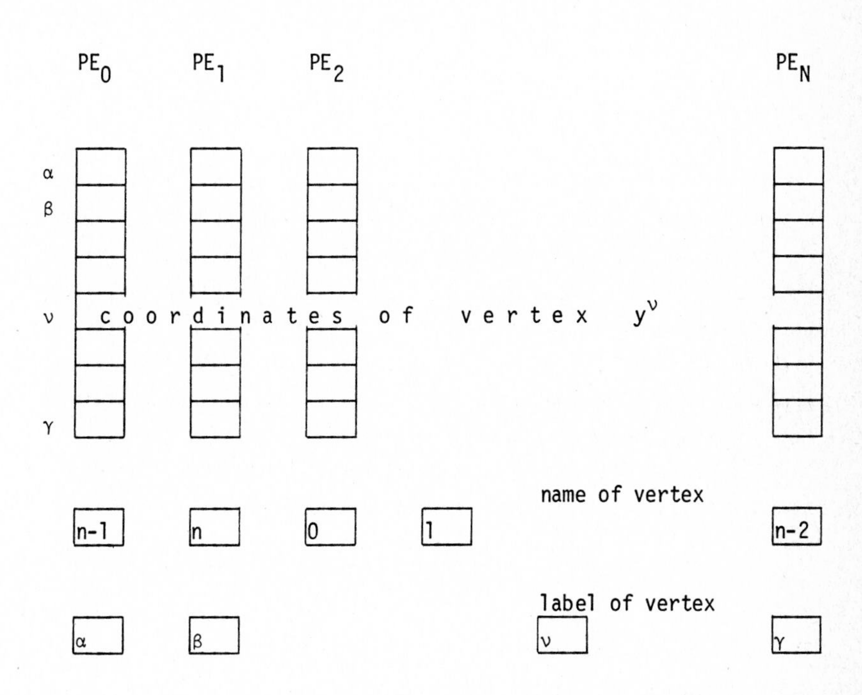

We use the first $n+1$ storage locations for the coordinates of the vertices y^μ (storage locations $\alpha, \beta, \ldots, \nu, \ldots, \gamma$), the (n+2)nd storage locations for the names of the vertices (e.g. 3 means y^3, etc.), the (n+3)rd storage location to store the label of the vertex which is stored in the (n+2)nd storage location.
In our example in storage location α of the memory of all $n+1$ PE's the coordinates of y^{n-1} are stored, in storage location β of the memory of all $n+1$ PE's the coordinates of y^n are stored, etc.
We assume e.g. that the coordinates stored in memory location α of all PE's corespond to the vertex with the name stored in the (n+2)nd storage location of PE_0, etc. Renaming of vertices then requires only transfer

of names and labels to neighbouring PE's and this is a task parallel processors are well suited for. Because we delete coordinates of vertices, compute the coordinates of new vertices and rename vertices, the data stored will change from step to step.

4. COMPUTING ROOTS OF POLYNOMIALS

In this section we sketch how an array processor can be used to speed up the search for roots of polynomials. If we consider polynomials with complex coefficients then we have a two-dimensional problem. The basic algorithm is due to Kuhn [9], [10]. We can use array processors in the following manner to improve the performance of the algorithm. We start searching the boundary (of the region given by Kuhn's estimates) for a completely labelled 1-simplex. Once we have found such a completely labelled 1-simplex we start searching the interior for a completely labelled 2-simplex. On an array processor we can at the same time continue to search the boundary for further completely labelled 1-simplices. In this manner we achieve some overlap in the search for the roots. If we assume that at some stage in the search we will have found all completely labelled 1-simplices on the boundary, then from this stage on all roots of the polynomial are searched for at the same time. This implies that a considerable saving in computation time can be achieved.

5. CONCLUSIONS

Utilizing advanced computer structures it is possible to speed up fixed point algorithms by a factor of n, where n is the dimension of the problem. For special types of problems, e.g. root-finding of polynomials, array processors can be applied with considerable savings in computation time. In the two examples discussed, we have assumed that we use integer labelling. The savings in computation time achieved in the first example (i.e. the parallel machine equivalent for one pivot step is independent of n) does not seem possible for vector labelling.

REFERENCES

1. Saigal, R., Investigations into the Efficiency of the Fixed Point Algorithms, in Fixed Points Algorithms and Applications, S. Karamardian (Ed.), Academic Press, New York 1977, pp. 203-223.

2. Kojima, M., On the Homotopic Approach to Systems of Equations with Separable Mappings, Mathematical Programming Study 7, 170-184 (1978).

3. Saigal, R.; Todd, M.J., Efficient Acceleration Techniques for Fixed Point Algorithms, to appear in SIAM J. Numer. Anal.

4. Todd, M.J., Exploiting Structure in Fixed Point Algorithms, to appear in the Proceedings of a Summerschool and Conference on "Functional Differential Equations and Approximation of Fixed Points, Bonn, July 1978.

5. Mara, P.S., Triangulations for the Cube, J. of Comb. Theory (A) 20, 170-176 (1976).

6. Bouknight, W.J.; Denenberg, S.A.; McIntyre, D.E.; Randall, J.M.; Sameh, A.H.; Slotnick, D.L., The Illiac IV System, Proceedings of the IEEE 60, 369-388 (1972).

7. Todd, M.J., The Computation of Fixed Points and Applications, Springer, Berlin-Heidelberg-New York 1976.

8. Kuhn, H.W.; MacKinnon, J.G., The Sandwich Method for Finding Fixed Points, J. Optimization Theory and Applications 17, 189-204 (1975).

9. Kuhn, H.W., A New Proof of the Fundamental Theorem of Algebra, Mathematical Programming Study 1, 148-158 (1974).

10. Kuhn, H.W., Finding Roots of Polynomials by Pivoting, in Fixed Points Algorithms and Applications, S. Karamardian (Ed.), Academic Press, New York 1977, pp. 11-39.

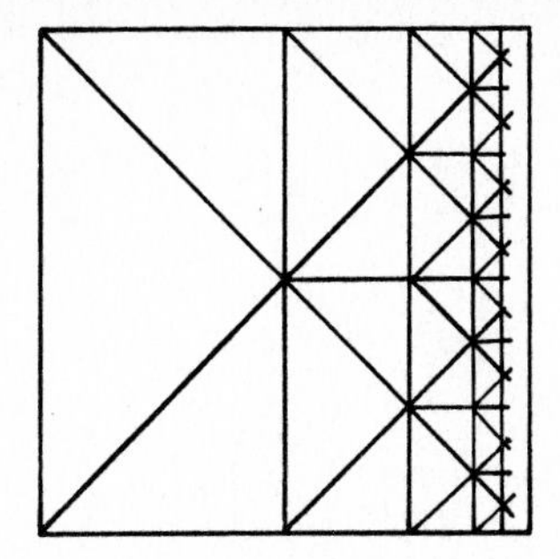

In the first paper of this section E. Sperner gives a generalization of his famous lemma. The other papers in this section either deal with Sperner's new lemma or are related to the classical Sperner Lemma.

Numerical Solution of Highly Nonlinear Problems
W. Forster (ed.)
© North-Holland Publishing Company, 1980

FIFTY YEARS OF FURTHER DEVELOPMENT OF A COMBINATORIAL LEMMA. PART A.

Emanuel Sperner

Universität Hamburg,
Hamburg, West Germany

EDITOR'S NOTE:
The following paper is a transcription of the slides E. Sperner used for his Southampton lecture. The reader of the following pages has to keep this in mind. The material is very condensed and includes an excellent summary of combinatorial lemmas and the main steps leading to the generalized version of Sperner's Lemma.

1. COMBINATORIAL LEMMA

$S = \{a_0, a_1, a_2, \ldots, a_n\}$ = n-dimensional simplex in R^n ,

a_i = vertex for all $i = 0, 1, \ldots, n$;

$\{a_{\nu_0}, a_{\nu_1}, \ldots, a_{\nu_m}\}$ = m-dimensional face $(0 \leq \nu_i \leq m)$,

$Z(S)$ = simplicial decomposition of S ,

A = set of all vertices in $Z(S)$.

Labelling (valuation):

$\varphi : A \to \{0, 1, 2, \ldots, n\}$.

Boundary condition:

For all $a \in A \cap \{a_{\nu_0}, a_{\nu_1}, \ldots, a_{\nu_m}\}$

$\Rightarrow$ there exist $i : 0 \leq i \leq m : \varphi(a) = \nu_i$.

Then the lemma says:

<u>Lemma</u>: For each $Z(S)$ and φ there exists in $Z(S)$ an odd number of subsimplices $s' = \{a'_0, a'_1, \ldots, a'_n\}$ with $\varphi(s') = \{0, 1, \ldots, n\}$.

□

Corollary (Covering theorem):

M_i closed point-sets in R^n ,

s_i = (n-1)-dimensional face of S opposite to a_i .

Corollary:

$$\left.\begin{array}{l} S = \bigcup_{i=0}^{n} M_i \\ M_i \cap s_i = \emptyset \quad \text{for all} \quad i = 0, 1, \ldots, n \end{array}\right] \Rightarrow \bigcap_{i=0}^{n} M_i \neq \emptyset .$$

□

First Applications:

1) Invariance of dimension (Sperner, 1928).
2) Invariance of region (Sperner, 1928).
3) Theorem of verification (Rechtfertigungssatz) in Menger's Theory of Dimension (K. Menger, 1928).
4) Brouwer's fixed point theorem (Knaster, Kuratowski, Mazurkiewicz, 1929).
5) Matrices with elements $\geqq 0$ (Ky Fan, 1958), theorems of Perron, Frobenius and others.

2. RELATED DEVELOPMENTS

Lattice in R^n = set of all points $I = (x_1, x_2, \ldots, x_n)$, $x_k \in Z$,

in particular: $i = (i, i, \ldots, i)$, $i \in Z$.

Definition:

$I \leqq I' :\Leftrightarrow x_k \leqq x'_k$; $I < I' :\Leftrightarrow x_k < x'_k$ for all $k = 1, \ldots, n$.

$I \leq I' :\Leftrightarrow I \leqq I' \wedge I \neq I'$.

1-cube: $\Gamma_1^{I_0} = \{I : I_0 \leqq I \leqq I_0 + 1\}$, I_0 = initial point .

p-cube: $\Gamma_p^0 = \{I : 0 \leqq I \leqq p\}$, 0 = initial point .

I = "interior" point of $\Gamma_p^0 :\Leftrightarrow 0 < I < p$;

otherwise: "boundary" point .

Simplicial subdivision (standard form):

$$Z(\Gamma_p^0) = \{S \subset \Gamma_p^0 : S = \{I_0 , I_1 , \ldots , I_n\} \text{ with}$$

$$I_0 \in \Gamma_{p+1}^0 \wedge I_0 \leq I_1 \leq I_2 \leq \ldots \leq I_0 + 1\} .$$

<u>Lemma</u> (A.W. Tucker, 1945):

A labelling φ of Γ_p^0 , such that

$\varphi : \Gamma_p^0 \to \{\pm 1 , \pm 2 , \ldots , \pm n\}$ and

1) $(I) + \varphi(I') \neq 0$ for all I , I' with $I \leqq I' \leqq I + 1$,

2) $\varphi(I) + \varphi(I') = 0$ for boundary points I , I'

with $I + I' = p$ ("antipodal" !)

does not exist.

□

<u>Ky Fan's Lemma</u> (1951):

For each labelling φ of Γ_p^0 such that

$\varphi : \Gamma_p^0 \to \{\pm 1 , \pm 2 , \ldots , \pm m\}$, and with properties 1), 2),

there exists in the subdivision $Z(\Gamma_p^0)$ an odd number of simplices

$S = \{I_0 , I_1 , \ldots , I_n\}$ with

$\varphi(S) = \{k_0 , -k_1 , k_2 , -k_3 , \ldots , (-1)^n k_n\} \wedge 1 \leqq k_0 < k_1 < \ldots < k_n \leqq m$.

□

<u>Corollary</u>: $m \geqq n + 1$, i.e. Tucker's lemma !

Definition:

"Faces" of Γ_p^0 : $s_{i0} = \{I \in \Gamma_p^0 : x_i = 0\}$,

$s_{i1} = \{I \in \Gamma_p^0 : x_i = p\}$, $\quad i = 1, 2, \ldots, n$.

H.W. Kuhn's "Cubical Sperner Lemma" (1960):

For each labelling $\varphi : \Gamma_p^0 \to \{0, 1, \ldots, n\}$

with $\varphi(I) \neq i - k$ for all $I \in s_{ik}$ {boundary condition}

there exists $S \in Z(\Gamma_p^0)$ with $\varphi(S) = \{0, 1, 2, \ldots, n\}$.

□

Applications:

Tucker-Fan: Antipodal theorems !

1. Lusternik-Schnirelmann (Covering of the n-sphere S^n by $(n+1)$ closed sets).
2. Borsuk-Ulam (Transformation of S^n into R^n).

Kuhn: 3. Derivation of Sperner's lemma from Brouwer's fixed point theorem ?

(3. finally solved by M. Yoseloff, 1972).

Another recent application: Constructive Proof of the Fundamental Theorem of Algebra (H.W. Kuhn, 1974, 1977).

3. SEARCH FOR A COMMON GENERALIZATION

1. Ky Fan (1960): Theorems on pseudomanifolds, "Combinatorial Properties of Certain Simplicial and Cubical Vertex Maps" (Arch. Math. 11, 368-377 (1960)).

2. P. Mani (1967): "Zwei kombinatorisch-geometrische Sätze vom (H. Hadwiger) Typus Sperner-Tucker-Ky Fan" (Monatshefte f. Math. 71, 427-435 (1967)).
3. E. Sperner (1970/71): "Kombinatorik bewerteter Komplexe" (Hamb. Abh. 39, 21-43 (1973)).

Main result:

$$\Phi^{\Delta C^m}_{K^{m+1}} \equiv \Phi^{C^m}_{\partial K^{m+1}} \qquad (\text{mod } 2)$$

(will be explained later).

4. DEVELOPMENT OF RELATED TOPICS ON ORIENTED COMPLEXES

1. F. Bagemihl (1953): "An Extension of Sperner's Lemma, ... " (Fund. Math. 40, 3-12 (1953)).
2. A.B. Brown (1961): "Oriented Version" of Sperner's Lemma in: S.S. Cairns: "Introductory Topology" "... a strengthening of Sperner's Lemma, which permits a substantial simplification of earlier proofs of the topological invariance of homology properties."
3. Ky Fan (1967): Oriented version of Fan's generalization of Tucker's Lemma; main formula on an oriented n-pseudomanifold M with labelling $\varphi : M \to \{\pm 1, \pm 2, \ldots, \pm m\}$ (with a certain restrictive property):

$$\sum_{0<k_0<k_1<\ldots<k_{n-1}\leqq m} \beta\,(k_0, -k_1, k_2, -k_3, \ldots, (-1)^{n-1} k_{n-1}) =$$

$$= \sum_{0<k_0<k_1<\ldots<k_n\leqq m} [\alpha\,(-k_0, k_1, -k_2, \ldots, (-1)^{n+1} k_n) +$$

$$+ (-1)^n \alpha\,(k_0, -k_1, k_2, -k_3, \ldots, (-1)^n k_n)] .$$

Explanation: Oriented version of Sperner's Lemma:

Let

p(i j) = number of boundary edges, which carry labels (i , j) ,

q(i j k) = number of triangles, which carry labels (i , j , k) ,

then there is always

$$p(i\ j) - p(j\ i) = q(i\ j\ k) - q(i\ k\ j)$$

for fixed $i \neq j$ and $k \neq i, j$.

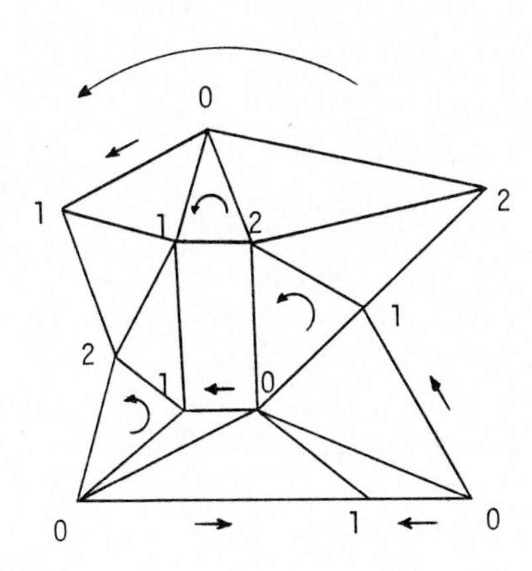

Problem: If C is any system of labelling n-sets, does there always exist a corresponding system ΔC of labelling (n+1)-sets, such that for each oriented and labelled complex containing "boundary" n-sets and "interior" (n+1)-sets, the algebraic sums of

1) those boundary n-sets, which carry labelling sets of C , and
2) those interior (n+1)-sets, which carry labelling sets of ΔC

are equal ?

5. SET - ORIENTATION

Given R = ring with 1 ,

$A = \{a_i\}_{i \in J}$ = a set , (later: basis set) .

Consider $S^m \in A^m$, i.e. $S^m = (a_1, a_2, \ldots, a_m) =$

= ordered m-tuple with $a_i \in A$.

Call S^m $\begin{cases} \text{regular, if } a_i \neq a_k \text{ for all } i, k : 1 \leq i < k \leq m , \\ \text{irregular: otherwise} . \end{cases}$

Denote by $*A^m$ the set of all regular elements of A^m .

Choose $A_m \subset *A^m$, say $A_m = \{S^m_\rho\}_{\rho \in P_m}$.

Call A_m representative, if for each $S^m \in *A^m$ there exists

exactly one $\rho \in P_m$ such that S^m is a permutation of S^m_ρ .

Put $S^m = \varepsilon S^m_\rho$: $\varepsilon \in R$ and

$$\varepsilon = \begin{cases} +1 & \text{if } S^m \text{ is an even permutation of } S^m_\rho \\ -1 & \text{if } S^m \text{ is an odd permutation of } S^m_\rho . \end{cases}$$

<u>Result (5)</u>: We have obtained a transformation

$$\begin{bmatrix} *A^m \to P_m \\ S^m \to \rho(S^m) \end{bmatrix} \quad \text{such that} \quad S^m = \varepsilon S^m_\rho$$

with $\varepsilon = \pm 1 \in R$, $\rho \in P_m$, both ε , ρ uniquely determined.

□

Suppose A_m fixed for all $m = 1, 2, 3, \ldots$.

Then A is called oriented by $A = (A_1, A_2, A_3, \ldots)$.

<u>Application</u>: Construct a vector space K_m , $m = 1, 2, 3, \ldots$,

with A_m as basis and R as the set of scalars, such that the

elements of K_m are $K^m = \{\alpha_\rho\}_{\rho \in P_m}$, $\alpha_\rho \in R$,

with component-wise addition and scalar (left-) multiplication.

Another way of writing this:

Put

$$S^m_{\rho_0} \equiv \{\alpha_\rho\}_{\rho \in P_m} , \text{ if } \begin{cases} \alpha_{\rho_0} = 1 \\ \alpha_\rho = 0 \text{ for } \rho \neq \rho_0 \end{cases} ,$$

then we obtain $K^m = \sum_{\rho \in P_m} \alpha_\rho S^m_\rho$.

Important subspace: $K^0_m = \{K^m : |\{\rho : \alpha_\rho \neq 0\}| < \infty\}$.

6. THE Δ - OPERATOR

Suppose $B = \{b_i\}_{i \in J}$ = set of "labels" ,

oriented by $B = (B_1, B_2, B_3, \ldots)$ as in section 5.

Notation: $B_m = \{T^m_\sigma\}_{\sigma \in \Gamma_m}$ [i.e. T^m instead of S^m

σ instead of ρ

Γ_m instead of P_m] .

Result (6): By B and the accompanying definitions (as in section 5.)

a transformation $\left[\begin{array}{ccc} {}^*B^m & \to & \Gamma_m \\ T^m & \to & \sigma(T^m) \end{array}\right]$ is determined

for all $m = 1, 2, 3, \ldots$

such that $T^m = \eta\, T^m_\sigma$ with $\eta = \pm 1 \in R$ (η, σ unique).

□

Corresponding vector space L_m

with elements $C^m = \{\beta_\sigma\}_{\sigma\in\Gamma_m} = \sum_{\sigma\in\Gamma_m} \beta_\sigma T^m_\sigma$; $(\beta_\sigma \in R)$.

Definition of Δ: Given $C^m = \{\beta^{(m)}_\sigma\}_{\sigma\in\Gamma_m}$.

For each $\sigma \in \Gamma_{m+1}$ with $T^m_\sigma = (b_0, b_1, \ldots, b_m)$

determine for all $k = 0, 1, \ldots, m$:

$$T^m_k = (b_0, \ldots, b_{k-1}, b_{k+1}, \ldots, b_m)$$

and furthermore with $\sigma(T^m_k) = \sigma_k \in \Gamma_m$:

$$T^m_k = \eta_k\, T^m_{\sigma_k} , \quad k = 0, 1, \ldots, m .$$

Then take β_{σ_k} out of C^m and form the sum

$$\beta^{(m+1)}_\sigma := \sum_{k=0}^{m} (-1)^k\, \eta_k\, \beta_{\sigma_k} .$$

With these $\beta^{(m+1)}_\sigma$ define

$$\Delta C^m = \{\beta^{(m+1)}_\sigma\}_{\sigma\in\Gamma_{m+1}} \quad \left(= \sum_{\sigma\in\Gamma_{m+1}} \beta^{(m+1)}_\sigma T^{m+1}_\sigma \right) .$$

□

Notice: 1) Δ is a transformation of L_m into L_{m+1} .

2) Δ is independent of the choice of B_m , B_{m+1} .

3) Δ is additive in C^m , i.e. $\Delta C_1^m + \Delta C_2^m = \Delta(C_1^m + C_2^m)$.

7. EFFECT OF LABELLING

From now on $\left[\begin{array}{l} A = \text{basis set} \\ \\ B = \text{set of labels} \end{array}\right]$ both oriented as in section 5. and 6.

Boundary operator ∂ (operating on A):

1) Definition of ∂S_ρ^m for all $S_\rho^m \in A_m$.

Suppose $S_\rho^m = (a_1, a_2, \ldots, a_m)$

and define
$$\partial S_\rho^m = \sum_{k=1}^{m} (-1)^{k+1} (a_1, \ldots, a_{k-1}, a_{k+1}, \ldots, a_m)$$
$$= \sum_{k=1}^{m} (-1)^{k+1} \varepsilon_k S_{\rho_k}^{m-1} ,$$

if according to Result (5) :

$$(a_1, \ldots, a_{k-1}, a_{k+1}, \ldots, a_m) = \varepsilon_k S_{\rho_k}^{m-1} .$$

2) Definition of ∂K^m for all $K^m \in K_m^0$.

Suppose $K^m = \{\alpha_\rho\}_{\rho \in P_m}$

and define
$$\partial K^m = \sum_{\rho \in P_m} \alpha_\rho \partial S_\rho^m = \sum_{\rho \in P_{m-1}} \alpha'_\rho S_\rho^{m-1} .$$

□

Notice: $\partial : K_m^0 \to K_{m-1}^0$.

Labelling (valuation): = mapping $\varphi : A \to B$,

especially $\varphi(S^m) = (\varphi(a_1), \varphi(a_2), \ldots, \varphi(a_m))$

for $S^m = (a_1, a_2, \ldots, a_m)$.

Definition of derived mappings: $\varphi'_m : P_m \to \Gamma_m$, $\overline{\varphi}_m : P_m \to R$.

1) $\varphi(S^m_\rho) \in {}^*B^m$ => there exists one $\sigma_\rho \in \Gamma_m$:

$$\varphi(S^m_\rho) = \eta^{(m)}_\rho T^m_{\sigma_\rho} \qquad \text{(by Result (6))} .$$

Then put $\varphi'_m(\rho) = \sigma_\rho$, $\overline{\varphi}_m(\rho) = \eta^{(m)}_\rho$ ($= \pm 1 \in R$) .

2) $\varphi(S^m_\rho) \notin {}^*B^m$, then put $\varphi'_m(\rho) = \sigma^m_0$, $\overline{\varphi}_m(\rho) = 0$ ($\in R$)

with a fixed $\sigma^m_0 \in \Gamma_m$.

□

Effect of a labelling (valuation) φ :

For any $K^m \in K^0_m$ and $C^m \in L_m$,

$$K^m = \{\alpha_\rho\}_{\rho \in P_m} , \quad C^m = \{\beta_\sigma\}_{\sigma \in \Gamma_m} ,$$

define

$$\Phi^{C^m}_{K^m} := \sum_{\rho \in P_m} \alpha_\rho \, \beta_{\varphi'_m(\rho)} \, \overline{\varphi}_m(\rho) .$$

□

Notice: 1) Φ is independent of the choice of A_m , B_m .

2) Φ is additive in K^m and C^m , i.e.

$$\Phi^{C^m}_{K_1^m} + \Phi^{C^m}_{K_2^m} = \Phi^{C^m}_{K_1^m + K_2^m} \quad , \quad \Phi^{C_1^m}_{K^m} + \Phi^{C_2^m}_{K^m} = \Phi^{C_1^m + C_2^m}_{K^m}$$

8. MAIN RESULT

For all C^m , K^{m+1} , φ we have

$$\Phi^{C^m}_{\partial K^{m+1}} = \Phi^{\Delta C^m}_{K^{m+1}} \quad .$$

Interpretation: Purely combinatorial identity. Reduction in m .

Mode of application:

1) Specialization of A , B , R , C^m , K^{m+1} .

2) Repeated application of the reduction (used in mathematical induction).

3) Additional conditions on φ (e.g. boundary conditions).

Examples:

1) $B = \mathrm{R} = Z$,

$$C^m = T^m_{\sigma_0} = (1 , 2 , \ldots , m) \; ,$$

$$\Delta C^m = \sum_{a \neq 1 , 2 , \ldots , m} (a , 1 , 2 , \ldots , m) \; ,$$

→ oriented version of Sperner's Lemma .

2) $B = R = Z$,

$$C^m = \sum_{0 < b_1 < -b_2 < \dots < (-1)^{m-1} b_m} (b_1 , b_2 , \dots , b_m) ,$$

$$\Delta C^m = \sum_{\substack{0 < -x < b_1 \\ (b_i \text{ as in } C^m)}} (x , b_1 , \dots , b_m) +$$

$$+ \sum_{\substack{(-1)^{m-1} b_m < (-1)^m x \\ (b_i \text{ as in } C^m)}} (x , b_1 , \dots , b_m) +$$

$$+ \sum_{\substack{x= -b_i \\ (1 \leq i \leq m) \\ (b_i \text{ as in } C^m)}} (x , b_1 , \dots , b_m) .$$

Proper restriction of φ eliminates third sum →

→ Theorem of Ky Fan .

3) $B = Z$, R = prime field of characteristic 2 .

a) A = simplex in R^n , C^m as in 1) →

→ Sperner's Lemma (without orientation) .

b) A = n-dimensional lattice , C^m as in 2) →

→ Lemma of Tucker - Ky Fan .

4) A = B = all lattice points in R^n , R = real field, etc. →

→ Integral-Theorem of Gauß (or Stokes) .

Numerical Solution of Highly Nonlinear Problems
W. Forster (ed.)
© North-Holland Publishing Company, 1980

FIFTY YEARS OF FURTHER DEVELOPMENT OF A COMBINATORIAL LEMMA. PART B.

Emanuel Sperner

Universität Hamburg,
Hamburg, West Germany

EDITOR'S NOTE:
This paper was presented at a Summerschool and Conference on "Functional Differential Equations and Approximation of Fixed Points" held in Bonn form 17th to 22nd July 1978. Emanuel Sperner presented for the first time his new generalized version of his well known lemma. The talk was first announced with the title "A Few Remarks On Labelling Theorems", but in the lecture E. Sperner used the title "Kombinatorische Eigenschaften bewerteter Komplexe mit Orientierung". The material of the following pages is a translation and transcription of the slides used in the lecture. The reader has to keep this in mind. The material, however, is important and as yet unpublished. The following pages contain the core of the proof of the generalized version of Sperner's Lemma. In the words of E. Sperner, this new version "reaches out to the very limits of combinatorial topology".

1. HISTORICAL BACKGROUND

1928 E. Sperner: Simplex Lemma.
Applications: Invariance of dimension and domain; L.E.J. Brouwer.
Menger's definition of dimension, Theorem of Justification ("Rechtfertigungssatz").

1929 B. Knaster, C. Kuratowski, S. Mazurkiewicz: Simplex lemma.
Application: Brouwer's fixed point theorem.

1945 A.W. Tucker: Lemma for n-Cube (n-Polyhedron, n-Sphere).
Applications: Theorems of Lusternik-Schnirelmann, Borsuk-Ulam.

1949 S. Lefschetz: Lemma for n-cube.
Applications: as above.

1952 Ky Fan: Generalization and Lemma for n-Disc.
Applications: as above.

1957 Ky Fan: Applications to matrices with positive elements,
Theorems on eigenvalues of such matrices by Perron, Frobenius, etc.

1960 H.W. Kuhn: Cubical Sperner Lemma.
Application: "Equivalence" with Brouwer's fixed point theorem.

1967 H. Scarf: Application to the approximation of fixed points.

1969 H.W. Kuhn: Application as above.

1971 E.L. Allgower, C.L. Keller, etc.: Application as above.

1974 M. Yosiloff: Approach in reverse direction: Derivation of combinatorial lemmas from appropriate topological theorems.

Unification

1960 Ky Fan: Theorems on pseudomanifolds.

1968 P. Mani (H. Hadwiger): Combinatorial-geometrical theorems of Sperner-Tucker-Ky Fan type.

1971 ff. E. Sperner: Labelling of complexes and their combinatorial properties.

2. MOTIVATION

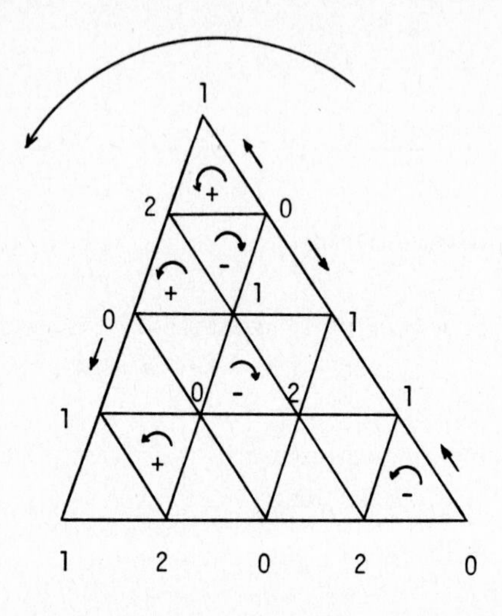

Fig. 1.

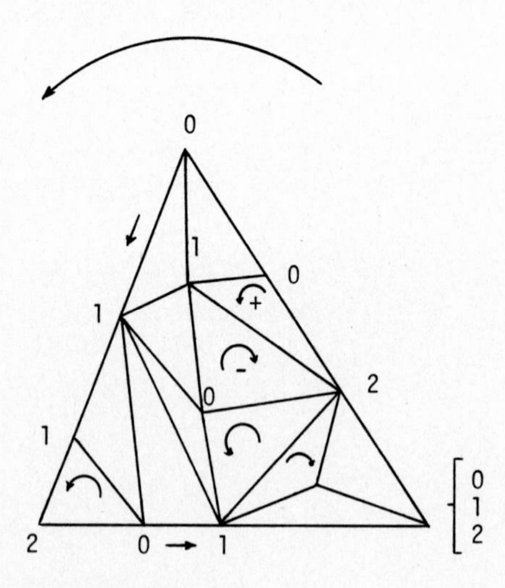

Fig. 2.

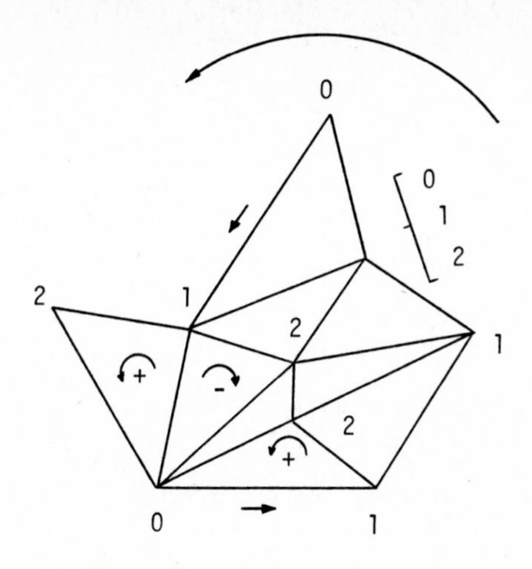

Fig. 3.

If we go round the figures counterclockwise ("positive orientation") and denote the number of boundary elements with vertices (i, j) with $p(i, j)$ and similarly denote the number of triangles with vertices (i, j, k) with $q(i, j, k)$,
<u>then the following holds</u>:

$$p(0, 1) - p(1, 0) = q(0, 1, 2) - q(1, 0, 2) .$$

<u>Corollary</u>:

$$p(0, 1) - p(1, 0) = p(1, 2) - p(2, 1) =$$

$$= p(2, 0) - p(0, 2) .$$

3. COMPLEXES WITH ORIENTATION AND THEIR LABELLINGS

Assumptions: Given

Basis set: $A = \{a_i\}_{i \in J}$ ("points") ,

Labelling set: $B = \{b_i\}_{i \in J}$ ("labels") ,

Ring of coefficients: $R = \{\alpha_i , \beta_k\}$

with addition commutative and associative,
multiplication associative and with unit element,
and both distributive laws hold.

Definitions:

D 1: ordered m-tuple: $S^m = (a_1 , a_2 , a_3 , . . . , a_m)$

$(a_i \neq a_k \text{ for } i \neq k)$

$$= \text{sign} (\nu_1 , \nu_2 , . . . , \nu_m) (a_{\nu_1} , a_{\nu_2} , . . . , a_{\nu_m}) ,$$

$$\text{sign} (\nu_1 , .. , \nu_m) = \pm 1 .$$

D 2: marking of the set of all $S^m \subset A$ with a set of indices P^m , such that:

D 2.1: for all S^m => there exist $\rho \in P^m$: $S^m = \pm S^m_\rho$,

D 2.2: $\rho , \rho' \in P^m$, $\rho \neq \rho'$ => $S^m_\rho \neq \pm S^m_{\rho'}$.

D 3: m-complex:

$$K^m = \sum_{\rho \in P^m} \alpha_\rho S_\rho^m \quad : \quad \left[\begin{array}{ll} \text{D 3.1:} & \alpha_\rho \in R \ , \\ \text{D 3.2:} & |\{\rho : \alpha_\rho \neq 0\}| < \infty \\ & (\text{i.e. } \alpha_\rho = 0 \text{ for} \\ & \text{almost all } \rho) \ . \end{array} \right.$$

D 4: addition of two m-complexes:

$$\sum_\rho \alpha_\rho S_\rho^m + \sum_\rho \alpha'_\rho S_\rho^m = \sum_{\rho \in P^m} (\alpha_\rho + \alpha'_\rho) S_\rho^m \quad .$$

D 5: multiplication of K^m by $\alpha \in R$:

$$\alpha \sum_\rho \alpha_\rho S_\rho^m = \sum_\rho (\alpha\, \alpha_\rho) S_\rho^m \quad .$$

D 6: boundary map ∂ :

for $S^{m+1} = (a_0 , a_1 , \ldots , a_m)$

define $\partial S^{m+1} = \sum_{\rho \in P^m} \varepsilon_\rho S_\rho^m$,

$$\text{with} \quad \varepsilon_\rho = \left[\begin{array}{l} \varepsilon : \text{there exist } k \ , \ 0 \le k \le m \ : \\ \qquad \varepsilon S_\rho^m = \\ \qquad = (-1)^k (a_0 , \ldots , a_{k-1} , a_{k+1} , \ldots , a_m) \ , \\ 0 \ \text{otherwise} \ . \end{array} \right.$$

For

$$K^m = \sum_{\rho \in P^m} \alpha_\rho S^m_\rho$$

define $\partial K^m = \sum_{\rho \in P^m} \alpha_\rho \, \partial S^m_\rho$.

D 7: ordered m-tuple of labels:

$$z^m = (b_1 , b_2 , . . . , b_m) \qquad (b_i \in B)$$

$$= \text{sign} (\nu_1 , \nu_2 , . . . , \nu_m) (b_{\nu_1} , b_{\nu_2} , . . . , b_{\nu_m}) .$$

D 8: marking of the set of m-labels with a set of indices T^m ,

such that:

- D 8.1: for all z^m => there exist τ ,
- D 8.2: $\tau \neq \tau'$, $\tau , \tau' \in T^m$ => $z^m_\tau \neq \pm z^m_{\tau'}$.

D 9: labelling m-complex:

$$C^m = \sum_{\tau \in T^m} \beta_\tau z^m_\tau$$

with

- D 9.1: $\beta_\tau \in R$,
- D 9.2: $\beta_\tau = 0$ if $z^m_\tau = -z^m_\tau$ and characteristic of $R \neq 2$.

D 10: labelling: $=$ map $\varphi : A \to B$,

in particular $\varphi(S^m_\rho) = \pm z^m_{\tau_\rho}$.

D 11: effect of labelling φ for arbitrary

$$K^m = \sum_{\rho\in P^m} \alpha_\rho S^m_\rho \quad , \qquad C^m = \sum_{\tau\in T^m} \beta_\tau z^m_\tau \quad .$$

For the map $\rho \to \tau_\rho$, given by $\varphi(S^m_\rho) = \varepsilon_{\tau_\rho} z^m_{\tau_\rho}$,

define

$$\Phi^{C^m}_{K^m} = \sum_\rho \alpha_\rho \beta_{\tau_\rho} \varepsilon_{\tau_\rho} \quad .$$

Note: $\Phi^{C^m}_{K^m}$ is independent of the normalization of S^m_ρ and z^m_τ

D 12: coboundary map Δ : $=$ map $\Delta : \{C^m\} \to \{C^{m+1}\}$.

Given $C^m = \sum_{\tau\in T^m} \beta^{(m)}_\tau z^m_\tau$,

we have to define the coefficients $\beta^{(m+1)}_\tau$ in

$$\Delta C^m = \sum_{\tau\in T^{m+1}} \beta^{(m+1)}_\tau z^{m+1}_\tau \quad .$$

For each $z^{m+1} = (b_0 , b_1 , ... , b_m)$ $(b_i \in B)$

form (for all k , $0 \le k \le m$) :

$$(-1)^k (b_0 , . . . , b_{k-1} , b_{k+1} , . . . , b_m) = \varepsilon_{\tau_k} z^m_{\tau_k}$$

and define $\beta^{(m+1)}_\tau = \sum_{k=0}^m \beta^{(m)}_{\tau_k} \varepsilon_{\tau_k}$.

Note: ΔC^m is independent of the normalization of

$$z^m_\tau \text{ and } z^{m+1}_\tau .$$

4. SIMPLE CONCLUSIONS

4.1. Additivity of ${}_\Phi C^m_{K^m}$.

Given
$$K^m = \sum_{\rho\in P^m} \alpha_\rho S^m_\rho ,$$

$$\overline{K}^m = \sum_{\rho\in P^m} \overline{\alpha}_\rho S^m_\rho ,$$

$$C^m = \sum_{\tau\in T^m} \beta_\tau z^m_\tau ;$$

Proposition:

$$(1) \qquad {}_\Phi C^m_{K^m} + {}_\Phi C^m_{\overline{K}^m} = {}_\Phi C^m_{K^m+\overline{K}^m}$$

$$\sum_\rho \alpha_\rho \beta_{\tau_\rho} \varepsilon_{\tau_\rho} + \sum_\rho \overline{\alpha}_\rho \beta_{\tau_\rho} \varepsilon_{\tau_\rho} = \sum_\rho (\alpha_\rho + \overline{\alpha}_\rho) \beta_{\tau_\rho} \varepsilon_{\tau_\rho}$$

□

Calculation of

4.2.

$$(2) \qquad {}_{\Phi}C^{m}_{\alpha_\rho \, \partial S^{m+1}_{\rho}} \quad ,$$

with $S^{m+1}_{\rho} = (a_0 , a_1 , \ldots , a_m)$

and $C^m = \sum_{\tau \in T^m} \beta_\tau z^m_\tau \quad .$

Utilizing D 6. and the independence of equation (2) of the normalization of S_ρ , we can write

$$\partial S^{m+1}_{\rho} = \sum_{k=0}^{m} S^m_{\rho_k} \quad ,$$

with

$$S^m_{\rho_k} = (-1)^k (a_0 ,\ldots, a_{k-1} , a_{k+1} ,\ldots, a_m) \quad .$$

4.3.

$$(3) \qquad {}_{\Phi}\Delta C^{m}_{\alpha_\rho \, S^{m+1}_{\rho}} \quad .$$

The sum we have to form according to D 11. for equation (3) consists of only one term, namely

$$(3) \qquad = \alpha_\rho \, \beta^{(m+1)}_{\tau_\rho} \, \varepsilon_{\tau_\rho} \quad .$$

The only coefficient $\beta^{(m+1)}_{\tau_\rho}$ of interest in ΔC^m is the one with index τ_ρ from

$$\varphi(S^{m+1}_{\rho}) = \varepsilon_{\tau_\rho} z^{m+1}_{\tau_\rho} \quad .$$

Let $\qquad \varphi(a_i) = b_i \quad .$

According to D 11. we form

$$\varphi(S^m_{\rho_k}) =$$

$$= (-1)^k \, (\varphi(a_0), \ldots, \varphi(a_{k-1}), \varphi(a_{k+1}), \ldots, \varphi(a_m)) =$$

$$= \varepsilon_{\tau_{\rho_k}} \, z^m_{\tau_{\rho_k}} \quad .$$

From D 11. we obtain:

$$_{\Phi}C^m_{\alpha_\rho \, \partial S^{m+1}_\rho} = \sum_{k=0}^{m} \alpha_\rho \, \beta_{\tau_{\rho_k}} \, \varepsilon_{\tau_{\rho_k}} \quad .$$

Utilizing the independence of equation (3) of the normalization of z^{m+1}_τ, we can write

$$(b_0, b_1, \ldots, b_m) = z^{m+1}_{\tau_\rho}$$

or

$$\varepsilon_{\tau_\rho} = 1 \quad .$$

According to D 12. we have to form

$$(-1)^k \, (b_0, \ldots, b_{k-1}, b_{k+1}, \ldots, b_m) =$$

$$= \varepsilon_{\tau_{\rho_k}} \, z^m_{\tau_{\rho_k}} \quad .$$

From D 12. follows

$$\beta^{(m+1)}_{\tau_\rho} = \sum_{k=0}^{m} \beta_{\tau_{\rho_k}} \, \varepsilon_{\tau_{\rho_k}} \quad .$$

Substitution into equation (3) then gives

$$ {}_{\Phi}\Delta C^{m}_{\alpha_\rho S^{m+1}_\rho} = \alpha_\rho \sum_{k=0}^{m} \beta_{\tau_{\rho_k}} \varepsilon_{\tau_{\rho_k}} . $$

4.4. Comparison of the results of 4.2. and 4.3. leads immediately to:

$$ (4) \qquad {}_{\Phi}\Delta C^{m}_{\alpha_\rho S^{m+1}_\rho} = {}_{\Phi} C^{m}_{\alpha_\rho \partial S^{m+1}_\rho} $$

for all φ , $\alpha_\rho S^{m+1}_\rho$, C^m .

4.5. Given an arbitrary (m+1)-complex

$$K^{m+1} = \sum_{\rho \in P^{m+1}} \alpha_\rho S_\rho^{m+1}$$

and therefore

$$\partial K^{m+1} = \sum_{\rho \in P^{m+1}} \alpha_\rho \partial S_\rho^{m+1} ,$$

we obtain from equation (4) by summing over ρ and utilizing equation (1) the following

$$(5) \qquad \Phi_{K^{m+1}}^{\Delta C^m} = \Phi_{\partial K^{m+1}}^{C^m} \qquad \text{for all } K^m , C^m , \varphi .$$

5. EXAMPLES

5.1. The special case R = prime field of characteristic 2 allows the following interpretation:

$$K^m = \sum_{\rho \in P^m} \alpha_\rho S_\rho^m \quad : \quad S_\rho^m \in K^m \ :\Longleftrightarrow \ \alpha_\rho = 1$$

$$(S_\rho^m \notin K^m \ :\Longleftrightarrow \ \alpha_\rho = 0) ,$$

$$C^m = \sum_{\tau \in T^m} \beta_\tau z_\tau^m \quad : \quad z_\tau^m \in C^m \ :\Longleftrightarrow \ \beta_\tau = 1 .$$

$$\Phi_{K^m}^{C^m} = \sum_{\rho \in P^m} \alpha_\rho \beta_{\tau_\rho} \varepsilon_{\tau_\rho} = |\{S_\rho^m \in K^m : \varphi(S_\rho^m) \in C^m\}| ,$$

$\Delta C^m =$

$= \{z^{m+1} : |\{k \in Z : 0 \le k \le m ,$

$(b_0 , .. , b_{k-1} , b_{k+1} , .. , b_m) \in C^m\}| \equiv 1 \pmod 2) \}$

with $z^{m+1} = (b_0 , b_1 , ... , b_m)$.

For $B = Z$ this special case reduces to a known theorem mod 2 .

5.2. Return to the starting point

We consider a special case of equation (5) (theorem (5)), now m-dimensional.

In equation (5) we set

$A = R^m$,

$B = \{0 , 1 , . . . , m\}$,

$R = Z$,

K^{m+1} = simplicial complex in R^m (homogeneous in dimension m) ,

$c^m_\nu = (-1)^\nu (0 , ... , \nu-1 , \nu+1 , ... , m)$ $(= z^m_\nu)$,

$\varphi : R^m \to B$,

and we obtain

$$\Delta C_\nu^m = (0 , . . . , \nu-1 , \nu , \nu+1 , . . . , m) = C^{m+1} ,$$

$\Phi_{K^{m+1}}^{C^{m+1}}$ = difference between the number of simplices with positive orientation and the number of simplices with negative orientation; the simplices are labelled with $(0 , 1 , . . . , m)$;

$\Phi_{\partial K^{m+1}}^{C_\nu^m}$ = "algebraic sum" of the simplices on the boundary labelled with $(0 , . . . , \nu-1 , \nu+1 , . . . , m)$.

<u>Theorem</u>: $\Phi_{K^{m+1}}^{C^{m+1}} = \Phi_{\partial K^{m+1}}^{C_\nu^m}$, for all $\nu = 0 , 1 , ... , m$.

<u>Corollary</u>: $\Phi_{\partial K^{m+1}}^{C_\nu^m} = \Phi_{\partial K^{m+1}}^{C_\mu^m}$, for all ν , μ with $0 \leq \nu \leq \mu \leq m$.

Numerical Solution of Highly Nonlinear Problems
W. Forster (ed.)
© North-Holland Publishing Company, 1980

FIFTY YEARS OF FURTHER DEVELOPMENT OF A COMBINATORIAL LEMMA. PART C.

EDITOR'S COMMENTS:

The boundary operator ∂ used in the paper is well known from homology theory, the coboundary operator Δ is well known from cohomology theory. The modern abstract approach utilizes the natural duality between the two theories. An invariant similar to the invariant called "the effect of the labelling" is given e.g. in

P.J. Hilton and S. Wylie, Homology Theory, Cambridge University Press, Cambridge 1967,

and is called "Kronecker product" (it is a special case of the cap product). It has to be pointed out that Sperner's approach does not use any abstract machinery from algebraic topology and that this makes his new result easily accessible to a wide audience.

We supplement the material by a list of references.

REFERENCES

1. Brouwer, L.E.J., Über eineindeutige, stetige Transformationen von Flächen in sich, Math. Ann. 69, 176-180 (1910).

2. Sperner, E., Neuer Beweis für die Invarianz der Dimensionszahl und des Gebietes, Abh. math. Sem. Hamburg 6, 265-272 (1928).

3. Menger, K., Dimensionstheorie, B.G. Teubner, Leipzig und Berlin 1928.

4. Knaster, B.; Kuratowski, K.; Mazurkiewicz, S., Ein Beweis des Fixpunktsatzes für n-dimensionale Simplexe, Fundamenta Math. 14, 132-137 (1929).

5. Borsuk, K., Drei Sätze über die n-dimensional Euklidische Sphäre, Fundamenta Math. 20, 177-190 (1933).

6. Alexandroff, P.; Hopf, H., Topologie I, Springer, Berlin 1935 (see page 487 for Lusternik - Schnirelmann - Borsuk).

7. Tucker, A.W., Some Topological Properties of Disk and Sphere, Proc. First Can. Math. Congress, Montreal 1945, pp. 285-309.

8. Lefschetz, S., Introduction to Topology, Princeton University Press, Princeton, New Jersey 1949.

9. Ky Fan, A Generalization of Tucker's Combinatorial Lemma with Topological Applications, Ann. of Math. 56, 431-437 (1952).

10. Ky Fan, Topological Proofs for Certain Theorems on Matrices with Non-Negative Elements, Monatshefte für Mathematik 62, 219-237 (1958).

11. Ky Fan, Combinatorial Properties of Certain Simplicial and Cubical Vertex Maps, Archiv d. Math. 11, 368-377 (1960).

12. Ky Fan, A Covering Property of Simplexes, Math. Scand. 22, 17-20 (1968).

13. Ky Fan, Simplicial Maps from an Orientable n-Pseudomanifold into S^m with the Octahedral Triangulation, J. Comb. Theory 2, 588-602 (1967).

14. Ky Fan, A Combinatorial Property of Pseudomanifolds and Covering Properties of Simplices, J. of Math. Analysis and Applications 31, 68-80 (1970).

15. Bagemihl, F., An Extension of Sperner's Lemma, with Applications to Closed-set Coverings and Fixed Points, Fund. Math. 40, 3-12 (1953).

16. Kuhn, H.W., Some Combinatorial Lemmas in Topology, I.B.M. Journ. Res. Develop. 4, 518-524 (1960).

17. Kuhn, H.W., Simplicial Approximation of Fixed Points, Proc. Nat. Acad. Sc. 61, 1238-1242 (1968).

18. Kuhn, H.W., Approximate Search for Fixed Points, in: Computing Methods in Optimization Problems 2, Academic Press, New York 1969, pp. 199-211.

19. Kuhn, H.W., A New Proof of the Fundamental Theorem of Algebra, Mathematical Programming Study 1, 148-158 (1974).

20. Cairns, S.S., Introductory Topology, The Ronald Press, New York 1961 (includes A.B. Brown's contribution).

21. Scarf, H., The Approximation of Fixed Points of a Continuous Mapping, SIAM J. Applied Math. 15, 1328-1343 (1967).

22. Mani, P., Zwei kombinatorisch-geometrische Sätze vom Typus Sperner - Tucker - Ky Fan, Monatshefte für Mathematik 71, 427-435 (1967).

23. Allgower, E.L.; Keller, Ch.L., A Search Routine for a Sperner Simplex, Computing 8, 157-165 (1971).

24. Sperner, E., Über die kombinatorischen Grundlagen gewisser simplizialer Sätze, Atti del Convegno di Geometria Combinatoria e sue Applicazioni, Perugia 1970, pp. 385-401.

25. Sperner, E., Kombinatorik bewerteter Komplexe, Abh. math. Sem. Hamburg 39, 21-43 (1973).

26. Sperner, E., Ein kombinatorischer Umschliessungssatz nebst Anwendungen, to be published in Mathematical Results (1979).

27. Yoseloff, M., Topological Proofs of some Combinatorial Theorems, J. Comb. Theory (A) 17, 95-111 (1974).

Numerical Solution of Highly Nonlinear Problems
W. Forster (ed.)
© North-Holland Publishing Company, 1980

AN APPLICATION OF THE GENERALIZED SPERNER LEMMA TO THE COMPUTATION OF FIXED POINTS IN ARBITRARY COMPLEXES

Walter Forster

University of Southampton,
Southampton, England

1. INTRODUCTION

Fixed point theorems have long been used in applied mathematics to show that systems of equations have at least one solution, i.e. to give existence proofs. The capability to turn these nonconstructive proofs into constructive numerical methods for obtaining solutions to highly nonlinear problems (since Scarf's first constructive proof of Brouwer's fixed point theorem in 1967 [1]) has made fixed point methods and related methods one of the fastest growing areas of constructive mathematics. A number of algorithms has been developed and implemented.

In this paper we utilize a recently given generalization of Sperner's lemma to give a new type of fixed point theorem. In order to explain the new character of this theorem let us briefly look at classical fixed point theorems. Consider e.g. Brouwer's fixed point theorem [2], or the Leray-Schauder fixed point theorem [3], or the Lefschetz fixed point theorem [4], or the Kakutani fixed point theorem [5], etc. For all these fixed point theorems we have to check certain conditions (usually by paper and pencil) on the map, the region, etc. in order to ensure the existence of a fixed point. This approach is important for classical existence proofs, but in many cases the conditions which have to be satisfied for a fixed point to exist are difficult to verify. Therefore for computational purposes a different approach is desirable. What we want is a numerical method which either finds a fixed point (or an approximate fixed point) if one exists, or alternatively gives an indication that it has not found a fixed point. Such an approach allows all the calculations to be executed on the computer (without having to check complicated conditions by paper and pencil). In the following sections we will give such a theorem.

2. A COMPUTATIONAL FIXED POINT THEOREM

2.a. In 1978 Sperner gave a new version of his well known lemma [6]. It reads as follows

$$\Phi^{\Delta C^n}_{K^{n+1}} = \Phi^{C^n}_{\partial K^{n+1}} \quad .$$

For details see Sperner's paper [7].

The importance of this new lemma lies in the fact that it connects an "algebraic count" of e.g. labelled (n-1)-simplices C^n on the boundary ∂K^{n+1} of a given simplicial complex K^{n+1} with an "oriented count" of e.g. labelled n-simplices ΔC^n in the interior of the complex K^{n+1}. Interpreted for computational purposes, this allows us to construct a pivotal path from the boundary to the interior.

2.b. Let us consider the special case of a complex K^{n+1} which is homogeneously n-dimensional, i.e. every simplex is a face of some n-simplex of the complex. Using the notation of Sperner's paper (Part B) we have

$$A = R^n \quad ,$$

$$B = \{0, 1, \ldots, n\} \quad ,$$

$$R = Z \quad ,$$

K^{n+1} = simplicial complex (homogeneously n-dimensional) in R^n ,

$$C_\nu^n = (-1)^\nu (0 , \ldots , \nu-1 , \nu+1 , \ldots , n) \quad ,$$

and we obtain

$$\Delta C^n = (0 , 1 , \ldots , \nu-1 , \nu , \nu+1 , \ldots , n) = C^{n+1} \quad .$$

This gives the following

Lemma (Sperner, 1978):

$$\Phi_{K^{n+1}}^{C^{n+1}} = \Phi_{\partial K^{n+1}}^{C_\nu^n}$$

for all $\nu = 0 , 1 , \ldots , n$.

□

Corollary (Sperner, 1978):

$$\Phi_{\partial K^{n+1}}^{C_\nu^n} = \Phi_{\partial K^{n+1}}^{C_\mu^n}$$

for all ν , μ , with $0 \le \nu \le \mu \le n$.

□

If we take the special case $\nu = n$, then we can state the lemma in the following more familiar form.

Lemma: Let K be a complex homogeneously n-dimensional in R^n. If the vertices of K are labelled arbitrarily from the set $\{0, 1, ..., n\}$, then the number of completely labelled n-simplices in K (i.e. the vertices carry all the labels from 0 to n) with positive orientation minus the number of completely labelled n-simplices with negative orientation is equal to the "algebraic sum" of completely labelled (n-1)-simplices on the boundary of K.

□

Corollary: The "algebraic sum" of completely labelled (n-1)-simplices on the boundary of K is equal to the "algebraic sum" of the (n-1)-simplices on the boundary of K labelled with $(0, 1, ..., \nu-1, \nu+1, ..., n)$, where $0 \leq \nu \leq n$.

□

In other words, the "algebraic sum" of the (n-1)-simplices on the boundary of K does not depend on the missing label ν.

Utilizing the usual arguments (see e.g. [8]) for pivoting algorithms, we can construct a chain of simplices which starts at the boundary of the n-dimensional complex with an (n-1)-simplex with n different labels and

(i) either ends with an n-simplex with $n+1$ different labels, or
(ii) alternatively ends on the boundary with another (n-1)-simplex with the same n different labels.

We give a few details [9]. The (n-1)-simplex with n different labels on the boundary of the complex is a face of an n-simplex with $n+1$ labels. If the (n+1)st label is different from all the other n labels of the simplex, then we have found an n-simplex with $n+1$ different labels and the path ends here.
If the (n+1)st label is one of the n labels of the (n-1)-simplex we started with, then there is another face of this simplex which has the same n different labels and we leave the n-simplex through this face and enter another n-simplex.

We continue in this manner and the path will end
(i) either with an n-simplex with $n+1$ different labels, or
(ii) with an $(n-1)$-simplex with the same n different labels on the boundary of the complex K
(subject to the remarks about cycling in section 3.).

2.c. It is well known (see e.g. [10]) that a general n-dimensional complex can be embedded in an R^{2n+1}. This case requires a different version of the generalized Sperner lemma. In order to be able to embed an arbitrary n-dimensional complex K^{n+1} in an R^{2n+1} (or R^{2n+2}, see later) and label with the usual labelling function

$$i = \min_j \{j \mid f_j(y) \leq x_j > 0\} ,$$

we utilize the following form of the generalized Sperner lemma (oriented version of Sperner's lemma, see Sperner's paper [7], Part A).
We take (notation as in Sperner's paper [7], Part B)

$$A = R^{2n+1} ,$$

$$B = \{0, 1, \ldots, N\} , \quad N \geq n \quad \text{(here e.g. } N = 2n+1) ,$$

$$R = Z ,$$

K^{n+1} = n-dimensional simplicial complex in R^{2n+1} (or R^{2n+2}, see later), not necessarily homogeneous of dimension n,

$$C^n = (1, 2, \ldots, n) ,$$

and we obtain

$$\Delta C^n = \sum_{\substack{a \neq 1, 2, \ldots, n \\ a \in B}} (a, 1, 2, \ldots, n) .$$

Lemma (Sperner, 1979):

$$\Phi^{\Delta C^n}_{K^{n+1}} = \Phi^{C^n}_{\partial K^{n+1}} ,$$

with $C^n = (1 , 2 , . . . , n)$
and

$$\Delta C^n = \sum_{\substack{a \neq 1, 2, \ldots , n \\ a \in B}} (a , 1 , 2 , . . . , n) .$$

□

Here again we have a statement about an "algebraic count" of (n-1)-dimensional simplices on the boundary of the n-dimensional complex and an "oriented count" of n-dimensional simplices in the interior of the complex. This allows us to construct a pivotal path from the boundary to the interior.
The lemma is obviously true for arbitrary m , $0 \leq m \leq n$.
Let K^{m+1} be an m-dimensional subcomplex of K^{n+1} such that the m-dimensional simplices of K^{m+1} have all their faces but are not faces of $m+1$ or higher dimensional simplices. If we now consider subcomplexes which are homogeneous of dimension m , then we can start a pivotal path with an (n-1)-simplex on the boundary of the m-dimensional subcomplex K^{m+1} (we assume this (m-1)-dimensional simplex has m different labels) and proceed with pivoting steps in the usual way until we

(i) reach an m-dimensional simplex with $m+1$ different labels, or
(ii) alternatively end on the boundary of the m-dimensional complex with an (m-1)-dimensional simplex with the same m different labels we started with.

In the case of a general complex this path does not have to be unique and we have to take precautions to avoid cycling (see section 3).

Note: We can assume without loss of generality that the m different labels of the (m-1)-simplex on the boundary of the m-dimensional subcomplex K^{m+1} are $(1, 2, . . . , m)$. If the m different labels of a particular simplex on the boundary consist of other labels from the set $\{0, 1, . . . , N\}$, then we relabel in a suitable manner so that we obtain the labels $(1, 2, . . . , m)$. This is done to be in accordance with the Lemma (Sperner, 1979) stated above. For the actual implementation of the

algorithm this relabelling is not necessary (m different labels on an (m-1)-simplex on the boundary will do).
For implementing the algorithm we do not have to calculate the invariants

$$\Phi^{\Delta C^m}_{K^{m+1}}$$

or

$$\Phi^{C^m}_{\partial K^{m+1}} \quad .$$

We use the generalized Sperner lemma only as a statement which connects a count of certain simplices on the boundary with a count of certain simplices in the interior. This in turn allows us to make statements about pivotal paths.

2.d. We now state these results as a theorem.

Theorem (pivotal form): Starting with an (m-1)-simplex (with m different labels) on the boundary of a suitably labelled m-dimensional simplicial complex K^{m+1} (homogeneously m-dimensional) at least one of the following holds:

(i) there is a pivotal path to an m-dimensional simplex with $m+1$ different labels, or

(ii) there is a pivotal path to another (m-1)-simplex on the boundary of K^{m+1} with the same m different labels.

□

For avoidance of cycling see section 3. What to do with complexes not homogeneously m-dimensional see section 4.

If we use the labelling function given above, then an m-simplex with $m+1$ different labels is an approximation to a fixed point (if a fixed point exists). If we subdivide the simplices ("big simplices") of the complex and let the diameter tend to zero, then we can utilize a similar proof to the one given in [8] to show that an m-simplex (which has all its faces and is not a face of an $m+1$ or higher dimensional simplex), which has $m+1$

different labels for arbitrarily small diameter is an approximation to a fixed point (in the case of a general complex there is no guarantee that there is a fixed point),
or alternatively we can use an argument given in [11] to show that the path of (m-1)-simplices starting on the boundary of an m-dimensional subcomplex and ending on the boundary of the m-dimensional subcomplex is a compact path for the diameter of the simplices tending to zero.
This can be shown for all m , $0 \leq m \leq n$.

Theorem (limit form) [12], [13]:
Given a continuous map from an arbitrary (finite) complex K to itself, then at least one of the following holds
(i) there exists at least one fixed point in K ,
(ii) there exists at least one compact path in K .

Note: We assume the complex K is of dimension at most n , but not necessarily homogeneously of dimension n .

Proof: For part (i) embed the n-dimensional complex K in R^{2n+2} so that all coordinates are positive and add to one. Use the labelling function given above. Using the complex K (instead of a simplex) the proof given in [8] page 11 carries over with minor modifications. We have to take the limit of m-dimensional simplices (diameter of the simplices tends to zero) with m+1 different labels. These m-dimensional simplices belong to m-dimensional subcomplexes of K .
This procedure is repeated for all m , $0 \leq m \leq n$.
For part (ii) use part (ii) of the theorem (pivotal form) and the argument in [11]. Repeat for all m , $0 \leq m \leq n$.

□

3. AVOIDANCE OF CYCLING

We assume that the complex K consists of a finite number of simplices, which we call "big simplices". For the computation of fixed points we subdivide these "big simplices" in a suitable manner and we call the simplices obtained by subdivision "small simplices". The subdivision of each "big simplex" gives a pseudomanifold and we do not have to worry

about cycling of the algorithm, because the Lemke and Howson argument [14] applies to each "big simplex" individually. When we consider the complex as a whole things are more complicated. Unless we take precautions cycling can occur.
Consider e.g. the complex consisting of 1-dimensional simplices ("big simplices")

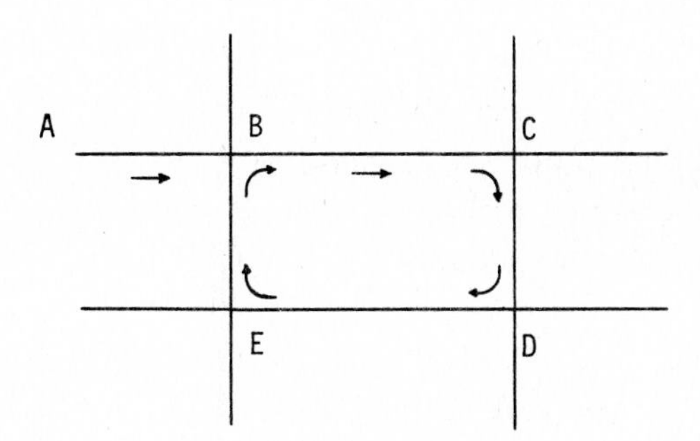

With appropriate orientation of the 1-dimensional simplices ("big simplices") the complex will have no boundary at the junctions B, C, D, E. Let us assume that subdivision of the "big simplices" <A, B>, <B, C>, <C, D>, <D, E>, <E, B> into "small simplices" and labelling leads to labels 0 on all the vertices of the "small simplices". Unless we remember which big simplex we leave and which big simplex we enter at the junctions B, C, D, E the algorithm could cycle.
Therefore we have to add the following rule to the usual pivoting rules. If more than two "big simplices" (m-dimensional) have an (m-1)-face in common then at such a junction one has to remember the "big simplex" one has left and the "big simplex" one has entered in order to avoid cycling. When the algorithm reaches such a junction one can then apply simple rules to avoid following the same path (or parts of an earlier path) twice. Otherwise there are no changes in the pivoting algorithm as usually applied.

4. COMPLEXES NOT HOMOGENEOUSLY OF DIMENSION N

We assume we have embedded our n-dimensional complex in an R^{2n+2} order to use the labelling function mentioned earlier). We leave the complex as it is, but consider only connected subcomplexes consisting of m-dimensional "big simplices", i.e. the m-dimensional simplices considered have all their faces but are not a face of an (m+1)-dimensional or higher dimensional simplex.

Example:
Given e.g. the following complex (2-dimensional).

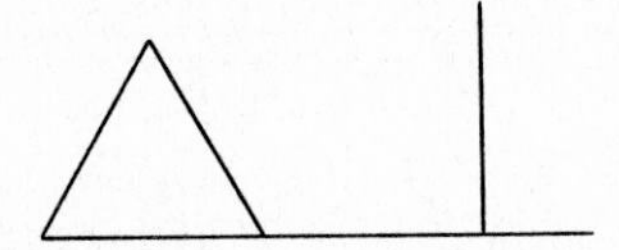

For computational purposes we consider the following two subcomplexes.

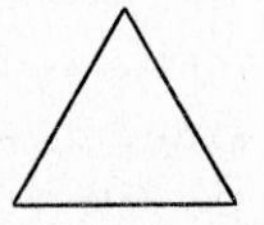

2-dimensional subcomplex.

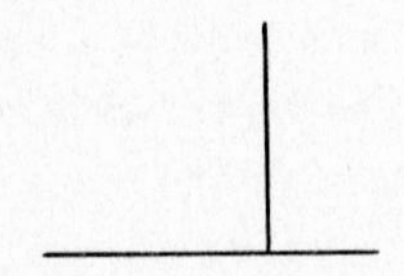

1-dimensional subcomplex.

The generalization to higher dimensional cases is obvious.
This process does not require any changes in the embedding mentioned earlier and does not change any labels. We only assume that the simplices of the various m-dimensional subcomplexes $(0 \leq m \leq n)$ have all their faces and are not a face of a higher dimensional simplex, and that the original complex is the union of these subcomplexes. The pivoting algorithm is then applied to each of the connected pieces of such an m-dimensional subcomplex (and to each starting simplex with m different labels on the boundary of the m-dimensional subcomplex).

5. STARTING PROCEDURES

Starting procedures are not crucial to the main ideas developed in this paper, but starting procedures can have a considerable influence on pivoting algorithms. We have to find (m-1)-dimensional simplices on the boundary of m-dimensional subcomplexes $(0 \leq m \leq n)$, i.e. we look for starting simplices.
We could start with a zero-dimensional simplex and search for a 1-dimensional simplex with two different labels. Then search for a two-dimensional simplex with 3 different labels, and proceed inductively until an (m-1)-dimensional simplex with m different labels is found. This starting procedure is very similar to the variable dimension approach (see e.g. [8]), or can be considered a variant of [15].
Another possibility is to hold (m-1) coordinates constant in the m-dimensional subcomplex and search for 1-dimensional simplices with 2 different labels, then hold certain (m-2) coordinates constant and search for simplices with 3 different labels, etc.
Another possibility is to consider each "big simplex" separately and search each big simplex in the usual way.
Another possibility is to consider an inductive application of the generalized Sperner Lemma.
Obviously there are many more variants of starting procedures. For a general complex none of these procedures can guarantee the existence of an m-dimensional simplex with $m+1$ different labels.

6. CONCLUSIONS

In this paper two special versions of the generalized Sperner Lemma have been exploited to give algorithms for finding m-dimensional simplices with m+1 different labels starting from (m-1)-dimensional simplices with m different labels on the boundary of an m-dimensional subcomplex. The novel feature of the generalized Sperner Lemma is that it connects an "oriented count" of m-dimensional simplices in the interior of an m-dimensional subcomplex labelled with m+1 different labels, with an "algebraic count" of (m-1)-dimensional simplices with m different labels on the boundary of the subcomplex. This fact can then be used to construct a pivotal path. (Other combinatorial lemmas stated e.g. in [7] Part A can be treated similarly).

In connection with the computation of zeros of polynomials a similar feature has been exploited in [16], [17] for the special case of a convex region. The generalized version of Sperner's lemma also sheds new light on algorithms for finding more than one fixed point, e.g. relabelling algorithms [18]. Another example to be mentioned in this context is [19].

The novel result of the present paper is a new type of fixed point theorem for general complexes embedded in R^{2n+1} (or R^{2n+2}). As opposed to classical fixed point theorems, no conditions on the region, boundary, etc. have to be verified. On the other hand no fixed point is guaranteed. If we assume that we have found by a starting procedure all (m-1)-dimensional starting simplices (for $0 \le m \le n$), then the theorem is constructive in the sense that the algorithm will either find an (approximate) fixed point (i.e. the algorithm will e.g. find one or more completely labelled simplices), or the algorithm will give an indication that it has not found an (approximate) fixed point (i.e. the algorithm will end on the boundary of the complex).

This theorem subsumes a number of classical fixed point theorems and related theorems.

REFERENCES

1. Scarf, H.E., The Approximation of Fixed Points of a Continuous Mapping, SIAM J. Appl. Math. 15, 1328-1343 (1967).

2. Brouwer, L.E.J., Über eineindeutige, stetige Transformationen von Flächen in sich, Math. Ann. 69, 176-180 (1910).

3. Leray, J.; Schauder, J., Topologie et équations fonctionnelles, Ann. Sci. Ecole Norm. Sup. 51, 45-78 (1934).

4. Lefschetz, S., Intersections and Transformations of Complexes and Manifolds, Trans. A.M.S. 28, 1-49 (1926).

5. Kakutani, S., A Generalization of Brouwer's Fixed Point Theorem, Duke Math. J. 8, 457-459 (1941).

6. Sperner, E., Neuer Beweis für die Invarianz der Dimensionszahl und des Gebietes, Abh. Math. Sem. Univ. Hamburg 6, 265-272 (1928).

7. Sperner, E., Fifty Years of Further Development of a Combinatorial Lemma, in this volume.

8. Todd, M.J., The Computation of Fixed Points and Applications, Springer, Berlin-Heidelberg-New York 1976.

9. Forster, W., Computing Lefschetz Fixed Points, Faculty of Mathematical Studies, University of Southampton, Report No. 21, January 1979.

10. Hilton, P.J.; Wylie, S., Homology Theory, Cambridge University Press, Cambridge 1967.

11. Eaves, B.C.; Saigal, R., Homotopies for Computation of Fixed Points on Unbounded Regions, Mathematical Programming 3, 225-237 (1972) (see Theorem 4.3).

12. Forster, W., A Constructive Fixed Point Theorem, Mitteilungen der GAMM, Heft 1, 136-137, March 1979.

13. Forster, W., A Fixed Point Theorem for Arbitrary Complexes, Tenth International Symposium on Mathematical Programming, Abstracts, 25-26, Montreal 1979.

14. Lemke, C.E.; Howson, J.T., Equilibrium Points of Bimatrix Games, SIAM J. Appl. Math. 12, 413-423 (1964).

15. Van der Laan, G.; Talman, A.J.J., A New Algorithm for Computing Fixed Points, Interfaculteit der Actuariele Wetenschappen en Econometrie, Vrije Universiteit, Amsterdam, Report, Maart 1978.

16. Kuhn, H.W., A New Proof of the Fundamental Theorem of Algebra, Math. Programming Study 1, 148-158 (1974).

17. Kuhn, H.W., Finding Roots of Polynomials by Pivoting, in S. Karamardian (Ed.), Fixed Points Algorithms and Applications, Academic Press, New York 1977, pp. 11-39.

18. Jeppson, M.M., A Search for the Fixed Points of a Continuous Mapping, Ph.D. Thesis, Colorado State University, 1972.

19. Garcia, C.B., A Fixed Point Theorem including the Last Theorem of Poincaré, Mathematical Programming 9, 227-239 (1975).

Numerical Solution of Highly Nonlinear Problems
W. Forster (ed.)
© North-Holland Publishing Company, 1980

ON THE SHAPLEY-SPERNER LEMMA

Harold W. Kuhn

Princeton University,
Princeton, New Jersey, U.S.A.

Shapley's elegant proof [1] of Scarf's theorem [2] (asserting that the core of a balanced game without side payments is nonempty) is based on an interesting generalization of Sperner's Lemma [3]. To state this result, which we shall call the Shapley-Sperner Lemma, let $N = \{1, \dots, n\}$ and denote by $\mathcal{N}$ the family of all nonempty proper (i.e., $\neq N$) subsets of N. Let M be the simplex $\{x \mid x = (x_1, \dots, x_n) \geq 0 \text{ and } \Sigma_i x_i = 1\}$. A <u>Shapley labelling</u> of a triangulation of M is a function L defined on the vertices of the triangulation with $L(x) \in \mathcal{N}$ and $x_i > 0$ whenever $i \in S = L(x)$. A subfamily $\mathcal{B}$ of $\mathcal{N}$ is said to be <u>balanced</u> if there exist nonnegative weights w_S for $S \in \mathcal{B}$ such that

$$\sum_{S,\, i \in S} w_S = 1 \qquad \text{for } i = 1, \dots, n .$$

<u>Shapley-Sperner Lemma</u>: For any Shapley labelling there exists a simplex of the triangulation whose labels form a balanced family.

Note that Sperner's Lemma is an obvious special case of this result. (If the labels are singletons $\{i\}$, then the only balanced family is

---

Research supported, in part, by National Science Foundation Grant MCS 76-19670.

$\{\{1\}, \ldots, \{n\}\}$. However, as Shapley has observed, the strong form of Sperner's Lemma (asserting the existence of an <u>odd</u> number of simplices of the triangulation with labels $\{\{1\}, \ldots, \{n\}\}$) does not follow from this result. The following two figures illustrate the difficulty.

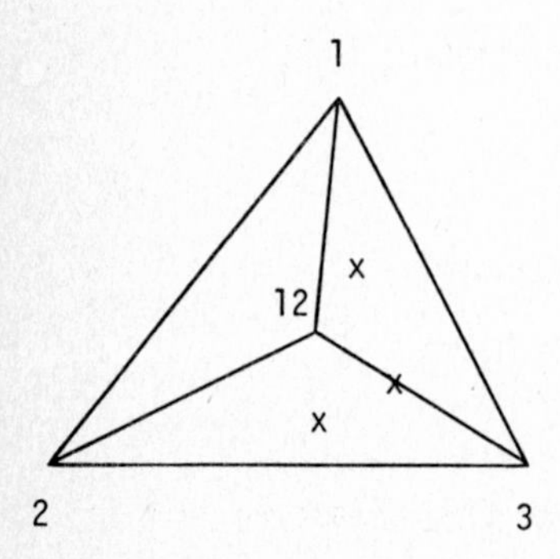

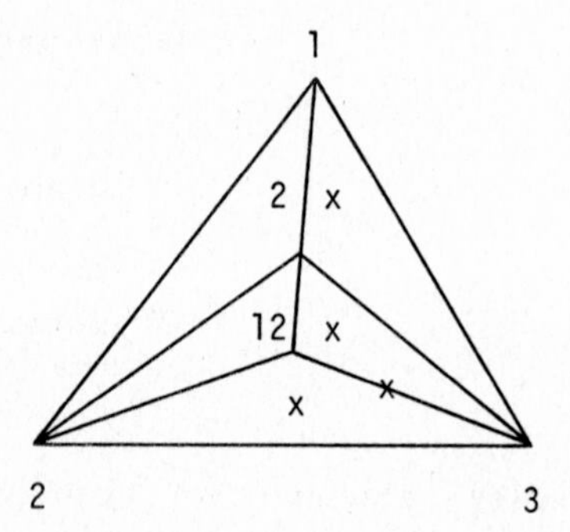

Figure 1.

In the figure at the left, if we count full dimensional simplices (with a balanced family as labels) the count is 2. If we include simplices of all dimensions, then the count is 3. In the figure at the right, the situation is reversed, the count of full dimensional simplices is odd and the count of all simplices is even.

The purpose of this note is to approach the Shapley-Sperner Lemma via a combinatorial theorem which asserts the "odd count" property for certain proper subsets of the full family of balanced sets.

<u>Definition 1</u>: A <u>Shapley labelling with labels in $N' \subset N$</u> of a triangulation of M is a function L defined on the vertices x of the triangulation with $L(x) \in N'$ and $L(x) = S$ implies $x_i > 0$ for all $i \in S$.

Definition 2: A class P of balanced families B is called pivotal with respect to N' if (1) every $B \in P$ contains n sets $S \in N$;
(2) $\{\{1\}, \{2\}, \ldots, \{n\}\} \in P$;
(3) if $B \in P$, $T \in N'$, $T \notin B$ then there exists a unique $S \in B$ such that $(B - \{S\}) \cup \{T\} \in P$.

Theorem: For any Shapley labelling with labels in N' and any class P that is pivotal with respect to N' , the number of simplices whose labels are in P is odd.

(Note that the strong form of Sperner's Lemma follows as a corollary by taking $N' = \{\{1\}, \ldots, \{n\}\}$ and $P = \{\{\{1\}, \ldots, \{n\}\}\}$.)

Proof: Without loss of generality, we may assume that the triangulation of M has no vertices in the faces $x_i = 0$ other than the unit vectors. If this is not the case, we can use a familiar imbedding (illustrated below; see [4], p. 48 , for formal details) of the triangulation of M in the triangulation of a larger simplex without adding any simplices whose labels form a balanced family. (In Figure 2 , the triangulation of M is shown by heavy lines; the new vertices are "opposite" old vertices i and are labelled i-1 (mod n).) We now construct a triangulation of a simplex $\bar{M}$ of dimension n by joining one additional vertex (outside the plane of M) labelled 1 to all of the vertices of the given triangulation. (To visualize this, imagine the extra vertex above the plane of Figure 2 .) As a direct consequence of properties (1) and (3) of Definition 2 , the n-dimensional simplices of this triangulation enjoy the familiar property: if an (n-1)-dimensional face has as labels a family in P then exactly two such faces have as labels a family in P . Since only one boundary face of $\bar{M}$ has a balanced family ($\{\{1\}, \ldots, \{n\}\} \in P$) from P as labels, the theorem follows. (Algorithmically, pivoting leads from the boundary face to one of the simplices of the triangulation of M labelled by a family from P ; the other such simplices are joined in pairs by pivotal paths.)

□

The proof of the Shapley-Sperner Lemma is completed by the following preliminary result.

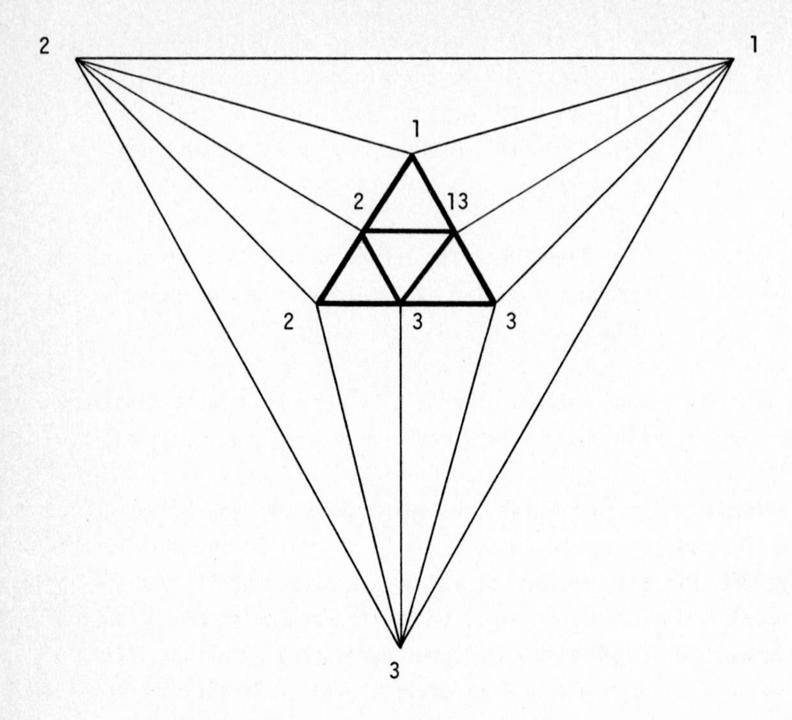

Figure 2.

<u>Lemma</u>: For every n, there exists a class P that is pivotal with respect to N.

<u>Proof of the Lemma</u>: Consider the convex polytope defined by

$$\sum_{S,\ i \in S} w_S = 1 \qquad \text{for} \quad i = 1, \ldots, n$$

and $w_S \geq 0$ for $S \in N$. The extreme points of this polytope correspond to the <u>minimal</u> balanced families. Precisely, each extreme point $(\bar{w}_S)$ specifies $B = \{S \mid \bar{w}_S > 0\}$. If all extreme points had n positive components, we would have properties 1, 2, and 3 immediately. However, since there are balanced families B with fewer than n sets (degeneracy, in linear programming jargon), we must perturb the system to obtain the desired result. To be explicit, if we replace the system of equations by

$$\sum_{S,\ i \in S} w_S = 1 + \varepsilon^i \qquad \text{for} \quad i = 1 , \ldots , n$$

and $\varepsilon > 0$ small enough, then all the extreme points will have n positive components. The $\mathcal{B}$ so determined comprise $\mathcal{P}$.

□

(For $N = \{1 , 2\}$ the only class $\mathcal{P}$ that is pivotal with respect to N is $\mathcal{P} = \{\{1 , 2\} , \{1 , 12\}\}$, up to a permutation of $(1 , 2)$. The diagrams below present two classes $\mathcal{P}$ that are pivotal with respect to N for $N = \{1 , 2 , 3\}$; they can be shown to be unique up to a permutation of $(1 , 2 , 3)$. They should be "viewed" as convex diamonds with a base labelled 13 in the plane of the paper and facets above that plane with labels as shown. An element of $\mathcal{P}$ corresponds to a vertex in the diagram and is the <u>complement</u> of the set of labels on facets incident at the vertex. To verify that they are pivotal with respect to N you need only verify that each family is balanced. Property 3 , the pivoting property, follows from the structure of the simple convex polytopes from which they are formed. Note that the two families both give a count of 1 for the left hand triangulation of Figure 1 but $\mathcal{P}_1$ gives a count of 3 and $\mathcal{P}_2$ gives a count of 1 to the right hand triangulation of Figure 1.

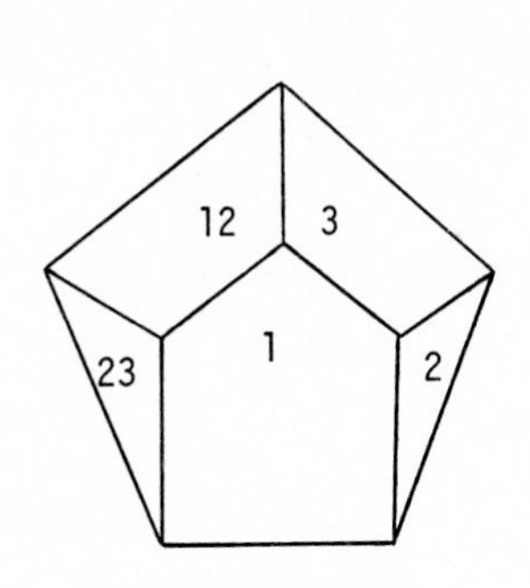

$\mathcal{P}_1$: {1 , 2 , 3}
{2 , 3 , 12}
{2 , 3 , 13}
{1 , 2 , 23}
{1 , 12 , 23}
{2 , 13 , 23}
{3 , 12 , 23}
{12 , 13 , 23}

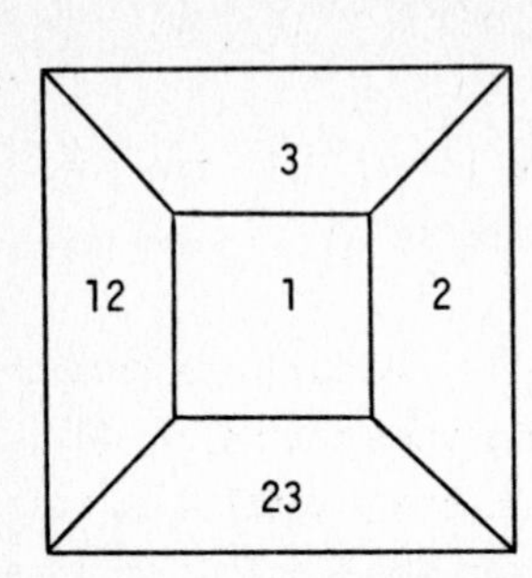

P_2 : {1 , 2 , 3}
{1 , 3 , 12}
{2 , 3 , 13}
{1 , 2 , 23}
{1 , 12 , 23}
{2 , 13 , 23}
{3 , 12 , 13}
{12 , 13 , 23}

For N = {1 , 2 , 3 , 4} , the situation is even more complicated. Classes P that are pivotal with respect to N have been constructed with 157, 181, and 185 members. No simple structure has yet been discovered for any of these classes. A purely combinatorial construction of P seems to be a difficult problem.)

REFERENCES

1. Shapley, L.S., On Balanced Games Without Side Payments, in Mathematical Programming, T.C. Hu; S.M. Robinson (Eds.), Academic Press, New York-London 1973, pp. 261 - 290.

2. Scarf, H., The Core of an N Person Game, Econometrica 35, 50 - 69 (1967).

3. Sperner, E., Neuer Beweis für die Invarianz der Dimensionszahl und des Gebietes, Abh. Math. Sem. Hamb. Univ. 6, 265 - 272 (1928).

4. Todd, M.J., The Computation of Fixed Points and Applications, Springer, Berlin-Heidelberg-New York 1976.

Numerical Solution of Highly Nonlinear Problems
W. Forster (ed.)
© North-Holland Publishing Company, 1980

ON THE KAKUTANI FIXED POINT THEOREM

Daniel I.A. Cohen

The Rockefeller University,
New York, U.S.A.

The Kakutani Fixed Point Theorem, an elegant generalization of the Brouwer Fixed Point Theorem [1], is usually proven using approximation by piecewise linear functions. The Brouwer Theorem is then applied to these functions allowing for delicate algorithmic calculation [2, 3]. Below we will show that this theorem can be proven directly from Sperner's Lemma.

Theorem: Let S be an n-simplex with vertices labelled $0, 1, \ldots, n$. Let C be the collection of compact convex subsets of S. Let $f: S \to C$ be an upper semicontinuous point-to-set map. Then there exists a point $x \in S$ such that $x \in f(x)$, i.e. a fixed point for f.

Proof: Let us assume that no such fixed point exists. Let T be any triangulation of S. Let t be any vertex of T. There must exist a vertex w of S such that i) w is a vertex of the carrier of t, ii) there is a hyperplane dividing the simplex such that w and t are on one side and $f(t)$ is on the other.
Take any such w and label t with the label of w.

If we apply this process to every vertex of T, we will obtain a labelled triangulation which is proper in the Sperner sense. Therefore, there will exist an n-simplex of T with a complete set of labels.

Taking finer and finer triangulations will produce smaller and smaller complete n-simplices in S and these will have at least one limit point x.

Let m be any vertex of S .

Let v_1 , v_2 , ... be a sequence of vertices of the complete triangles all of which have the label m and which converge to x .

Let x_m be the point of $f(x)$ closest to m . If x_m is 3ε closer to m than x is, then there is a $0 < \delta < \varepsilon$ such that for all v_i within δ of x , $f(v_i)$ is within ε of $f(t)$. But this means that $f(v_i)$ is closer to m than v_i is. This contradiction shows that $f(x)$ cannot be closer to m than x is.

But x cannot be closer (or as close) to each vertex of S than $f(x)$ is.

□

The ability to use this proof for calculation depends on the feasibility of labelling the triangulation T and the algorithms for finding complete Sperner Triangles [3, 4, 5].

REFERENCES

1. Kakutani, S., A Generalization of Brouwer's Fixed Point Theorem, Duke Math. J. 8, 457-459 (1941).

2. Eaves, B.C., Computing Kakutani Fixed Points, SIAM J. of Appl. Math. 21, 2, 236-244 (1971).

3. Todd, M.J., The Computation of Fixed Points and Applications, Springer, Berlin-Heidelberg-New York 1976.

4. Cohen, D.I.A., On the Sperner Lemma, J. Comb. Theory 2, 585-587 (1967).

5. Kuhn, H.W., Simplicial Approximation of Fixed Points, Proceedings of the National Academy of Sciences, USA, 61, 1238-1242 (1968).

Numerical Solution of Highly Nonlinear Problems
W. Forster (ed.)
© North-Holland Publishing Company, 1980

FIXED POINT ALGORITHMS AND SOME GRAPH PROBLEMS

Lidia Filus

Central School of Planning and Statistics,
Warsaw, Poland

In this paper we show the relationship between fixed point algorithms and the problem of obtaining (in a non-oriented graph $\Gamma = (W, Q)$) a path in the set $V \subset W$ with the property $\#\{y \in W \mid xQy\} \leq 2$ for every $x \in V$ ending in $W \backslash V$.
We give necessary and sufficient conditions for the existence of an odd number of such paths.

1. INTRODUCTION

When we look at some fixed point algorithms (e.g. at those based on simplicial subdivision or on primitive sets) from the combinatorial point of view (see e.g. [4], [5]), then instead of considering the family of primitive sets or the family of n-simplices in the simplicial subdivision we can consider an arbitrary finite family P of $(n+1)$-element sets.
In the family P we have a binary replacement relation R defined. It is defined in such a way that two sets Z and Z' are related by this

binary replacement relation if and only if one set can be obtained from the other by replacing exactly one element.

We have also the labelling function ℓ which maps the union

$$\cup P = \bigcup_{Z \in P} Z$$

into the set $N = \{0, 1, \ldots, n\}$. This allows us to treat various procedures used in fixed point algorithms as a property of the relation R (see [6]) and to look at fixed point algorithms as a procedure for obtaining a special sequence of almost complete sets (with respect to certain $i = 0, 1, \ldots, n$) ending with a complete set in P .
Since the replacement relation R is symmetric and nonreflexive we can treat the members of P as vertices of a non-oriented graph and R as the set of edges of the graph.
Similar to [6] let us denote

$W = \{ Z \in P \mid \ell(Z) \supset N\backslash\{i\} \}$,
$V = \{ Z \in P \mid \ell(Z) = N\backslash\{i\} \}$,
$Q = \{ (Z, Z') \in R \mid \ell(Z) \cap \ell(Z') = N\backslash\{i\} \}$,
for certain fixed $i = 0, 1, \ldots, n$.
Obtaining in P a sequence of almost complete sets (with respect to that fixed $i = 0, 1, \ldots, n$) ending with a complete set is then equivalent to obtaining in the subgraph (W, Q) of the graph (P, R) a path with vertices in V and ending in $W\backslash V$.
In fixed point algorithms uniqueness of the replacement of multiple elements in
$V = \{ Z \in P \mid \ell(Z) = N\backslash\{i\} \}$ is assumed (see [5], [6]). Moreover, the first element Z_0 of the sequence of almost complete sets has the property that exactly one multiple element is replaceable in Z_0 . This guarantees that for every $Z \in V$ the set
$W(Z) = \{ Z' \in W \mid ZQZ' \}$ has at most two elements and that the set $W(Z_0)$ has exactly one element (see [6], section 3).
This graph description makes it possible to formulate a somewhat more general problem.

2. THE GRAPH PROBLEM

Given a finite non-oriented graph $\Gamma = (W, Q)$. W is a finite set of vertices and Q is a set of edges ($Q \subset W^2$ is a symmetric and nonreflexive relation) of the graph Γ . For an arbitrary set $V \subset W$ and $x \in W$ $V(x) = \{ y \in V \mid xQy \}$. We denote by $\#V$ the cardinality of V . By a path in Γ we mean an arbitrary sequence $(x^0, \ldots, x^k)$ of vertices such that either $k = 0$ or $(x^j, x^{j+1}) \in Q$ for $j = 0, \ldots, k-1$.

The graph description of section 1 can be formulated as the following somewhat more general problem:
There is a subset V of the set of vertices W with the property $\#W(x) \leq 2$ for every $x \in V$. What can we say about existence and the number of all paths $(x^0, \ldots, x^k)$ in Γ such that
1) $x^j \in V$ for $j = 0, \ldots, k-1$,
2) $\#W(x^0) = 1$,
3) $x^k \in W \setminus V$,
4) $x^{j-1} \neq x^{j+1}$ for $j = 1, \ldots, k-1$.
The following theorem gives an answer to this question.

Theorem 2.1: If $\#W(x) \leq 2$ for every $x \in V \subset W$, then the number of all paths $(x^0, \ldots, x^k)$ in Γ which satisfy the conditions 1), 2), 3), 4) is not greater than $\#\{ x \in V \mid \#W(x) = 1 \}$ and the number of all such paths is odd if and only if $\#\{ x \in V \mid \#W(x) = 1 \}$ is odd.

□

This theorem follows directly from the following lemmas, which are true under the assumption that $\#W(x) \leq 2$ for every $x \in V$. They are similar to the lemmas 2.2 to 2.4 for chains in [5].

Lemma 2.2: If $(x^0, \ldots, x^k)$ is a path in Γ such that $x^j \in V$ for $j = 0, \ldots, k$ and $x^{j-1} \neq x^{j+1}$ for $j = 1, \ldots, k-1$, then there exists at most one $x^{k+1} \in W$ such that $(x^0, \ldots, x^k, x^{k+1})$ is a path in Γ and $x^{k+1} \neq x^{k-1}$.

Proof: Assume the contrary, that there exist two different elements $\overline{x^{k+1}}$, x in W such that $x^{k+1} \neq x^{k-1}$, $x \neq x^{k-1}$ and that $(x^0, \ldots, x^k, x^{k+1})$ and $(x^0, \ldots, x^k, x)$ are the paths in Γ .

From this it follows that $W(x^k) \supset \{x^{k-1}, x^{k+1}, x\}$ which contradicts the assumption that $\#W(x) \le 2$ for every $x \in V$.

□

Lemma 2.3: If $(x^0, \ldots, x^k)$ is a path in Γ such that
$x^j \in V$ for $j = 0, \ldots, k$,
$\#W(x^0) = 1$ and
$x^{j-1} \ne x^{j+1}$ for $j = 1, \ldots, k-1$,
then the path $(x^0, \ldots, x^k)$ does not cycle.

Proof: Assume the contrary, that for certain $j = 1, \ldots, k-1$ there exists $i < j$ such that $x^i = x^j$. Let j^* be the smallest index from the set $\{1, 2, \ldots, k-1\}$ with this property. Then for $i \ne 0$ we have $W(x^i) = \{x^{i-1}, x^{i+1}\} = \{x^{j^*-1}, x^{j^*+1}\} = W(x^{j^*})$. From this follows that either $x^{j^*-1} = x^{i-1}$ or $x^{j^*-1} = x^{i+1}$. Since the relation Q is nonreflexive we have $i \ne j^*-1$. From the assumption $x^{j-1} \ne x^{j+1}$ we obtain that $i \ne j^*-2$. So for j^*-1 we can find $i' = i+1$ or $i' = i-1$ such that $i' < j^*-1$ and $x^{i'} = x^{j^*-1}$, which contradicts the assumption that j^* is the smallest index for which there exists an i such that $x^i = x^{j^*}$. From this contradiction it follows that none of the vertices $x^1, \ldots, x^k$ can be the beginning of a cycle. And since the relation Q is nonreflexive and $\#W(x^0) = 1$ we have also that x^0 cannot be the beginning of a cycle; which completes the proof of Lemma 2.3.

□

We will say that the path $(x^0, \ldots, x^k)$ with $x^{j-1} \ne x^{j+1}$ $(j = 1, \ldots, k-1)$ can be extended in V if there exists $x^{k+1} \in V$, $x^{k+1} \ne x^{k-1}$ such that $(x^0, \ldots, x^k, x^{k+1})$ is a path in Γ.

Lemma 2.4: If $(x^0, \ldots, x^k)$ and $(y^0, \ldots, y^k)$ are two paths which cannot be extended in V with $x^j \in V$, $y^i \in V$ for $j = 0, \ldots, k$; $i = 0, \ldots, \ell$,
$\#W(x^0) = \#W(y^0) = 1$ and at least one of them can be extended in W, and if for some $j_0 = 0, \ldots, k$ and $i_0 = 0, \ldots, \ell$ we have

$x^{j_0} \ne y^{i_0}$, then we have $x^j = y^i$ for each $j = 0, \ldots, k$ and each $i = 0, \ldots, \ell$.

Proof: Let us assume that $(y^0, \ldots, y^\ell)$ can be extended in W. Suppose that there exist $j = 0, \ldots, k$ and $i = 0, \ldots, \ell$ such that $x^j = y^i$. Let $j^* \in \{0, \ldots, k\}$ be the smallest index such that for certain $i \in \{0, \ldots, \ell\}$ we have $x^j = y^i$. We will show that $j^* = 0$.

Suppose that $j^* \neq 0$ and $j^* \neq k$. Then we have $W(x^{j^*}) = \{x^{j^*-1}, x^{j^*+1}\} = W(y^i) = \{y^{i-1}, y^{i+1}\}$ and $i \neq 0$. From this it follows that either $x^{j^*-1} = y^{i-1}$ or $x^{j^*-1} = y^{i+1}$ which contradicts the assumption that j^* is the smallest index for which there exists $i \in \{0, \ldots, \ell\}$ such that $x^{j^*} = y^i$.

We obtain the analogous contradiction in the case $j^* = k$. In this case if $\#W(x^k) = 2$ and $x^k = y^i$ for certain $i = 0, \ldots, \ell$, then $W(x^k) = \{x^{k-1}, x^{k+1}\} = W(y^i) = \{y^{i-1}, y^{i+1}\}$ and we have either $x^{k-1} = y^{i-1}$ or $x^{k-1} = y^{i+1}$.

If $\#W(x^k) = 1$ and $x^k = y^i$ for certain $i = 0, \ldots, \ell$, then $W(x^k) = \{x^{k-1}\}$ and since $\#W(y^i) = 2$ for $i = 1, \ldots, \ell$ we have $x^k = y^0$ and $W(x^k) = \{x^{k-1}\} = \{y^1\} = W(y^0)$. The equality $x^{k-1} = y^1$ contradicts the assumption that $j^* = k$. Therefore $j^* = 0$ and for certain $i = 0, \ldots, \ell$ we have $x^0 = y^i$ and $\#W(x^0) = \#W(y^i) = 1$. Since $\#W(y^i) = 2$ for each $i = 1, \ldots, \ell$ we have $x^0 = y^0$ and from Lemma 2.2 we obtain that $k = 1$ and $x^j = y^j$ for each $j = 0, \ldots, k$. So Lemma 2.4 is true.

□

Lemma 2.5: If $(x^0, \ldots, x^k)$ and $(y^0, \ldots, y^k)$ are paths which cannot be extended in W such that $x^j \in V$, $y^i \in V$ for each $j = 0, \ldots, k$ and $i = 0, \ldots, \ell$,
and $\#W(x^0) = \#W(y^0) = 1$,
and $x^j = y^i$ for some $j = 0, \ldots, k$ and $i = 0, \ldots, \ell$,
then $k = \ell$ and for each $j = 0, \ldots, k$ we have either $x^j = y^j$ or $y^j = x^{k-j}$.

Proof: On the basis of Lemma 2.2 it is enough to show that either $x^0 = y^0$ or $x^0 = y^\ell$. Similar to the proof of Lemma 2.4 let j^* be the smallest index from the set $\{0, \ldots, k\}$ with the property that there exist $i = 0, \ldots, \ell$ such that $x^{j^*} = y^i$. If we assume that $j^* \neq 0$ and $j^* \neq k$, then we obtain a contradiction in the same way as in the proof of Lemma 2.4. So we have either $j^* = 0$ or $j^* = k$.

If $j^* = 0$ and $x^0 = y^i$ for certain $i = 0, \ldots, \ell$, then $\#W(x^0) = \#W(y^i) = 1$. Then we obtain that either $x^0 = y^0$ or $x^0 = y^\ell$.

The case $j^* = k$ is identical to the case $j^* = 0$.

This completes the proof of Lemma 2.5.

□

If two paths $(x^0, \ldots, x^k)$ and $(y^0, \ldots, y^k)$ have the property that $x^j = y^{k-j}$ for $j = 0, \ldots, k$, then we say that one of the paths is the inverse of the other.

Proof of Theorem 2.1: From Lemma 2.2 and Lemma 2.3 it follows that each element x^0 from the set $\{x \in V \mid \#W(x) = 1\}$ uniquely determines a path $(x^0, \ldots, x^{k-1})$ with its elements in V and which cannot be extended in V.

From Lemma 2.4 and Lemma 2.5 it follows that every two such paths determined by different elements of the set $\{x \in V \mid \#W(x) = 1\}$ are either disjoint or one is the inverse of the other. So there is an even number of elements in the set $\{x \in V \mid \#W(x) = 1\}$ determining the paths which cannot be extended in W. The number of all paths $(x^0, \ldots, x^k)$ such that $\#W(x^0) = 1$ and $x^i \in V$ for each $i = 0, \ldots, k-1$ which can be extended in $W \setminus V$ is not greater than $\#\{x \in V \mid \#W(x) = 1\}$ and the number of all such paths is odd if and only if $\#\{x \in V \mid \#W(x) = 1\}$ is an odd number.

From Lemma 2.2 it follows that the number of all paths $(x^0, \ldots, x^{k-1}, x^k)$ which satisfy conditions 1), 2), 3), 4) is equal to the number of all paths $(x^0, \ldots, x^{k-1})$ with $\#W(x^0) = 1$, $x^i \in V$ for $i = 0, \ldots, k-1$ and which can be extended in $W \setminus V$.

This completes the proof of Theorem 2.1.

□

From Lemmas 2.2 to 2.5 we obtain

Proposition 2.6: If $\#W(x) \leq 2$ for every $x \in V$, then the set A consisting of all elements $x \in W \setminus V$ for which there exists a path $(\xi^0, \ldots, \xi^{r-1}, \xi^r)$ satisfying

1) $\xi^i \in V$ for $i = 0, \ldots, r-1$,

2) $\#W(\xi^0) = 1$,

3') $\xi^r = x$,

has not more elements than the set B consisting of all paths $(x^0, \ldots, x^{k-1}, x^k)$ satisfying 1), 2), 3), 4). Moreover, if $\#V(x) \leq 1$ for every $x \in W\setminus V$, then $\#A = \#B$ and the paths from B are disjoint.

Proof: First note that if for certain $x \in W\setminus V$ there exists a path $(\xi^0, \ldots, \xi^{r-1}, \xi^r)$ which satisfies 1), 2), 3'), then there exists also a path $(x^0, \ldots, x^k)$ in B such that $x^k = x$ and $x^0 = \xi^0$. From this and from Lemme 2.2 we obtain that the set A has not more elements than the set B. From Lemma 2.4 we have that if $(x^0, \ldots, x^k)$ and $(y^0, \ldots, y^\ell)$ are paths in B with $x^0 \neq y^0$, then the paths $(x^0, \ldots, x^{k-1})$ and $(y^0, \ldots, y^{\ell-1})$ are disjoint. Then to prove Proposition 2.6 it is enough to show that if $\#V(x) \leq 1$ for every $x \in W\setminus V$, then $x^k \neq y^\ell$.
Suppose that $x^k = y^\ell$. Then $\{x^{k-1}, y^{\ell-1}\} \subset V(x^k)$ and since $x^{k-1} \neq y^{\ell-1}$ we have $\#V(x^k) \geq 2$, which contradicts the assumption that $\#V(x) \leq 1$ for every $x \in W\setminus V$.
This completes the proof of Proposition 2.6.

□

3. REMARKS

Theorem 2.1 and Proposition 2.6 can be used to prove a theorem (related to the fixed point algorithms mentioned in section 1) which gives necessary and sufficient conditions for the existence of an odd number of sequences $(Z_0, Z_1, \ldots, Z_k)$ in the family $\mathcal{P}$ described in section 1 (see [6], Theorem 3.2), and to prove a theorem which gives necessary and sufficient conditions for the existence of an odd number of complete sets in the family $\mathcal{P}$ (see [6], Theorem 3.1). This is related to Sperner's Lemma in the sense that from this theorem, using the same method of proof as in the proof of Theorem 3.1 in [5] (Sperner's Lemma for polyhedra), we can obtain a stronger version of Sperner's Lemma for pseudomanifolds.

Let T be a triangulation of an n-pseudomanifold K and let ℓ be a labelling function defined on the set of all vertices of this triangulation into the set $N = \{0, 1, \ldots, n\}$.

Theorem 3.1: If for some $i \in \{0, 1, \ldots, n\}$ there is an odd number of $(n-1)$-simplices in T which lie on the boundary of K and for which the set of labels is $N \setminus \{i\}$, then there exists an odd number of complete n-simplices in T and the number of $(n-1)$-simplices in T which lie on the boundary of K and for which the set of labels is $N \setminus \{i\}$ is odd for each $i = 0, \ldots, n$.

REFERENCES

1. Alexandroff, P.S., Combinatorial Topology, Vol. I, Graylock Press, Rochester, New York 1956.

2. Cohen, D.I.A., On the Sperner Lemma, Journal of Combinatorial Theory 2, 585-587 (1967).

3. Eaves, B.C., Homotopies for Computation of Fixed Points, Mathematical Programming 3, 1-22 (1972).

4. Filus, L., A Combinatorial Lemma Related to the Search of Fixed Points, Bulletin de L'Academie Polonaise des Sciences, Sêrie de sciences math., astr. et phys., Vol. 25, No. 7, 615-616 (1977).

5. Filus, L., A Combinatorial Lemma for Fixed Point Algorithms, in Nonlinear Programming 3, O.L. Mangasarian; R.R. Meyer; S.M. Robinson (Eds.), Academic Press, New York 1978, pp. 407-427.

6. Filus, L., Combinatorial Fixed Point Algorithms, Proceedings of the Seminar on Game Theory and Related Topics (September 1978 Bonn/Hagen), North Holland, to appear.

7. Knaster, B.; Kuratowski, C.; Mazurkiewicz, S., Ein Beweis des Fixpunktsatzes für n-dimensionale Simplexe, Fundamenta Mathematicae 14, 132-137 (1929).

8. Kuhn, H.W., Simplicial Approximation of Fixed Points, Proc. Nat. Acad. Sci. 61, 1238-1242 (1968).

9. Kuhn, H.W.; MacKinnon, J.G., Sandwich Method for Finding Fixed Points, Journal of Optimization Theory and Applications,17, 189-204 (1975).

10. Scarf, H.E., The Approximation of Fixed Points of a Continuous Mapping, SIAM j. Appl. Math 15, 1328-1343 (1967).

11. Sperner, E., Neuer Beweis für die Invarianz der Dimensionszahl und des Gebietes, Abh. Math. Sem. Univ. Hamburg 6, 265-272 (1928).

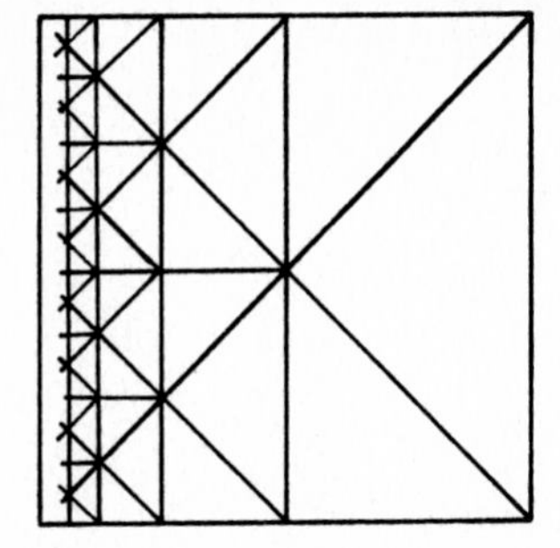

In this section we have collected papers with a mainly theoretical content.

Numerical Solution of Highly Nonlinear Problems
W. Forster (ed.)
© North-Holland Publishing Company, 1980

HOMOTOPY METHODS FOR APPROXIMATING SEVERAL SOLUTIONS TO NONLINEAR SYSTEMS OF EQUATIONS

Eugene L. Allgower

Colorado State University,
Fort Collins, Colorado, U.S.A.

and

Kurt Georg

Universität Bonn,
Bonn, West Germany

1. INTRODUCTION

The problem which we consider here is that of numerically approximating additional zeros of $g : R^n \to R^n$ after one zero has already been approximated or determined. Our approach will be to construct homotopies which deform g to some auxiliary function e.g. a constant. The numerical aspect will then consist of tracing the induced homotopy paths via continuation or simplicial algorithms according as g is

---

This paper was written while the first author was a guest of the Deutsche Forschungsgemeinschaft at SFB 72, University of Bonn.

Sponsored in part by AFOSR Grant 76-3019.

sufficiently smooth or not.

If g is smooth, then in theory, Newton's method is adequate to approximate all of the zeros of g, provided the proper starting values are chosen. This proviso is however a practical difficulty, because in practice it usually results that some of the zeros of g have domains of attraction relative to Newton's method which are very difficult to hit with randomly chosen starting values. A remedial measure which may be taken to combat this difficulty is to perform a "norm deflation" by multiplying g by $||x - \tilde{x}^*||^{-1}$, where x^* is a zero of g which has already been satisfactorily approximated by $\tilde{x}^*$. However, although this measure is frequently effective, it may also frequently fail or result in slow convergence, or lose effectiveness with repeated application [8]. The global convergence property of homotopy methods makes it possible to overcome some of these difficulties.

Homotopy methods have recently been applied by several authors [10], [11], [12], [16], [17] to the problem of approximating all zeros of complex polynomial maps $P : C^n \to C^n$. This approach consists of choosing an appropriate polynomial $Q : C^n \to C^n$ having $d = \Pi d_j$ ($d_j = \deg P_j$) known zeros, and forming a homotopy H such that $H(0, z) = Q(z)$ and $H(1, z) = P(z)$. For the type of Q used in [10], it has been shown that for almost all possible Q's, $H^{-1}(0)$ consists of smooth paths c_j, $j = 1, \ldots, d$, that each c_j eventually comes arbitrarily close to $\{1\} \times C^n$, and that each isolated zero of P can be obtained by a limit point in $\{1\} \to C^n$ of some c_j. Since each c_j can be followed via a continuation or simplicial algorithm, the problem of approximating all zeros of P seems to be well in hand at least for moderate degrees d.

The present problem differs from the above case in that it is not generally possible to determine the number of zeros of g a priori. Even if this number is known, an appropriate choice of initial map g_0 in a deformation homotopy between g and g_0 so as to connect the zeros of g_0 and g one-to-one with smooth paths generally requires too much a priori knowledge of the structure of g, such as the local Brouwer degrees of the zeros of g, [2].

One of our aims here will be to determine homotopies h so that $h^{-1}(0)$ will connect several zeros of g. It is also desirable to determine homotopy paths which will remain bounded for all values of the homotopy parameter, and if possible connect several or all zeros of g.

If the homotopy paths remain bounded for all values of the homotopy parameter, they consist of loops and hence practical stopping criteria are possible.

In our present discussions we shall not dwell upon details of the continuation or simplicial methods for following homotopy paths. These have been adequately outlined elsewhere, see [2] and the relevant references cited therein. It might be noted before proceeding however, that if $h^{-1}(0)$ has bifurcations, it might be necessary to make some modifications in the case of continuation algorithms. For details of this aspect, see [2], [24], [32].

In the last section of our paper we indicate an application of our methods to nonlinear eigenvalue problems. In particular, it is possible to establish the accessibility of a second solution despite the fact that the two solutions lie on separated components of the "bifurcation diagram". This investigation has been motivated by recent numerical results reported by H.O. Peitgen and M. Prüfer [27] and by H. Jürgens and D. Saupe [21].

2. REVIEW OF HOMOTOPY METHODS FOR SOLVING SYSTEMS OF EQUATIONS

2.1. Let us begin to make definite the problem we will consider and the methods to be used. Let $g : R^n \to R^n$ be a continuous map. The following lemma can be shown by a simple application of the Brouwer degree:

<u>Lemma</u>: Suppose that there exists a "starting" point $c \in R^n$ and a bounded open neighbourhood V of c so that

(a) $h(t, x) = A(x - c) + t\, g(x) \neq 0$ for $t \geq 0$ and $x \in \partial V$.

Then g has a zero-point x^* in V.

□

In fact, using a more sophisticated degree argument [29] or a "constructive" simplicial pivoting argument [14], [27], [28], one can assure that the component C of $h^{-1}(0) \cap [0, \infty) \times R^n$ which contains $(0, c)$ is unbounded and contained in $[0, \infty) \times V$. Hence any limit point $x^* = \lim x_i$ with $(t_i, x_i) \in C$ and $\lim t_i = \infty$ is a zero-point of g. There are essentially two different numerical methods for "following" C in order to approximate x^*. We sketch the two methods very briefly below.

2.2. A CONTINUATION METHOD

Here we suppose that g is sufficiently smooth. Let us make the assumption that zero is a regular value of h, i.e. that the Jacobian (h_t, h_x) of h has maximal rank n for all zero-points (t, x) of h. It can be shown [9] that this is true for almost all starting points $c \in R^n$. Then C is a smooth curve which can be parametrized by arc length s [26] and which can be "followed" by numerically integrating the associated Davidenko differential equation

(a)
$$\begin{cases} h_t(t, x) \frac{dt}{ds} + h_x(t, x) \frac{dx}{ds} = 0 \\ (\frac{dt}{ds})^2 + (\frac{dx}{ds})^T (\frac{dx}{ds}) = 1 \\ (t(0), x(0)) = (0, c) \end{cases}$$

by predictor-corrector methods tracing back at least to [18]. For details and a survey of other recent work on continuation homotopy methods, see e.g. [2], [9], [15], [37].

2.3. A SIMPLICIAL HOMOTOPY METHOD [13], [14]

Simplicial methods for approximating zero-points require technical machinery which we now sketch for later reference.

(a) Let T be a triangulation of $R \times R^n$ with continuous refinement of grid size, e.g. J_3 [35]. In particular, the nodes T^0 of T consist of points (k, x) such that $k \in Z$ (= set of integers) and $2^k x \in Z^n$, i.e. on level k we have the grid size 2^{-k}. We shall not further describe here the simplices in T, nor their pivoting rules. For the details we refer the reader to [35].

(b) The labelling $\lambda : T^0 \to R^n$ is defined by

$$\lambda(t, x) = \begin{cases} g(x) & \text{for } t \geq k_0 \text{ and } x \in V \\ A(x - c) & \text{elsewhere} \end{cases}$$

where $k_0 \in Z$ is a chosen level (grid size).

(c) An n-face $\tau \in T^n$ with vertices $v_0 , \dots , v_n$ is said to be completely labelled if the convex hull $co\{\lambda(v_0) , \dots , \lambda(v_n)\}$ contains the vector $(\varepsilon , \dots , \varepsilon^n)$ for all small enough $\varepsilon > 0$.

(d) Each simplex in T contains none or exactly two completely labelled n-faces.

(e) On level $\{k_0\} \times R^n$ there is exactly one completely labelled n-face, say τ_0 . Let σ_1 be the simplex of T which contains τ_0 and is contained in $[k_0 , \infty) \times R^n$.

(f) There is a unique infinite chain having no repetitions, $\tau_0 , \sigma_1 , \tau_1 , \sigma_2 , \tau_2 , \dots$ of n-faces τ_i and simplices σ_i , such that $\tau_i = \sigma_i \cap \sigma_{i+1}$ is completely labelled, $i = 1 , 2 , \dots$

(g) Any simplex of (f) intersects $[k_0 , \infty) \times V$, hence the chain posesses at least one "limit point" $x^* \in \bar{V}$, i.e. for any $k \in Z$ and any neighbourhood U of x^* some simplex σ_{j_k} of (f) is contained in $[k , \infty) \times U$.

(h) If x^* is a limit point of (f) , then by the boundary condition 2.1.(a) : $x^* \in V$ and thus $g(x^*) = 0$.

3. HOMOTOPY METHODS FOR APPROXIMATING ADDITIONAL SOLUTIONS

3.1. Let us suppose that a zero-point x^* of g has already been approximated by some method or is known via some other considerations. Analogously to 2. we formulate a boundary condition which not only assures the existence of a second solution, but can also be developed into a numerical homotopy method for approximating it.

Lemma: Let x^* be an isolated zero-point of g which is nondegenerate in the sense that the local Brouwer degree differs from zero. Suppose that there exists a "direction" $d \in R^n\backslash\{0\}$ and a bounded open neighbourhood U of x^* such that

(a) $h_d(t , x) = g(x) + t d \neq 0$ for $t \geq 0$ and $x \in \partial U$.
Then g has a second zero-point $y^* \in U \setminus \{x^*\}$.

□

Again, as in 2., it can be shown that the component C_d of $h_d^{-1}(0) \cap [0 , \infty) \times R^n$ which contains $(0 , x^*)$ is compact, contained in $[0 , \infty) \times U$ and contains some point $(0 , y^*)$ with $x^* \neq y^*$.

3.2. A CONTINUATION HOMOTOPY METHOD

If g is sufficiently smooth and zero is a regular value of g , then zero is a regular value of h_d for almost all choices of d [9] , and as in 2.2. a predictor-corrector method can be used for "running" along the curve C_d , starting at $(0 , x^*)$ and stopping at $(0 , y^*)$.

3.3. A SIMPLICIAL HOMOTOPY METHOD

(a) We use the triangulation T as described in 2.3.(a).

(b) Given $k_0 \in Z$ (e.g. $2^{-k_0} \sim$ diam U), we define a labelling $\lambda_d : T^0 \to R^n$ by

$$\lambda_d(t , x) = \begin{cases} g(x) & \text{for } t > k_0 \text{ and } x \in U , \\ d & \text{elsewhere} . \end{cases}$$

(c) For a start, we need a completely labelled n-face τ_0 in a level $\{k\} \times U$ $(k > k_0)$ which "approximates" x^* in the following sense: if σ_1 is the simplex in $[k , \infty) \times R^n$ which contains τ_0 , then the simplicial algorithm $\tau_0 , \sigma_1 , \ldots$ remains in $[k , \infty) \times U$ and converges to x^* .

(d) If $\tilde{\sigma}_1$ is the simplex in $(-\infty , k] \times R^n$ which contains τ_0 , then the simplicial algorithm $\tau_0 , \tilde{\sigma}_1 , \ldots$ cannot cycle by (c), never has empty intersection with $(k_0 , \infty) \times U$ by (b) and thus possesses a limit point $y^* \in \bar{U}$. By the boundary condition 3.1.(a): $y^* \in U$ and hence $g(y^*) = 0$.

(e) In practice, one will get a starting n-face τ_0 in (c) by a performance of algorithm 2.2. Note that τ_0 is allowed to be quite large.

Also the labelling (b) is given in such a way that the algorithm (d) will come near y^* by using large steps.

(f) The assumptions in (c) do not assure however, that $x^* \neq y^*$, though this will be extremely unlikely in practice and to us seems to be an academic question. In any event, by the assumptions in 3.1. on x^* and the fact that Brouwer degree and simplicial algorithms are linked [2], [27], [28], the existence of a start τ_0 such that $x^* \neq y^*$ can be proven. A more concrete condition to guarantee this is that $0 \notin \text{co}\, \lambda(\rho)$ for all (n-1)-faces ρ of τ_0. For example, if g is differentiable at x^*, with nonsingular Jacobian, and if $k \in Z$ is sufficiently large, then this condition is true whenever (k, x^*) is the barycenter of an n-face τ_0 in $\{k\} \times R^n$.

3.4. We now indicate some simple examples in which it is easy to specify an appropriate pair (U, d) for the boundary condition 3.1.(a) so that the methods 3.2. or 3.3. may be applied (according as g is smooth) to approximate a second solution.

Suppose that there exists a $\Psi \in R^n$ such that $\Psi^T g(x) > 0$ for $||x|| \geq r > 0$. Choose any d so that $\Psi^T d > 0$. Then clearly $h_d(t, x) \neq 0$ for $||x|| \geq r$ and $t \geq 0$.

To be more specific, suppose that $g_1(x) > 0$ for $||x|| \geq r > 0$. Then clearly $\Psi = (1, 0, \ldots, 0)^T$ in 3.4. will do, and $h_d(t, x) \neq 0$ for $t \geq 0$ and $||x|| \geq r$ whenever the first component of d is positive. Simple examples are:

(i) $g_1(x) = x^T A_1 x + x^T b_1 + c_1$ is a quadratic form with positive definite real $n \times n$-matrix A_1, $b_1 \in R^n$, $c_1 \in R$.

(ii) More generally, let g_1 be a convex function with bounded level sets, $\{x \in R^n : g_1(x) \leq \alpha\}$, $\alpha \in R$. Actually, it is routine to see that if one level set is bounded and nonempty, then all are bounded. The uniformly convex functions viz.

$$y^T g_1''(x) y \geq \varepsilon y^T y \quad \text{for some } \varepsilon > 0 \text{ and all } x, y \in R^n$$

have bounded level sets. For a more specific discussion on convex functions with bounded level sets, see e.g. [34].

3.5. We have already noticed that if a pair U , d exists such that 3.1. Lemma holds, then the component C_d of $h_d^{-1}(0) \cap [0, \infty) \times R^n$ which contains $(0, x^*)$ is compact, contained in $[0, \infty) \times U$ and "bends back" to a second solution $(0, y^*)$. By choosing d slightly more restrictively, it is possible to attain the stronger condition

(a) $h_d(t, x) = g(x) + t\,d \neq 0$ for $t \in R$ and $x \in \partial U$.

In this case the larger component $\tilde{C}_d$ of $h_d^{-1}(0)$ which contains $(0, x^*)$ is compact, contained in $R \times U$ and consists, loosely speaking, of a closed "loop". In particular, if g is sufficiently smooth and zero is a regular value of h , then $\tilde{C}_d$ is a smooth, closed curve.

The advantage that this fact may offer is that $\tilde{C}_d$ may contain more than two zero-points. Several examples are given below. Theoretical results giving sufficient conditions that more than two zeros will lie on the loop remain to be developed.

3.6. Suppose that x^* is an isolated zero-point of g with nonvanishing local Brouwer degree, and that there exists an open bounded neighbourhood U of x^* and two linearly independent vectors $\Phi, \Psi \in R^n$ so that

(a) $\Phi^T g(x) > 0$, $\Psi^T g(x) > 0$ for $x \in \partial U$.

Then for any d such that

(b) $(\Phi^T d)(\Psi^T d) < 0$,

the stronger boundary condition 3.5.(a) holds. Of course, if one such Φ has been found, $\Psi \neq \alpha\Phi$ may be obtained by a sufficiently small perturbation of Φ . Then a d satisfying (b) exists and can usually be easily found.

Example: Assume that $\min_{||x||=r} g_1(x) \geq m > 0$, and that

$\max_{||x||=r} |g_2(x)| < M$ for some $r > 0$. Then the choices

$\Phi = (M, -m, 0, \dots, 0)^T$, $\Psi = (M, m, 0, \dots, 0)^T$

are possible, and hence $h_d(t, x) \neq 0$ for $t \in R$ and $||x|| = r$, whenever $\delta_2 \neq 0$ and

$\delta_1 \in (-\frac{m}{M}|\delta_2|, \frac{m}{M}|\delta_2|)$ where $d = (\delta_1, \delta_2, \dots, \delta_n)^T$.

3.7. As the preceding example suggests, the choice of parameters in d is still rather large. The absolute and relative magnitudes of the parameters in d influence the length, steepness and curvature of the loop $\tilde{c}_d$, and this is still a practical numerical consideration. We conclude this section with two concrete examples which have been used as simple numerical tests for a continuation predictor-corrector method.

(i) Example: Consider the map $g : R^2 \to R^2$ given by

$$g(x) = \begin{bmatrix} \xi_1^2 + \xi_2^2 - 2 \\ \xi_1^2 - \xi_2^2 \end{bmatrix} , \quad s = (\xi_1 , \xi_2)^T \in R^2 .$$

The zeros of g are the four points $x = (\pm 1 , \pm 1)^T$.

For $d = (\delta_1 , \delta_2)^T$ and $(t , x) \in h_d^{-1}(0)$, we have the parametrization

$$\begin{bmatrix} \xi_1^2 \\ \xi_2^2 \end{bmatrix} = \begin{bmatrix} 1 - \frac{\delta_2 + \delta_1}{2} t \\ 1 + \frac{\delta_2 - \delta_1}{2} t \end{bmatrix} ,$$

which implies immediately that for $\delta_2^2 > \delta_1^2$, the zero set $h_d^{-1}(0)$ is bounded and consists of one single loop.

(ii) Example: Consider the map $g : R^3 \to R^3$ given by

$$g(x) = \begin{bmatrix} \xi_1^2 + \xi_2^2 + \xi_3^2 - 3 \\ \xi_1^2 + \xi_2^2 - \xi_3^2 - 1 \\ \xi_1^2 - \xi_2^2 + \xi_3^2 - 1 \end{bmatrix} , \quad x = (\xi_1 , \xi_2 , \xi_3)^T \in R^3 .$$

The zeros of g are the eight points $x = (\pm 1 , \pm 1 , \pm 1)^T$.

For $d = (\delta_1, \delta_2, \delta_3)^T$ and $(t, x) \in h_d^{-1}(0)$ we have the parametrization

$$\begin{bmatrix} \xi_1^2 \\ \xi_2^2 \\ \xi_3^2 \end{bmatrix} = \frac{1}{2} \begin{bmatrix} 0 & 1 & 1 \\ 1 & 0 & -1 \\ 1 & -1 & 0 \end{bmatrix} \begin{bmatrix} 3 - t\,\delta_1 \\ 1 - t\,\delta_2 \\ 1 - t\,\delta_3 \end{bmatrix} = \begin{bmatrix} 1 - \frac{1}{2}(\delta_2 + \delta_3)\,t \\ 1 + \frac{1}{2}(\delta_3 - \delta_1)\,t \\ 1 + \frac{1}{2}(\delta_2 - \delta_1)\,t \end{bmatrix} ,$$

which implies immediately, that for $0 < \delta_1 < \delta_2 < \delta_3$ the zero set $h_d^{-1}(0)$ is bounded and consists of exactly two loops, each of which contains four solutions $x = (\pm 1, \pm 1, 1)^T$ resp. $x = (\pm 1, \pm 1, -1)^T$.

4. APPROXIMATING TWO POSITIVE SOLUTIONS FOR A CLASS OF NONLINEAR EIGENVALUE PROBLEMS

4.1. In this section we apply the methods of the preceding section to find a second positive solution to a class of nonlinear eigenvalue problems. The class of problems considered here is more general than the discretized boundary value problem treated in [2]. It also includes a class of discretized analogues of some nonlinear elliptic boundary value problems as treated in [4], [5], [6], [19]. Other simplicial methods have previously been used to obtain several solutions to discretizations of boundary value problems [1], [3], [20], [21], [27], but there does not yet seem to be any means for proving why these methods seem to work so surprisingly well for finding additional solutions. Although the present problem may be regarded as a nonlinear eigenvalue problem per se, it generally arises via a discretization of some nonlinear operator-eigenvalue problem. Thus there would remain a question of convergence of solutions of the discretized problem to solutions to the operator problem. This consideration will not come into our present scope. For discussions of this aspect, see e.g. [23].

4.2. Let M be an $n \times n$-matrix, $F : R^n \to R^n$ a map, and denote by $K = [0, \infty)^n$ the usual cone in R^n. The problem which we will consider is the following: Suppose that the system

$$\text{(a)} \quad g(\mu, x) = Mx - \mu F(x) = 0 \ , \quad \mu > 0 \ , \quad x \in K \ ,$$

has one solution. Find a second solution. We now discuss the assumptions on M and F, under which we shall consider this problem.

4.3. ASSUMPTIONS ON M :

$M = (m_{i,j})_{i,j=1, \ldots, n}$ is an irreducible weakly diagonally dominant Z-matrix [33], [36].

Let us briefly review the above terminology. Denote by $|M| = (|m_{i,j}|)_{i,j=1, \ldots, n}$ the matrix of absolute values. M is irreducible if the matrix

$$e^{|M|} = \sum_{i=0}^{\infty} (i!)^{-1} |M|^i \quad \text{has no zero entry.}$$

M is weakly diagonally dominant if $m_{i,j} \geq \sum_{\substack{j=1 \\ j \neq i}}^{n} |m_{i,j}| \ , \quad i = 1, \ldots, n \ ,$

where inequality holds for at least one $i \in \{1, \ldots, n\}$.
M is a Z-matrix if $m_{i,j} \leq 0$ for $i \neq j$.

4.4. For matrices M of the above form 4.3., the following results can be shown. For proofs and a thorough study of M-matrices, see [33]. M-matrices arise most frequently via finite difference discretizations of linear elliptic operators of second order, see [36].

(i) M has an inverse with strictly positive entries. For reasons of simplicity, let us normalize the spectral radius of M^{-1} to one.

(ii) There is an eigenvector $\Psi \in (0, \infty)^n$ with strictly positive entries such that $\Psi^T M = \Psi^T$.

(iii) Note that the assumptions 4.3. hold also for $M + D$ provided that D is a diagonal matrix with nonnegative entries.

4.5. For simplicity of discussion, let us introduce some definitions concerning the partial ordering of R^n .

(i) $x \le y \iff y \ge x \iff y - x \in K$,

(ii) $x < y \iff y > x \iff y \ge x$ and $y \ne x$,

(iii) $x \ll y \iff y \gg x \iff y - x \in \mathrm{int}(K) = (0 , \infty)^n$,

(iv) $x^+ = \sup (x , 0)$, $x^- = \sup (-x , 0)$, $|x| = x^+ + x^-$

where the sup is taken with respect to the partial ordering (i).

(v) F m (iv) it follows that $x = x^+ - x^-$.

4.6. ASSUMPTIONS ON F :

(i) $F(0) \ge 0$.

(ii)

$$\lim_{\substack{||x||\to\infty \\ x>0}} \frac{\Psi^T F(x)}{\Psi^T x} = +\infty .$$

(iii) There exists a diagonal matrix D with nonnegative entries such that $F(x) + Dx \ge 0$ for $x \ge 0$.

Since all norms on R^n are equivalent and $x \to \Psi^T|x|$ is a norm on R^n , the growth condition (ii) is implied e.g. by the following two growth conditions:

(iv)

$$\lim_{\substack{||x||\to\infty \\ x>0}} \frac{||F(x)^+||}{||x||} = +\infty ,$$

(v)

$$\lim_{\substack{||x||\to\infty \\ x>0}} \frac{||F(x)^-||}{||F(x)^+||} = 0 ,$$

which might be verified without explicit knowledge of the eigenvector Ψ .

4.7. Since we are only concerned with positive solutions in 4.2.(a), we may replace F by

(a) $\bar{F}(x) := F(x^+) + Dx^-$, $x \in R^n$,

and the system 4.2.(a) by

(b) $\bar{g}(\mu , x) = Mx - \mu\bar{F}(x) = 0$, $\mu > 0$.

From $\bar{F}(x) + Dx = F(x^+) + Dx^+$ and 4.6.(iii), it follows immediately that

(c) $\bar{F}(x) + Dx \geq 0$ for all $x \in R^n$.

We will make use of the following trick, cf. also [6]:

Lemma: Let $\mu > 0$ and $x \in R^n$. Then the following two statements are equivalent:

(i) $g(\mu , x) = 0$, $x \geq 0$,

(ii) $\bar{g}(\mu , x) = 0$.

In other words: Problems 4.2.(a) and 4.7.(b) are equivalent.

Proof: "(i) => (ii)". This follows from $F(x) = \bar{F}(x)$ for $x \geq 0$.
"(ii) => (i)". We have $(M + \mu D)x = (\bar{F} + D)x$. Since the right hand side is contained in K and $M + D$ has an inverse with positive entries, it follows that $x \geq 0$ and hence $F(x) = \bar{F}(x)$,
$g(\mu , x) = \bar{g}(\mu , x) = 0$.

□

4.8. We are now ready to prove a boundary 3.1.(a):

Lemma: Let $\mu > 0$ and $d \in R^n$, $d > 0$. Then there exists an $r > 0$ such that $\bar{g}(\mu , x) - td \neq 0$ for $||x|| \geq r$ and $t \geq 0$.

Proof: Suppose not. Then there exists a sequence
$t_i \geq 0$, $x_i \in R^n$, $||x_i|| \to \infty$ as $i \to \infty$ such that
$\bar{g}(\mu , x_i) - t_i d = 0$, hence

(a) $Mx_i = \mu\bar{F}(x_i) + t_i d$.

By adding μDx_i to both sides of (a), we see as in 4.7. (Proof), that $x_i \geq 0$. Hence $\bar{F}(x_i) = F(x_i)$, and multiplying (a) by the eigenvector Ψ^T gives

(b) $\Psi^T x_i = \mu \Psi^T F(x_i) + t_i \Psi^T d$.

This implies

(c) $\mu \Psi^T F(x_i) \le \Psi^T x_i \ , \quad \mu > 0 \ , \quad x_i \ge 0 \ , \quad ||x_i|| \to \infty \ ,$

which clearly contradicts the growth condition 4.6.(ii).

□

4.9. By using a degree technique essentially due to P.H. Rabinowitz [29], see also [4], [5], [6], [7], [19], [30], [31], it can be shown that the set $\Sigma = g^{-1}(0) \cap ((0 , \infty) \times K)$ of positive solutions contains unbounded components Σ_0 , Σ_∞ such that:

(i) there exists a sequence $(\mu_i , x_i) \in \Sigma_0$ with $\mu_i \to 0$, $x_i \to 0$ as $i \to \infty$,

(ii) there exists a sequence $(\mu_i , x_i) \in \Sigma_\infty$ with $\mu_i \to 0$, $||x_i|| \to \infty$ as $i \to \infty$.

By starting at (0 , 0) , the branch Σ_0 can easily be "followed" by a continuation or simplicial homotopy method similar to those in 2., using the parameter μ in 4.7.(b) as homotopy parameter, see e.g. [2], [9], [10], [11], [12], [13], [14], [15], [16], [17], [18], [21], [24], [25], [27], [28], [32], [35], [37]. Thus, if $\mu > 0$ occurs as parameter in Σ_0 , it is possible to approximate one solution $x^* \ge 0$ of 4.2.(a) . Supposing that this solution x^* is isolated with nonvanishing local Brouwer degree, by 4.8. , the methods described in 3. now are available to approximate a second solution $y^* \ge 0$, $y^* \ne x^*$.

4.10. The above method is of particular interest in case $\Sigma_0 \ne \Sigma_\infty$, since then, loosely speaking, y^* cannot be approximated by "further following" the branch Σ_0 . We conclude by giving a condition in the spirit of A. Ambrosetti and P. Hess [6] under which this occurs.

Proposition: Let F satisfy in addition to 4.6. the following assumptions:

(a) There exists $u_0 \in R^n$ such that $M u_0 > 0 \ge F(u_0)$,

(b) $F + D$ is monotone increasing on the order interval.
$[0 , u_0] = \{v \in R^n \ ; \ 0 \le v \le u_0\}$, i.e. $0 \le v \le w \le u_0$ implies $(F + D)v \le (F + D)w$.

Then Σ_0 is separated from Σ_∞ by u_0 , more precisely:
$\Sigma_0 \subset (0 , \infty) \times [0 , u_0]$, $\Sigma_\infty \cap ((0 , \infty) \times [0 , u_0]) = \Phi$.

Proof: By 4.4.(i) it follows that $u_0 \gg 0$, hence $[0 , u_0]$ is a neighbourhood of zero relative to K . By 4.9.(i) and (ii) , Σ_0 has points inside and Σ_∞ has points outside $(0 , \infty) \times [0 , u_0]$. Suppose that the conclusion of the proposition is not true. Then by the connectedness of Σ_0 (resp. Σ_∞) , one can deduce the existence of a solution $g(\mu , x) = 0$ with

(c) $\mu > 0$, $0 \le x \le u_0$, but $x \not\ll u_0$.

Let us show that this leads to a contradiction. By adding μDx to 4.2.(a) we get

(d) $(M + \mu D)x = \mu(F + D)x$.

Similarly, from (a) we get

(e) $(M + \mu D)u_0 > \mu(F + D)u_0$.

Subtracting (d) from (e) gives

(f) $(M + \mu D)(u_0 - x) > \mu[(F + D)u_0 - (F + D)x] \ge 0$,

where the last inequality follows from (b). Now by 4.4(i) and (ii) , it is seen that $u_0 - x \gg 0$ or $x \ll u_0$. But this is contradictory to (c) .

□

REFERENCES

1. Allgower, E.L., Application of a Fixed Point Search Algorithm, in Fixed Points: Algorithms and Applications, S. Karamardian (Ed.), Academic Press, New York 1977, pp. 87-111.

2. Allgower, E.L.; Georg, K., Simplicial and Continuation Methods for Approximating Fixed Points and Solutions to Systems of Equations, to appear in SIAM Review 1979.

3. Allgower, E.L.; Jeppson, M.M., The Approximation of Solutions of Nonlinear Elliptic Boundary Value Problems with Several Solutions, Springer Lecture Notes in Mathematics, 333, 1-20 (1973).

4. Amann, H., Fixed Point Equations and Nonlinear Eigenvalue Problems in Ordered Banach Spaces, SIAM Review 18, 620-709 (1976).

5. Amann, H., Nonlinear Eigenvalue Problems having Precisely Two Solutions, Math. Z. 150, 27-37 (1976).

6. Ambrosetti, A.; Hess, P., Positive Solutions to Asymptotically Linear Elliptic Eigenvalue Problems, Preprint (1979).

7. Brown, K.J.; Budin, H., Multiple Positive Solutions for a Class of Nonlinear Boundary Value Problems, J. Math. Anal. Appl. 60, 329-338 (1977).

8. Brown, K.M.; Gearhart, W.B., Deflation Techniques for the Calculation of Further Solutions of a Nonlinear System, Numer. Math. 16, 334-342 (1971).

9. Chow, N.; Mallet-Paret, J.; Yorke, J.A., Finding Zeros of Maps: Homotopy Methods that are Constructive with Probability One, Math. Comp. 32, 887-899 (1978).

10. Chow, N.; Malet-Paret, J.; Yorke, J.A., A Homotopy Method for Locating all Zeros of a System of Polynomials, Preprint 1979.

11. Drexler, F.J., Eine Methode zur Berechnung sämtlicher Lösungen von Polynomgleichungssystemen, Numer. Math., 29, 45-58 (1977).

12. Drexler, F.J., A Homotopy Method for the Calculation of all Zeros of Zero-Dimensional Polynomial Ideals, in Continuation Methods, H. Wacker (Ed.), Academic Press, New York 1978, pp. 64-94.

13. Eaves, B.C., Homotopies for Computation of Fixed Points, Mathematical Programming 3, 1-22 (1972).

14. Eaves, B.C.; Saigal, R., Homotopies for Computation of Fixed Points on Unbounded Regions, Mathematical Programming 3, 225-237 (1972).

15. Garcia, C.B.; Gould, F.J., Relations between Complementary Pivoting Algorithms and Local and Global Newton Methods, Preprint (1977).

16. Garcia, C.B.; Zangwill, W.I., Global Continuation Methods for Finding all Solutions to Polynomial Systems of Equations in n Variables, Preprint (1977).

17. Garcia, C.B.; Zangwill, H.I., Determining all Solutions to Certain Systems of Nonlinear Equations, Math. Op. Res. 4, 1-14 (1979).

18. Haselgrove, C., Solution to Nonlinear Equations and of Differential Equations with Two-Point Boundary Conditions, Comp. J. 4, 255-259 (1961).

19. Hess, P., Multiple Solutions of Asymptotically Linear Elliptic Boundary Value Problems, Proc. Equadiff IV, Prague 1977, to appear in Springer Lecture Notes in Mathematics.

20. Jeppson, M.M., A Search for the Fixed Points of a Continuous Mapping, in Mathematical Topics in Economic Theory and Computation, R.H. Day; S.M. Robinson (Eds.), SIAM, Philadelphia 1972, pp. 122-129.

21. Jürgens, H.; Saupe, D., Methoden der simplizialen Topologie zur numerischen Behandlung von nichtlinearen Eigenwert- und Verzweigungsproblemen, Diplomarbeit, University of Bremen, 1979.

22. Keener, J.P.; Keller, H.B., Positive Solutions of Convex Nonlinear Eigenvalue Problems, J. Diff. Equations 16, 103-125 (1974).

23. Keller, H.B., Accurate Difference Methods for Nonlinear Two-Point Boundary Value Problems, SIAM J. Num. Anal. 11, 305-320 (1974).

24. Keller, H.B., Numerical Solution of Bifurcation and Nonlinear Eigenvalue Problems, in Applications of Bifurcation Theory, P.H. Rabinowitz (Ed.), Academic Press, New York 1977, pp. 359-384.

25. Keller, H.B., Global Homotopies and Newton Methods, in Recent Advances in Numerical Analysis, C. de Boor; G.H. Golub (Eds.), Academic Press, New York 1978, pp. 73-94.

26. Milnor, J., Topology from the Differentiable Viewpoint, The University of Virginia Press, Charlottesville 1969.

27. Peitgen, H.O.; Prüfer, M., The Leray-Schauder Continuation Method is a Constructive Element in the Numerical Study of Nonlinear Eigenvalue and Bifurcation Problems, in Approximation of Fixed Points and Functional Differential Equations, H.O. Peitgen; H.O. Walther (Eds.), Springer Lecture Notes in Mathematics, 730, 326-409 (1979).

28. Prüfer, M., Simpliziale Topologie und globale Verzweigung, Dissertation, University of Bonn, 1978.

29. Rabinowitz, P.H., Some Global Results for Nonlinear Eigenvalue Problems, J. Functional Anal. 7, 487-513 (1971).

30. Rabinowitz, P.H., On Bifurcation from Infinity, J. Diff. Equations 14, 462-475 (1973).

31. Rabinowitz, P.H., Théorie du degré topologique et applications à des problèmes aux limites non linéaires. Notes University of Paris VI et CNRS, rédigés par H. Berestycki, 1975.

32. Rheinboldt, W.C., Numerical Methods for a Class of Finite Dimensional Bifurcation Problems, SIAM J. Numer. Anal. 15, 1-11 (1978).

33. Schröder, J., M-Matrices and Generalizations Using an Operator Theory Approach, SIAM Review 20, 213-244 (1978).

34. Stoer, J.; Witzgall, C., Convexity and Optimization in Finite Dimensions I, Springer, Berlin-Heidelberg-New York 1970.

35. Todd, M.J., The Computation of Fixed Points and Applications, Springer Lecture Notes in Economics and Mathematical Systems, 124, Springer, Berlin-Heidelberg-New York 1976.

36. Varga, R.S., Matrix Iterative Analysis, Prentice Hall, Englewood Cliffs, N.J. 1962.

37. Watson, L.T.; Fenner, D., Chow-Yorke Algorithms for Fixed Points or Zeros of C^2 Maps, to appear in ACM Trans. Math. Software, 1978.

Numerical Solution of Highly Nonlinear Problems
W. Forster (ed.)
© North-Holland Publishing Company, 1980

SOLVING EQUATIONS $0 \in f(x)$ UNDER GENERAL BOUNDARY CONDITIONS

Hoang Tuy

Institute of Mathematics,
Hanoi, Vietnam

0. INTRODUCTION

In this paper we are concerned with the problem of finding a zero of a multivalued mapping f from R^n into R^n [1] .

In order that such a zero exists in a given region C it is necessary to impose certain conditions on the values of f on the boundary of C (besides the usual assumptions of convexity, compactness and upper semi-continuity). A simple condition which has been long known is that at each boundary point x of C the set $f(x)+x$ must meet C . Other less familiar conditions are: The Leray-Schauder condition, stating that at each boundary point x the set $f(x)+x$ must not meet the halfline emanating from x in the direction $x-w$, where w is a given interior point of C ; the "inward" condition ([6], [1], [8]), requiring that at each boundary point x the set $f(x)$ meets the tangent cone of C at x ; the "nowhere outward" condition [8], requiring that at each boundary point x the set $f(x)$ be disjoint with the outward normal cone of C at x .

[1] The infinite-dimensional case will be treated in a subsequent paper (see [18] for preliminary results).

It has been proved by nonconstructive methods that each condition mentioned guarantees the existence of a zero of f in C. Our purpose is to show that such a zero, which theoretically exists, can be computed by combinatorial methods based an Scarf's fundamental algorithm [11]. In fact we shall develop more general conditions than those mentioned, which will secure the success of combinatorial methods when applied to our problem.

The paper is divided into three sections. In section 1 we shall describe the class of algorithms to be used in the sequel. The basic result of this section is a theorem pointing out a sufficient condition for termination of the algorithm. In section 2 we shall apply the algorithm to solve equations $0 \in f(x)$. A new boundary condition (condition 2.4) will be developed which will provide a generalization and unification of various known and new special conditions. In section 3 this condition 2.4 will be partially extended so as to serve as a basis for the application to the nonlinear complementary problem. We shall obtain here an existence theorem which contains as special cases a number of known propositions.

1. A CLASS OF RESTART ALGORITHMS

We first describe a general restart method for computing an approximate zero of a multivalued mapping from R^n into R^n. This method has the advantage that it does not involve an extra dimension and seems to require fewer artificial pivots than most others. A first particular variant of it, which corresponds to Example 1 below, has been presented in our paper [15] (see also [14] and [16]) and is merely a modified version of the original method of Scarf [11], [12]. A second variant, corresponding to Example 2 below, coincides essentially with the method by Van der Laan and Talman [20] as discussed in Todd [13]. In fact, the present generalization is an outgrowth of ideas developed earlier in [11], [15], [20], [13]. We also note that by a different approach and independently Van der Heyden [19] has come to an algorithm very close to the variant described in [15].

Let w be a prechosen point of R^n (a first guess of the solution to be sought). Select $n+1$ halflines $d^1, d^2, \ldots, d^{n+1}$ emanating from w so as to subdivide R^n into $n+1$ closed convex polyhedral solid cones $P_1, P_2, \ldots, P_{n+1}$, such that: 1) P_i contains all d^j with $j \neq i$;

2) the intersection of any k of these cones has dimension $n+1-k$.
Setting $P_{n+2} = R^n$, define for every nonempty set $I \subset \{1, 2, \dots, n+2\}$:

$$P_I = \bigcap_{i \in I} P_i \quad .$$

Consider a triangulation T of R^n which is consistent with the subdivision $P = \{P_1, P_2, \dots, P_{n+1}\}$, in the sense that for every simplex σ of T and every i the set $\sigma \cap P_i$ is a face (may be empty) of σ . So the restriction of T in P_I , $T_I = \{\sigma \cap P_I : \sigma \in T\}$, triangulates P_I .

Example 1: $S = [a^1, a^2, \dots, a^{n+1}]$ being an n-simplex containing w in its interior, let d^i denote the halfline from w through a^i , T the triangulation used in [15], [16]; see Fig. 1.

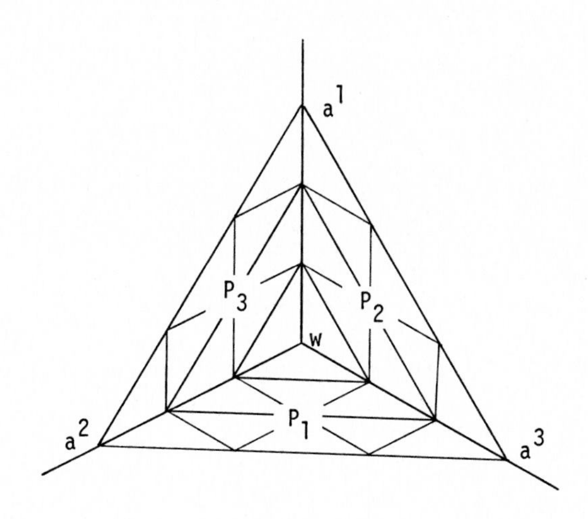

Figure 1 .

Example 2: S being an n-simplex as above, let N be a natural number, w a point of S such that Nw_i $(i = 1, . . . , n+1)$ are all integers (x_i denotes the i-th barycentric coordinate of x with respect to S),

$$d^i = \{x : x_i \leq w_i , x_{i+1} \geq w_{i+1} , x_j = w_j \ (j \neq i , i+1) \} ,$$

$$P_i = \{x : x_i \geq w_i , x_{i+1} \leq w_{i+1} \} ,$$

where $i+1 = 1$ if $i = n+1$; let T be the regular subdivision of degree N of S , as described for example in [7]; see Fig. 2.

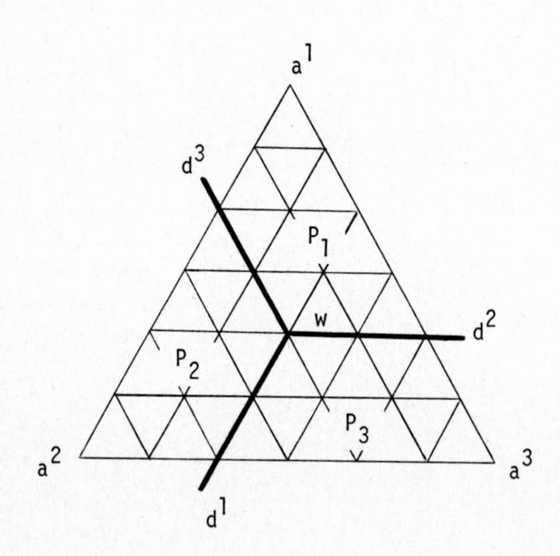

Figure 2 .

Let Q denote the set of vertices of the triangulation T, and let $\overline{Q} = Q \cup \{1, 2, \ldots, n+2\}$. A set U of $n+2$ distinct elements of $\overline{Q}$ is called a primitive set if $I = U \setminus Q$ is nonempty and if $U \cap Q$ is the vertex set of a simplex of maximal dimension of the triangulation T_I of P_I. Since $P_{n+2} = R^n$, it follows that a primitive set U contains the element $n+2$ only if $U \setminus \{n+2\} \subset Q$.

A subset W of a primitive set U is called a facet of it if $|W| = n+1$; the remaining element of U is then said to be opposite the facet W.

Now let f be the mapping of which we wish to compute an approximate zero. Denote by ℓ a selection of f, i.e. a single-valued mapping such that $\ell(x) \in f(x)$ for every x (we assume that $f(x) \neq \emptyset$ for every x). Select $n+1$ vectors of R^n, $b^1, b^2, \ldots, b^{n+1}$, satisfying

$$0 \in \text{int co } \{b^1, b^2, \ldots, b^{n+1}\}, \quad (1.1)$$

where co stands for "convex hull". We shall say that a set $W \subset \overline{Q}$ with $|W| = n+1$ is fair if $n+2 \notin W$ and if for all $\Theta > 0$ small enough :

$$[\Theta] \in \text{int co } \{\ell(W \cap Q) \; ; \; b^i, \; i \in W \setminus Q\}, \quad (1.2)$$

where $[\Theta]$ denotes the vector $(\Theta^1, \Theta^2, \ldots, \Theta^n) \in R^n$, Θ^i being the i-th power of Θ. Then using a familiar argument (which we omit) one can prove the following basic properties :

I) Whenever a primitive set U has a fair facet W, it has just one other fair facet if the element opposite to W is different from $n+2$, and none otherwise.

So a primitive set has at most two fair facets. If it has just one fair facet (the opposite element being necessarily $n+2$), we shall say that it is complete.

II) There is just one primitive set, namely $U_0 = \{1, \ldots, n+1, w\}$, with a fair facet $W_0 = \{1, \ldots, n+1\}$ entirely disjoint from Q.

III) If W is a fair facet of a noncomplete primitive set U, such that $W \neq W_0$, there is just one other primitive set having this facet in common with U.

From these properties it follows readily that a sequence Γ of primitive sets can be generated such that:

1) All primitive sets U_i in the sequence are distinct and meet Q ;
2) Any two consecutive U_i , U_{i+1} have a common fair facet W_{i+1} ;
3) The sequence starts from U_0 ; if it terminates, its last member is a complete primitive set.

The algorithm that generates Γ is the following.

Algorithm (w, P, T, b | f): Start with U_0 , W_0 , $x^0 = w$. Find $y^1 \in W_0$ such that $W_1 = (W_0 \setminus \{y^1\} \cup \{x^0\}$ is fair (using I) , then find $x^1 \in \overline{Q} \setminus U_0$ such that $U_1 = (U_0 \setminus \{y^1\}) \cup \{x^1\}$ is primitive (using III). Repeat the procedure with U_1 , W_1 , x^1 replacing U_0 , W_0 , x^0 . And so on. Stop when a primitive set U_k is reached which contains $n+2$ (and hence, is complete).

Thus, if the Algorithm terminates at step k , we have $W_k = U_k \setminus \{n+2\} \subset Q$, and from (1.2)

$$0 \in \text{co } f(W_k) \quad , \tag{1.3}$$

so that W_k provides an approximate fixed point of f .

Theorem 1.1: Let f be an u.s.c. (upper semi-continuous) multivalued mapping such that $f(x)$ is nonempty, compact, convex for every x . If there exists a compact set C containing w such that $\text{int } C \neq \emptyset$ and

$$\text{for all } x \in \partial C \qquad 0 \notin \text{co } \{f(x), b(x)\} \quad , \tag{1.4}$$

where $\partial C = C \setminus \text{int } C$, $b(x) = \text{co } \{b^i : x \in P_i \ , \ i \neq n+2\}$, then, provided the mesh of T is fine enough, Algorithm (w, P, T, b | f) will terminate in C yielding an approximate zero of f in C .

Proof: Since C is compact and $g(x) = co\{f(x), b(x)\}$ is an u.s.c. mapping, it is easily seen from (1.4) that there is $\delta > 0$ such that whenever a set V meets ∂C and has diam $V < \delta$ then $0 \notin co\, g(V)$. Indeed, for every $x \in \partial C$ there is a convex open neighbourhood $N(x)$ of $g(x)$ such that $0 \notin N(x)$. By the u.s.c. of g there is an open ball $B(x)$ around x such that $co\, g(B(x)) \subset N(x)$, hence $0 \notin co\, g(B(x))$. Let $B'(x)$ be an open ball around x, twice smaller than $B(x)$. Using the compactness of ∂C, we can find a finite number of balls $B'(x^i)$, $i = 1, \ldots, m$ covering ∂C. Let δ be the minimum of all radii of $B'(x^i)$, $i = 1, \ldots, m$. If V meets ∂C and diam $V < \delta$ then $V \subset B(x^i)$ for some i, hence $0 \notin co\, g(V)$.

Now assume that the mesh of T is $< \delta$. If the sequence Γ generated by Algorithm $(w, P, T, b \mid f)$ does not terminate in C, it must contain a primitive set U with $\partial C \cap co\,(U \cap Q) \neq \emptyset$. Let W be a fair facet of U. Because of Property III), we may assume $W \cap Q \neq \emptyset$. From (1.2) we get $0 \in co\,\{f(W \cap Q)\,;\, b^i, i \in W \setminus Q\} \subset co\, g(W \cap Q)$, since $\{b^i, i \in W \setminus Q\} \subset b(W \cap Q)$. But by the above $0 \notin co\, g(W \cap Q)$. This is a contradiction and completes the proof.

□

Note 1.2: Obviously, it is enough to suppose that $f(x)$ is nonempty, compact, convex for every $x \in C$ (and not for all x).

Corollary 1.3: Let f be an u.s.c. mapping such that $f(x)$ is nonempty, compact, convex for every x and $Im(f + id)$ is bounded. Then for w, P given such that every P_i is a pointed cone, one can choose $b^1, \ldots, b^{n+1}$ and a compact set C such that, whenever the mesh of T is fine enough, Algorithm $(w, P, T, b \mid f)$ will terminate in C, yielding an approximate zero of f in C.

(id denotes the identity mapping $x \mapsto x$; $Im\, g = g(R^n)$).

Proof: Since every P_i $(i = 1, \ldots, n+1)$ is a pointed cone, we can take a point $v^i \neq w$ in each halfline d^i so that $v^1, \ldots, v^{n+1}$ span an n-simplex S_1 containing w in its interior (indeed, the convex hull of $v^1, \ldots, v^{n+1}$ contains w in its boundary only if w is the convex combination of less than $n+1$ points $v^1, \ldots, v^{n+1}$). Choose b^i to be a nonzero normal vector to the face of S_1 opposite to v^i, directed inwards S_1. It is an easy matter to show that (1.1) holds.

Indeed, assuming the contrary, there is (by the separation theorem) a nonzero vector z such that $< z, b^i > < 0$ $(i = 1, . . . , n+1)$. Let $< z, v^{i_0} > = \min_i < z, v^i > = \min_{x \in S_1} < z, x >$. Then $< z, x - v^{i_0} > \geq 0$ for all $x \in S_1$, so that z is an inward normal to S_1 at v^{i_0} . It follows that $z \in \text{co} \{b^i : i \neq i_0\}$, conflicting with $< z, b^i > < 0$ $(i = 1, . . . , n+1)$. Hence, (1.1) holds. In particular, if $x \neq w$, then $u \neq 0$ for every $u \in b(x) = \text{co} \{b^i : x \in P_i \}$, since $\{i : x \in P_i\}$ is a proper subset of $\{1, . . . , n+1\}$.

Now let $t > 0$ be so large that $S_t = w + t(S_1 - w)$ contains $\text{Im}(f + \text{id})$ in its interior. Suppose that for some $x \in \partial S_t$ we have $0 \in \text{co} \{f(x), b(x)\}$. Then there exist $u \in b(x)$, $y \in f(x)$ and $\Theta \in [0 , 1]$ satisfying $-\Theta u = (1 - \Theta) y$. Since $u \neq 0$, it follows that $1 - \Theta > 0$. But u is obviously an inward normal to S_t at x , and since $y \in \text{int } S_t - x$, we can write $< u, y > > 0$, hence $< u, -\Theta u > = < u, (1 - \Theta) y > > 0$, which conflicts with $\Theta \geq 0$. Therefore, condition (1.4) holds for $C = S_t$. The conclusion now follows from the previous theorem.

□

Note 1.4: In practice, one needs not determine S_t in advance. If all conditions of the Corallary are fulfilled, and p, b are chosen as indicated above, then, whenever the mesh of T is fine enough, Algorithm $(w, P, T, b \mid f)$ necessarily terminates (in some bounded region independent of T).

2. EXISTENCE THEOREMS BASED ON RESTART ALGORITHMS

We now examine how the foregoing restart algorithms can be used to solve the equation

$$0 \in f(x) \quad , \qquad x \in C \quad , \qquad (2.1)$$

under various boundary conditions imposed upon the mapping f .

In the sequel, when we say that such an equation is solvable, or that a solution is computable, we mean that this can be done by some procedure $(w, P, T, b \mid g)$ with an appropriate choice of w, P, T, b and g . Therefore, the proofs which we shall provide are algorithmic in nature.

For a given closed set C in R^n, an u.s.c. mapping r from R^n into C is called a retraction (in the extended sense) if $r(y)$ is nonempty, compact (but not necessarily convex) for every $y \in R^n$ and $r(y) = \{y\}$ for every $y \in C$.

Theorem 2.1: Let C be a closed set in R^n, f an u.s.c. multivalued mapping from C into R^n such that $f(x)$ is nonempty, compact, convex for every $x \in C$ and $f(C)$ is bounded. Assume there exists a retraction r from R^n into C and a closed multivalued mapping s from R^n into R^n such that

$$\text{for all } y \notin C \qquad y \notin \text{co } (f(r(y)) \cap s(y)) \neq \emptyset \quad . \tag{2.2}$$

Then f has a computable fixed point in C.

Proof: Without loss of generality it may be assumed that $s(y) = R^n$ for every $y \in C$, because if we modify the values of s on C so as to satisfy this condition, we shall affect neither the closedness of s, nor condition (2.2). Consider now the mapping $g(y) = \text{co } (f(r(y)) \cap s(y) - y$. Clearly, g is u.s.c. and $g(y)$ is nonempty, compact, convex for every y. Furthermore, $\text{Im } (g + \text{id}) \subset f(C)$ is bounded. Therefore, by Corollary 1.3 a procedure $(w, P, T, b \mid g)$ can be chosen that computes a zero $\bar{x}$ of g, i.e. a point satisfying

$$\bar{x} \in \text{co } (f(r(\bar{x})) \cap s(\bar{x})) \quad . \tag{2.3}$$

From (2.2) this implies $\bar{x} \in C$. Hence, $r(\bar{x}) = \{\bar{x}\}$, and so $\bar{x} \in f(\bar{x})$.

□

Note 2.2: If r is single-valued, $r(y) \in \partial C$ for $y \notin C$ and $s(y)$ is convex for every y, then (2.2) has the form of a boundary condition:

$$\begin{aligned} &\text{for all } x \in \partial C \qquad \text{for all } y \in r^{-1}(x) \setminus \{x\} \\ &y \notin f(x) \cap s(y) \neq \emptyset \quad . \end{aligned} \tag{2.4}$$

We shall refer to this as the General Inward Boundary Condition (G.I.B.C.)

Note 2.3: If we do not require $r(y) = \{y\}$ for $y \in C$, all other conditions being unchanged, the previous proof shows that we can compute a point $\overline{x}$ satisfying (2.3).

It turns out that a variety of known as well as new fixed point propositions can be easily obtained as corollaries of the previous theorem by specializing the mappings r and s. Let us point out some of these propositions.

Corollary 2.4: Let C be a closed set in R^n, f an u.s.c. multivalued mapping from R^n into R^n such that $f(x)$ is nonempty, compact, for every $x \in C$ and $f(C)$ is bounded. Then the equation

$$y \in \text{co}\, f(\pi_C(y)) \tag{2.5}$$

is solvable, where

$$\pi_C(y) = \arg\min \{ ||y - x|| : x \in C\} . \tag{2.6}$$

Proof: It can easily be verified that $\pi_C(y)$ is a nonempty compact set for every y and that $\pi_C(y) = \{y\}$ for every $y \in C$. Moreover, π_C is an u.s.c. mapping, hence is a retraction from R^n into C. The result now follows from Theorem 2.1 where we take $s(y) = R^n$.

□

Corollary 2.5: Let C be a closed set in R^n, f an u.s.c. multivalued mapping from R^n into R^n such that $f(x)$ is nonempty, compact, convex for every $x \in C$, and $\text{Im}\,(\text{id} - f)$ is bounded. Let r be a continuous mapping from R^n into C. Then the equation

$$x \in C , \quad x \in f(x) + r^{-1}(x) \tag{2.7}$$

is solvable.

Proof: Applying Theorem 2.1 (see Note 2.3) to the mapping $\text{id} - f$, with $s(y) = R^n$, we can compute a point $\overline{y} \in r(\overline{y}) - f(r(\overline{y}))$. Setting $\overline{x} = r(\overline{y})$ yields a solution to (2.7).

□

Special case 2.5.1: If C is convex and $r(y) = \pi_C(y)$ is the projection of y on C, then, as is well known, $r^{-1}(x) - x = N_C(x)$ is the outward normal cone to C at the point x. In this case equation (2.7) becomes

$$x \in C \ , \qquad 0 \in f(x) + N_C(x) \ , \tag{2.8}$$

which may also be written as

$$\begin{aligned} &x \in C \ , \qquad y \in f(x) \ , \quad (\text{for all } u \in C) \\ &\qquad\qquad < y \, , u - x > \ \geq \ 0 \end{aligned} \tag{2.8'}$$

and is often termed a variational inequality.

Special case 2.5.2: More generally, if, C being convex,

$$r(y) = \text{prox}_C(y \mid \varphi) = \arg\min \{(1/2)||y - x||^2 + \varphi(x) \ : \ x \in C\} \tag{2.9}$$

where φ is a given closed proper convex function with $\text{dom}\,\varphi \cap C \neq \emptyset$, then $r^{-1}(x) - x = \partial\varphi(x)$, so that equation (2.7) has the form

$$x \in C \ , \qquad 0 \in f(x) + \partial\varphi(x) \tag{2.10}$$

and is equivalent to the variational inequality

$$\begin{aligned} &x \in C \ , \qquad y \in f(x) \ , \quad (\text{for all } u \in C) \\ &\qquad < y \, , u - x > + \varphi(u) - \varphi(x) \ \geq \ 0 \ . \end{aligned} \tag{2.10'}$$

Special case 2.5.3: More generally, if C is convex and

$$r = (\text{id} + T)^{-1} \ , \tag{2.11}$$

where T is a given maximal monotone multivalued mapping from R^n into

R^n , with dom $T \subset C$, then equation (2.7) becomes

$$x \in C \ , \qquad 0 \in f(x) + T(x) \qquad (2.12)$$

(it is known that (2.11) defines a single-valued mapping from R^n into C , satisfying $||r(y) - r(y')|| \leq ||y - y'||$) .

Corollary 2.6: Let C , f be as in Theorem 2.1 ; let r be a single-valued retraction from $R^n \backslash \text{int } C$ into ∂C . If

$$\begin{aligned} &\text{for all } x \in \partial C \ , \qquad \text{for all } y \in r^{-1}(x) \ , \\ &\text{there exist } v \in f(x) \ , \quad < y - x \ , v - x > \ \leq 0 \ , \end{aligned} \qquad (2.13)$$

then f has a computable fixed point in C .

Proof: Define $s(y) = \{v \ : \ < y - r(y) \ , v - r(y) > \ \leq 0\}$. Then for every $x \in \partial C$, $y \in r^{-1}(x)$ we have from (2.13) $f(x) \cap s(y) \neq \emptyset$. Since, on the other hand, $y \in s(y)$ implies $< y - r(y) \ , y - r(y) > \ \leq 0$, i.e. $y = r(y) = x$, G.I.B.C. holds. Hence the result.

□

Special case 2.6.1: If C is a convex set with a nonempty interior and $w : R^n \backslash \text{int } C \to \text{int } C$ is a continuous mapping, then

$$r(y) = \{w(y) + t(y - w(y)) \ : \ t \geq 0\} \cap \partial C \qquad (2.14)$$

defines a retraction from $R^n \backslash \text{int } C$ into ∂C . In that case, $r^{-1}(x) = \{y \ : \ x \in [y \ ; w(y)]\}$ (where $[a \ ; b]$ denotes the line segment joining a and b) ; hence condition (2.13) has the form:

$$\begin{aligned} &\text{for all } x \in \partial C \ , \quad \text{for all } y \in \{u \ : \ x \in [u \ ; w(u)]\} \ , \\ &\text{there exist } v \in f(x) \ , \quad < x - w(y) \ , v - x > \ \leq 0 \ . \end{aligned} \qquad (2.15)$$

This reduces to a well-known condition if f is single-valued and $w(y) \equiv w$ for all $y \notin \text{int } C$.

Special case 2.6.2: If C is convex and $r(y) = \pi_C(y)$, so that

$r^{-1}(x) = x + N_C(x)$, then (2.13) reduces to the so called "inward" condition which has been introduced (in an infinite-dimensional setting) by Halpern in [6]:

$$\begin{aligned} &\text{for all } x \in \partial C , \quad \text{for all } y \in N_C(x) , \\ &\text{there exist } v \in f(x) , \quad \langle y , v - x \rangle \leq 0 , \end{aligned} \tag{2.16}$$

or, equivalently,

$$\text{for all } x \in \partial C , \quad f(x) \cap (x + T_C(x)) \neq \emptyset , \tag{2.16'}$$

where $T_C(x) = \overline{\bigcup_{t \geq 0} t(C - x)}$ is the tangent cone to C at the point x.

Special case 2.6.3: If C is convex, with $w \in \operatorname{int} C$ and

$$r(y) = \{w + t(y - w) \;:\; t \geq 0\} \cap \partial C \qquad (y \notin C)$$

then it can easily be shown that condition (2.13) is implied by the following inward* condition introduced in [18]:

$$\text{for all } x \in \partial C , \quad f(x) \cap (x + N_C(x)) \neq \emptyset , \tag{2.17}$$

or, equivalently,

$$\begin{aligned} &\text{for all } x \in \partial C , \quad \text{for all } y \in T_C(x) , \\ &\text{there exist } v \in f(x) , \quad \langle y , v - x \rangle \leq 0 . \end{aligned} \tag{2.17'}$$

Corollary 2.7: Let C, f be as in Theorem 2.1. Assume there is a continuous mapping w from $R^n \setminus \operatorname{int} C$ into $\operatorname{int} C \neq \emptyset$, such that

$$\begin{aligned} &\text{for all } x \in \partial C , \quad \text{for all } t > 0 , \\ &x + t(x - w(x)) \notin f(x) . \end{aligned} \tag{2.18}$$

Assume, furthermore, that either C is convex or $w(R^n \setminus \text{int } C)$ is compact Then f has a computable fixed point in C.

Proof: Assume C is convex. Define r by (2.14) and let $s(y) \equiv R^n$. Condition (2.18) ensures that for every $y \notin C$: $y \notin f(r(y)) \cap s(y)$, since $y = x + t(x - w(x))$ with $x = r(y)$, $t > 0$.
Therefore G.I.B.C. holds and the result follows.

If C is not convex, but $w(R^n \setminus \text{int } C)$ is compact, we put

$$F(x) = \begin{cases} f(x) & x \in \text{int } C \\ w(x) & x \notin C \\ \text{co } \{f(x), w(x)\} & x \in \partial C \end{cases} .$$

Then $F(R^n)$ is bounded and since $\partial R^n = \emptyset$ it follows from the previous case that one can compute a point $\bar{x} \in F(\bar{x})$. Clearly, $\bar{x} \in C$. If $\bar{x} \in \partial C$, there is $\Theta \in [0, 1]$ satisfying $\bar{x} \in \Theta\, w(\bar{x}) + (1 - \Theta)\, f(\bar{x})$ which, in view of (2.18), is possible only if $\Theta = 0$. Therefore, $\bar{x} \in f(\bar{x})$.

□

Special case 2.7.1: If $w(x) \equiv w \in \text{int } C$, then (2.18) reduces to the well known Leray-Schauder condition:

$$\text{for all } x \in \partial C, \quad \text{for all } t > 0, \qquad x + t(x - w) \notin f(x). \tag{2.19}$$

Special case 2.7.2: More generally, if $w(x) = \text{prox}\,(x \mid \varphi)$, where φ is a closed proper convex function with $\text{dom}\, \varphi \subset \text{int } C$, then condition (2.18) has the form

$$\text{for all } x \in \partial C, \quad \text{for all } t > 0, \qquad x + t(x - \text{prox}\,(x \mid \varphi)) \notin f(x). \tag{2.20}$$

When φ is the indicator function of a closed convex set $A \subset \text{int } C$, this condition can be written as

$$\text{for all } x \in \partial C , \qquad \text{for all } t > 0 ,$$

$$x + t(x - \pi_A(x)) \notin f(x) .$$

<u>Corollary</u> 2.8: Let C , f be as in Theorem 2.1 . If there is a single-valued retraction r from $R^n \setminus \text{int } C$ into ∂C , such that

$$\text{for all } x \in \partial C , \qquad f(x) \cap r^{-1}(x) \subset \{x\} , \qquad (2.21)$$

then f has a computable fixed point in C .

<u>Proof</u>: For $x \in \partial C$, $y \in r^{-1}(x)$, if $y \in f(x)$, then (2.21) implies $y = x$. Thus G.I.B.C. holds with $s(y) \equiv R^n$. Hence the result.

□

<u>Special case</u> 2.8.1: If C is convex, $r(y) = \pi_C(y)$, then (2.21) reduces to the so called <u>nowhere outward condition</u>:

$$\text{for all } x \in \partial C , \qquad f(x) \cap (N_C(x) + x) \subset \{x\} , \quad (2.22)$$

which has been introduced in [6].

We have thus obtained a unifying view of various boundary conditions to be imposed upon the mapping f in order to ensure the solvability of (2.1).

Note 2.9: Since many of the above conditions involve a retraction r from $R^n \setminus \text{int } C$ into ∂C , it may be useful to indicate a general procedure for defining such a retraction.

C being a closed set in R^n , consider any closed multivalued mapping N from ∂C into $R^n \setminus \text{int } C$, such that:

a) $\cup\{N(x) : x \in \partial C\} = R^n \setminus \text{int } C$; $N(x) \cap N(x') = \emptyset$ $(x \neq x')$; $x \in N(x)$;

b) $\inf \{ ||y|| : y \in N(x)\} \to \infty$ as $||x|| \to \infty$.

Then, by a) there is for every $y \notin \text{int } C$ a unique $x = r(y) \in \partial C$ such that $y \in N(x)$; by the closedness of N this mapping r is closed and by b) it maps bounded sets into bounded sets. Hence, r is continuous and since $r(y) = y$ whenever $y \in \partial C$, r is a retraction. Obviously, $r^{-1}(x) = N(x)$.

3. SOME EXTENSIONS AND APPLICATIONS

We now partially generalize condition (2.4) in order to apply it to the nonlinear complementarity problem.

Theorem 3.1: Let C be a compact set in R^n containing 0, let f be an u.s.c. multivalued mapping from R^n into R^n such that $f(x)$ is nonempty, compact, convex for every x. Assume there are a compact set $D \subset C$ with $0 \in \text{int}_C D$ (interior of D relative to C) and a single-valued retraction r from R^n into D such that $r(y) \in \partial D$ for $y \notin D$, and for all $t \geq 1$:

$$\text{for all } x \in \partial C \cap \partial D_t , \quad \text{for all } y \in r_t^{-1}(x)\setminus\{x\} , \qquad (3.1)$$
$$y \notin f(x) ,$$

where $D_t = t D$, $r_t(y) = tr(y/t)$. Then f has a computable fixed point in C.

Proof: Consider an n-simplex $S = [a^1 , \ldots , a^{n+1}]$ in R^n containing 0 in its interior. Using the halflines d^i from 0 through a^i, subdivide R^n into $n+1$ closed convex polyhedral cones $P_1 , \ldots , P_{n+1}$ as was described in section 1. Let b^i be a nonzero normal to the face of S opposite to a^i, directed inwards S.

Now embed R^n into R^{n+1} by identifying each $x \in R^n$ with $(x, 1) \in R^n \times \{1\}$. Let $\tilde{b}^i = (b^i , 1)$, $\tilde{b}^0 = (0, \ldots , 0, -1) \in R^{n+1}$, so that

$$0 \in \text{int co} \{\tilde{b}^0 , \tilde{b}^1 , \ldots , \tilde{b}^{n+1}\} . \qquad (3.2)$$

Let $\tilde{P}_i = P_i \times (-\infty , 1]$ $(i = 1, \ldots , n+1)$, $\tilde{P}_0 = R^n \times [1 , \infty)$. Choose two integers α, β so large that $2C \subset D_\alpha$ and the simplex $S_\beta = \beta S$ contains $f(D_\alpha)$ in its interior. For each $z \in R^{n+1}$ write $z = (u , t)$ with $u \in R^n$, $t \in R$ and define

$$\varphi_t(u) = \begin{cases} f(r_t(u)) - u & \text{if} \quad t \geq 1 \\ \text{co } \{b^i \quad : \quad u \in P_i \quad , \quad i = 1, \ldots , n+1\} & \text{if} \quad t < 1 \end{cases}$$

(3.3)

$$F(z) = (\varphi_t(u) , 1) \quad . \tag{3.4}$$

N being a natural number, consider a triangulation T of R^{n+1} consistent with the subdivision $\tilde{P} = (\tilde{P}_0 , \tilde{P}_1 , \ldots , \tilde{P}_{n+1})$, such that every vertex of T lies in some $H_t = R^n \times \{t\}$ with tN integral, and the restriction of T in each H_t is a triangulation of H_t with a bounded mesh not increasing with $|t|$. Also mesh $T \to 0$ as $N \to \infty$. As previously, let Q denote the set of all vertices of T .

By applying the procedure $(w, \tilde{P}, T, \tilde{b} \mid F)$, with $w = (0 , 1) \in R^n \times \{1\}$, generate a chain Γ of primitive sets starting from $U_0 = (0 , 1 , \ldots , n+1 , w)$. Since $F(Q) \subset R^n \times \{1\}$, no primitive set U satisfies $0 \in F(U \cap Q)$, i.e. no primitive set is complete, so that Γ will never terminate. On the other hand, if Γ left $\tilde{P}_0$, it would contain a primitive set U with a fair facet W such that $W \cap Q \backslash \tilde{P}_0 \neq \emptyset$; then $0 \in \text{co } \{F(W \cap Q) \; ; \tilde{b}^i , \; i \in W \backslash Q\}$, which would contradict the fact that $F(W \cap Q) \subset R^n \times \{1\}$ (see (3.4)) and $W \backslash Q \subset \{1, \ldots , n+1\}$. Therefore, Γ will never leave $\tilde{P}_0$. Furthermore, provided N is large enough, Γ will never cross $\partial S_\beta \times [1 , \alpha]$. Indeed, for every $t \in [1 , \alpha]$, since by (3.3) $(\varphi_t + \text{id})(R^n) \subset f(D_\alpha) \subset \text{int } S_\beta$, we can show, by the same argument as that used in the proof of Corollary 1.3 , that

$$\text{for all} \quad u \in \partial S_\beta \quad , \qquad 0 \notin G(u , t) = \text{co } (\varphi_t(u) , b(u)) \quad ,$$

where $b(u) = \text{co } \{b^i \; : \; u \in P_i \; , \; i = 1, \ldots , n+1\}$. From the compactness of $\partial S_\beta \times [1, \alpha]$ and the u.s.c. of $G(u , t)$ it follows, by an argument similar to that used in the proof of Theorem 1.1 that for some $\delta > 0$ small enough we shall have

$$0 \notin \text{co } G(W) \tag{3.5}$$

whenever $W \subset R^n \times [1, \infty)$ meets $\partial S_\beta \times [1, \alpha]$ and has diam $W < \delta$. But every primitive set U of Γ has a fair facet and must satisfy

$$0 \in co\,\{F(U \cap Q)\ ;\ \tilde{b}^i\ ,\ i \in U \backslash Q\ ,\ i = 0, 1, \ldots, n+1\}\ . \tag{3.6}$$

Since for every $z = (u, t) \in U \cap Q$ we have $F(z) = (\varphi_t(u), 1)$ and $u \in P_i$ for $i \in U \backslash Q$, $i \neq 0$, since on the other hand $\tilde{b}^0 = (0, \ldots, 0, -1)$, it follows that

$$0 \in co\, G(U \cap Q)\ .$$

Then (3.5) shows that when N is so large that mesh $T < \delta$, we shall have for every primitive set U of Γ : $co(U \cap Q) \cap (\partial S_\beta \times [1, \alpha]) = \emptyset$. This proves our assertion that Γ never crosses $\partial S_\beta \times [1, \alpha]$.

Denote by Γ_α the part of Γ going from the starting primitive set U_0 to the first primitive set of Γ that meets $H_\alpha = R^n \times \{\alpha\}$. Note that, by the above,

$$co\,(U \cap Q) \subset S_\beta \times [1, \alpha] \quad \text{for all} \quad U \in \Gamma_\alpha\ .$$

Define for every $z = (u, t) \in R^n \times R : \rho(z) = r_t(u)$. Then it is easily seen that only three cases are possible:

a) For some $\Theta > 1$ there is a member U_Θ of Γ_α meeting H_Θ such that $\rho(co(U_\Theta \cap Q)) \cap \partial C \neq \emptyset$;

b) For $\Theta = \alpha$ there is a member U_Θ of Γ_α meeting H_Θ and such that $\rho(co(U_\Theta \cap Q)) \subset int\, C$;

c) For $\Theta = 1 + (1/N)$ there is a member U_Θ of Γ_α meeting H_Θ and such that $\rho(co(U_\Theta \cap Q)) \subset R^n \backslash C$.

Indeed, if case a) does not occur, the last member, U_α, of Γ_α must be such that: either $\rho(co(U_\alpha \cap Q)) \subset int\, C$, or $\rho(co(U_\alpha \cap Q)) \subset R^n \backslash C$; in the first alternative b) obviously holds (since U_α meets H_α), whereas in the second alternative the segment of Γ_α going from U_α to the nearest member U of Γ_α satisfying $\rho(co(U\cap Q)) \cap \partial C \neq \emptyset$ will contain a U_Θ satisfying c) (here we use the obvious line-connectedness of $\Delta = \cup\{\rho(co(U\cap Q)) : U \in \Gamma_\alpha\}$).

In each of the three listed cases, since U_Θ meets H_Θ with $\Theta > 1$, we have $U_\Theta \setminus Q = \{0\}$, hence, in view of (3.6),

$$0 \in co\ \{F(U_\Theta \cap Q) ; \tilde{b}^0\} ,$$

which, in view of (3.3) and (3.4), implies

$$0 \in co\ \{f(r_t(u)) - u : (u, t) \in U_\Theta \cap Q\} . \tag{3.7}$$

Thus, for each N we can find a set $V_N = U_\Theta \cap Q$ satisfying (3.7) and such that a), or b), or c) holds. Since $V_N \subset S_\beta \times [1, \alpha]$ and diam $V_N \to 0$ as $N \to \infty$, we may choose a subsequence along with $V_N \to (\overline{u}, \overline{t})$ and such that only one of the cases a), b), c) occurs for all N in this subsequence. We then get from (3.7)

$$\overline{u} \in f(\overline{x}) , \quad \overline{x} = r_{\overline{t}}(\overline{u}) . \tag{3.8}$$

In case a) we have $\overline{x} \in \partial C$; and
either $\overline{u} \in \overline{t}D$, which implies $\overline{x} = \overline{u}$;
or $\overline{u} \notin \overline{t}D$, so that $\overline{x} \in \partial D_{\overline{t}}$ and again $\overline{x} = \overline{u}$, by virtue of (3.8) and (3.1). Therefore in this case $\overline{x}$ is a fixed point of f in C .

In case b) we have for every

$$(u, t) \in U_\Theta \cap Q : r_t(u) \in \text{int } C \subset (1/2) \text{ int } D_\alpha ,$$

hence $(t/\alpha)r(u/t) \subset (1/2)$ int D . Since $t = \alpha$ or $t = \alpha - (1/N)$, this yields, for N large enough, $r_t(u) \in$ int D . Noting that $r(y) \in \partial D$ for $y \notin D$ we then get $(u/t) \in D$, hence $r(u/t) = (u/t)$, i.e.

$$r_t(u) = u \in C \quad \text{for all} \quad (u, t) \in U_\Theta \cap Q .$$

This implies $\overline{x} = \overline{u}$ and again (3.8) shows that $\overline{x}$ is a fixed point of f in C .

Finally, in case c) we have $\Theta \to 1$ as $N \to \infty$, so that $\bar{t} = 1$, $\bar{x} = r(\bar{u}) \in D \subset C$. Since, on the other hand, $r_t(u) \in R^n \backslash C$ for every $(u, t) \in U_\Theta \cap Q$, it follows that $\bar{x} \in \partial C$, which in turn implies $\bar{x} \in \partial C \cap \partial D$. From (3.8) and (3.1) we then conclude again $\bar{x} = \bar{u}$.

□

Note 3.2: Condition (3.1) can also be written as

$$\text{for all } x \in \partial C \cap \partial D_t \;, \quad f(x) \cap r_t^{-1}(x) \subset \{x\} \;. \tag{3.9}$$

The previous proof shows that if all conditions of Theorem 3.1, except possibly (3.1), are fulfilled, then one can compute a point $\bar{x} \in C$ such that either $\bar{x}$ is a fixed point of f, or for some $t \geq 1$:

$$\bar{x} \in \partial C \cap \partial D_t \;, \quad f(\bar{x}) \cap r_t^{-1}(\bar{x}) \neq \emptyset \;. \tag{3.10}$$

Corollary 3.3: Let K be a closed convex cone in R^n (with vertex at 0), f an u.s.c. multivalued mapping from R^n into R^n such that $f(x)$ is nonempty, compact, convex for every x. Assume there are a compact subset D of K such that $0 \in \text{int}_K D$, a retraction r from R^n into D such that $r(y) \in \partial D$ for every $y \notin D$ and

$$\text{for all } x \in \partial K \cap \text{int}_K D \;, \quad r^{-1}(x) \subset N_K(x) + x \tag{3.11}$$

and a compact set C such that $D \subset C \subset K$ and for every $t \geq 1$:

$$\text{for all } x \in \partial C \cap \partial_K D_t \;, \quad (x - f(x)) \cap r_t^{-1}(x) \subset N_K(x) + x \;, \tag{3.12}$$

where $D_t = tD$, $r_t(y) = tr(y/t)$. Then the complementarity problem

$$x \in K \;, \quad y \in K^+ \cap f(x) \;, \quad \langle x , y \rangle = 0 \tag{3.13}$$

has a computable solution in C.
(K^+ denotes the nonnegative polar cone of K, $N_K(x)$ is, as previously, the outward normal cone of K at x).

Proof: Consider the mapping $F(x) = x - f(x)$. By the above result (Note 3.2) we can compute a point $\bar{x}$ such that: either $\bar{x} \in C$, $\bar{x} \in F(\bar{x})$ (hence $0 \in f(\bar{x})$, so that $\bar{x}$ solves (3.13)) or for some $t \geq 1$:

$$\bar{x} \in \partial C \cap \partial D_t \quad , \quad F(\bar{x}) \cap r_t^{-1}(\bar{x}) \neq \emptyset \quad . \tag{3.14}$$

Let us distinguish two cases:

a) $\bar{x} \in \text{int}_K D_t$;

b) $\bar{x} \in \partial_K D_t$.

In the first case, since $\bar{x} \in \partial D_t$, we must have $\bar{x} \in \partial K$ and from (3.11) we get $r_t^{-1}(\bar{x}) \subset N_K(\bar{x}) + \bar{x}$, which together with (3.14) shows that

$$F(\bar{x}) \cap (N_K(\bar{x}) + \bar{x}) \neq \emptyset \quad , \text{ i.e. } \quad (-f(\bar{x})) \cap N_K(\bar{x}) \neq \emptyset \quad .$$

But the latter relation implies that $\bar{x}$ is a solution of (3.13), since we can take $\bar{y} \in f(\bar{x}) \cap (-N_K(\bar{x}))$ and since $-N_K(\bar{x}) \subset K^+$.

In the second case, we have from (3.12):

$$(-f(\bar{x})) \cap (r_t^{-1}(\bar{x}) - \bar{x}) \subset N_K(\bar{x}) \quad ,$$

but from (3.14): $(-f(\bar{x})) \cap (r_t^{-1}(\bar{x}) - \bar{x}) \neq \emptyset$, hence

$$(-f(\bar{x})) \cap N_K(\bar{x}) \neq \emptyset \quad ,$$

which, as previously, means that $\bar{x}$ solves (3.13).

□

Corollary 3.3 includes as special cases many known existence results on the complementarity problem. Note, moreover, that this Corollary contains more than an existence result, since, like the other results in this paper it asserts the effective solvability of the problem by combinatorial methods.

As an illustration, let us point out two special cases of Corollary 3.3.

Special case 3.3.1: Let K, f be as in Corollary 3.3 . Assume there are a compact convex set D and a compact set C such that $D \subset C \subset K$, $0 \in \text{int}_K D$ and for every $t \geq 1$:

$$\text{for all} \quad x \in \partial C \cap \partial_K D_t \quad , \qquad (-f(x)) \cap N_{D_t}(x) \subset N_K(x) \quad . \tag{3.15}$$

Then the complementarity problem (3.13) has a computable solution in C .

<u>Proof</u>: Consider $r(y) = \pi_D(y)$, i.e. r is the projection on the convex set D . Clearly $r_t^{-1}(x) = N_{D_t}(x) + x$, so it suffices to verify that (3.11) holds. But this follows from the fact that for every $x \in \partial K \cap \text{int}_K D$ we have $T_K(x) = T_D(x)$ (the tangent cones of K and D , respectively, at x), which implies $N_K(x) = N_D(x)$.

□

<u>Consequence</u>: Let K , f be as in Corollary 3.3 . Assume there are a compact convex set D and a compact set C such that $D \subset C \subset K$, $0 \in \text{int}_K D$ and for every $t \geq 1$:

$$\begin{aligned} &\text{for all} \quad x \in \partial C \cap \partial_K D_t \quad , \qquad \text{for all} \quad y \in f(x) \ , \\ &\text{there exist} \quad u \in \text{int}_K D_t \quad , \qquad < u - x , y > \ \leq \ 0 \quad . \end{aligned} \tag{3.16}$$

Then the complementarity problem (3.13) has a computable solution in C .

<u>Proof</u>: We first observe the following fact:

If D is a compact convex subset of K with $0 \in \text{int}_K D$ and if for some $y \in N_D(x)$ there is an $u \in \text{int}_K D$ satisfying $< u - x , y > \ \geq 0$, then $y \in N_K(x)$.

Indeed, consider an arbitrary $z \in K$. Since $u \in \text{int}_K D$ we have $\alpha z + (1 - \alpha)u \in D$ for all $\alpha > 0$ small enough. But since $y \in N_D(x)$ we have $< \alpha z + (1 - \alpha)u - x , y > \ \leq \ 0$, hence $\alpha < z - x , y > \ \leq \ -(1 - \alpha) < u - x , y > \ \leq \ 0$. Therefore, (for all $z \in K$) $< z - x , y > \ \leq \ 0$, which means that $y \in N_K(x)$ and proves our assertion.

Now, if $x \in \partial C \cap \partial_K D_t$ and $y \in (-f(x)) \cap N_{D_t}(x)$, then by the previous observation, (3.16) implies $y \in N_K(x)$, i.e. (3.15) holds.

□

Note that for the case where $f(x)$ is convex for every x the previous proposition is stronger than many analogous results obtained earlier by other authors (see e.g. [10]).

Special case 3.3.2: Let K, f be as in Corollary 3.3. Assume there is a compact subset C of K such that $0 \in \text{int}_K C$ and

$$\text{for all } x \in \partial_K C \ , \quad \text{for all } \Theta \geq 0 \ , \qquad (-f(x)) \cap (\Theta x + N_K(x)) \subset N_K(x) \ . \tag{3.17}$$

Then the complementarity problem (3.13) has a computable solution in C.

Proof: Consider any compact convex set $D \subset C$ such that $0 \in \text{int}_K D$. For every $y \in K$ let $\rho(y)$ denote the point where ∂D meets the halfline from 0 through y; for every $y \in R^n$ let $r(y) = \rho(\pi_K(y))$ if $\pi_K(y) \notin D$, $r(y) = y$ otherwise. Then r is a retraction from R^n into D, $r(y) \in \partial D$ for every $y \notin D$ and $r_t^{-1}(x) - x = \cup\{\Theta x + N_K(x) : \Theta \geq 0\}$ for every $x \in \partial_K D_t$. The latter relation shows that (3.17) implies (3.12). Since (3.11) is obvious, the result follows.

□

Note 3.4: In the previous propositions the set C plays the role of a set "separating D from ∞" - a notion which has become familiar in the recent literature on fixed points and complementarity problems. We have succeeded in avoiding this notion by a direct algorithmic proof of Theorem 3.1 (whereas the usual approach is to proceed through the intermediary of Browder's theorem).

REFERENCES

1. Browder, F., The Fixed Point Theory of Multivalued Mappings in Topological Vector Spaces, Math. Ann. 177, 283-301 (1968).

2. Eaves, B.C., Homotopies for Computation of Fixed Points, Math. Programming 3, 1-22 (1972).

3. Eaves, B.C.; Saigal, R., Homotopies for Computation of Fixed Points on Unbounded Regions, Math. Programming 3, 225-237 (1972).

4. Eaves, B.C.; Scarf, H., The Solution of Systems of Piecewise Linear Equations, Math. Oper. Res. 1, 1-28 (1976).

5. Fisher, M.L.; Gould, F.J.; Tolle, J.W., A New Simplicial Approximation Algorithm with Restarts: Relations between Convergence and Labelings, in Fixed Points, Algorithms and Applications, S. Karamardian (Ed.), Academic Press, New York, San Francisco, London 1977, pp. 41-58.

6. Halpern, B., Fixed Point Theorems for Set-Valued Mappings in Infinite-Dimensional Spaces, Math. Ann. 189, 87-98 (1970).

7. Kuhn, H.W.; MacKinnon, J.G., Sandwich Method for Finding Fixed Points, Journal of Optimization Theory and Applications 17, 189-204 (1975).

8. Ky Fan, Extensions of Two Fixed Point Theorems of Browder, Math. Z. 112, 234-240 (1969).

9. Merrill, O.H., Applications and Extensions of an Algorithm that Computes Fixed Points of Certain Upper Semi-Continuous Point to Set Mappings, University of Michigan, Ph.D. Thesis, 1972.

10. Saigal, R., Extensions of the Generalized Complementarity Problem, Math. Oper. Res. 1, 260-267 (1976).

11. Scarf, H., The Approximation of Fixed Points of a Continuous Mapping, SIAM J. Appl. Math. 15, 1328-1343 (1967).

12. Scarf, H., with the collaboration of Hansen, T., The Computation of Economic Equilibria, Yale University Press, New Haven, Connecticut 1973.

13. Todd, M.J., Fixed Point Algorithms that allow Restarting without an Extra Dimension, School of Operations Research and Industrial Engineering, College of Engineering, Cornell University, Ithaca, New York, Technical Report No. 379, 1978.

14. Hoang Tuy, Pivotal Methods for Computing Equilibrium Points: Unified Approach and New Restart Algorithm, Presented at IX. International Symposium on Mathematical Programming, Budapest 1976; Mathematical Programming 16, 210-227 (1979).

15. Hoang Tuy; Ng. v. Thoai; Le d. Muu, Un nouvel algorithme de point fixe, Comptes Rendus Acad. Sci. Paris 286, ser. A, 783-785 (1978).

16. Hoang Tuy; Ng. v. Thoai; Le d. Muu, A Modification of Scarf's Algorithm Allowing Restarting, Math. Operationsforsch. Statist. ser. Optimization 9, 353-369 (1978).

17. Hoang Tuy, Three Improved Versions of Scarf's Method Using Conventional Subsimplices and Allowing Restart and Continuation Procedures, Preprint, Institute of Mathematics, Hanoi 1979.

18. Hoang Tuy, Combinatorial Methods for Solving Nonlinear Equations, to appear in Acta Mathematica Vietnamica (1979).

19. Van der Heyden, Refinement Methods for Computing Fixed Points Using Primitive Sets, Ph.D. Thesis, Yale University, 1979.

20. Van der Laan, G.; Talman, A.J.J., A Restart Algorithm for Computing Fixed Points without an Extra Dimension, Mathematical Programming 17, 74-84 (1979).

Numerical Solution of Highly Nonlinear Problems
W. Forster (ed.)
© North-Holland Publishing Company, 1980

ON THE NUMBER OF SOLUTIONS OF PIECEWISE LINEAR EQUATIONS

Ruben Schramm

Tel-Aviv University,
Tel-Aviv, Israel

1. INTRODUCTION

The last decades have seen the emergence of new solution methods for nonlinear numerical equations in R^n. Of these the homotopy method [2], [3], [5] to [14], [17] to [22] and the bisection method [15], [25] seem to be the most important. The motivation for introducing the former, as well as the early formulations of this method, first came up in electric circuit theory [17]. In more recent investigations, however, the homotopy method was further developed, without any reference to extra-mathematical applications. As a result, many types of equations which previously defied attempts at a solution, can now be solved numerically. In some recent papers algorithms have even been presented, by which the *complete* solution set of a very comprehensive class of equations can be produced [13], [14], [19]. As a by-product of these investigations, a simple expression for the *number of solutions* of these equations is obtained.

In many of the new solution methods of nonlinear equations $f(x) = \gamma$ the functions $f: R^n \to R^n$ were first approximated by a *piecewise linear* (p.l.) *function* $F: R^n \to R^n$, and the actual solution process was then applied to the equation $F(x) = \gamma$. Consequently the development of a general theory of p.l. equations had become increasingly desirable.

Following this trend, Fujisawa, Kuh and Ohtsuki [11], Rheinboldt and Vandergraft [22], and Ohtsuki, Fujisawa and Kumagai [21] have provided sufficient conditions for the solvability of p.l. equations. Fujisawa and Kuh [10] have presented also sufficient conditions for a p.l. equation to have one solution at most. Their condition, however, is generally impractical from the computational point of view when the number of scalar unknowns is not sufficiently low. Eaves [4] was the first to provide a result on the cardinality of the solution set of p.l. equations. He proved that under certain conditions p.l. mappings possess an odd number of fixed points, a result which has been generalized by Charnes, Garcia and Lemke [2].

We shall here delineate further results of our own on the solution set of p.l. equations. The proofs, which can be found in [23], [24], are generally omitted.

After introducing our notation and some definitions in section 2, we shall in section 3 present necessary and sufficient conditions for the solvability of certain p.l. equations in *bounded domains*. Also, we shall set up simple conditions under which the number of solutions of p.l. equations $F(x) = \gamma$ in certain domains is "almost" independent of γ (Theorems 2 and 3). Theorems 2 and 3 contain also necessary and sufficient conditions for p.l. equations to have one solution at most in certain domains.

From Theorems 2 and 3 we shall in section 4 deduce further conditions for a p.l. equation to have one solution at most. These conditions too are necessary as well as sufficient (Theorems 4 to 6). Although generally Theorem 4 cannot be used in practice, it does lead on to other results, theoretical as well as practical.

Theorems 2(iv), 3(iv), 5 and 6 have been cast into the form of algorithms which can be applied successfully to equations actually occuring in practice (section 5).

The availability of an algorithm by which we can decide whether equations belonging to a certain family have one solution at most is of considerable interest in the treatment of some extra-mathematical problems. In electric circuit theory, e.g. we are very much interested to know whether the input of a certain system is uniquely determined by the output [20]. We can now avail ourselves of the relevant uniqueness results for the p.l. approximation of the equation under discussion. The same results have applications in mathematical economics as well [1].

In addition, we are now in a position to decide on some questions of stability for p.l. equations.

In section 7 we shall present upper bounds for the number of solutions of certain p.l. equations. In a certain sense, which we shall explain, these bounds are the best possible. The results of section 7 have been motivated by the study of *Riemann manifolds* of p.l. functions. In the theory of these functions Riemann manifolds play a role similar to the role played by Riemann surfaces in the theory of complex functions. Riemann manifolds of p.l. functions will be presented in section 6.

2. NOTATION AND DEFINITIONS

The closure of $A \subset R^n$ is $\overline{A}$. The Euclidean norm of $x \in R^n$ is $|x|$.

Definition 1: A continuous function $F: R^n \to R^n$ is piecewise linear, p.l. for short, when a finite set $H = \{H_1 , . . . , H_p\}$ of hyperplanes exists, such that $R^n \backslash \cup H$ is the disjoint union of finitely many open polyhedral sets $C_1 , . . . , C_q$, and for every $i = 1, ... , q$ the restriction F_i of F to $\overline{C}_i$ equals the restriction of an affine function to $\overline{C}_i$. Let $C = \{C_1 , . . . , C_q\}$. The values of F in $\overline{C}_i$ are computed from the expression $F(x) = F_i(x) = A_i x + b_i$, where A_i is a constant $n \times n$ matrix, and b_i a constant $n \times 1$ vector. The linear continuation of F_i to R^n , unique for nonsingular A_i , is $F_i^L = F|_{C_i}^L$.

Definition 2: Let E be a set of hyperplanes in R^n . The E^* is the set of points in R^n each of which belongs to at least two hyperplanes in E .

The importance of Definition 2 lies in the fact that eventual branch points of F are always in H^* .

The set of hyperplanes $F_j^L(H_i) = F|_{C_j}^L(H_i)$, with $H_i \in H$, $C_j \in C$, partitions the image space into a collection of polyhedral sets.

Definition 3: Let F_j^L be nonsingular for $j = 1, \ldots, q$, let $G = \{G_1, \ldots, G_s\}$ be the arbitrarily ordered set of hyperplanes

$\{F_j^L(H_i) \mid j = 1, \ldots, q, \; i = 1, \ldots, p\}$ and

$\Gamma = \{\Gamma_1, \ldots, \Gamma_t\}$ the set of open polyhedral sets which have

$R^n \backslash \cup G$ as their disjoint union. Thus $s \leq pq$. We shall also refer to Γ as "the set of polyhedral sets generated by H through F."

The relationship of the sets $\Gamma_k \in \Gamma$ and the sets $F(C_j)$, for $C_j \in C$, turns out to be particularly simple: whenever some $F(C_j)$ meets some Γ_k then Γ_k as a whole is contained in $F(C_j)$. It is because of this property that the set Γ plays a prominent role in the sequel.

Notation is required for certain subsets of H, C, Γ and R^n. It is expedient to define generally.

Definition 4: Let $E = \{E_1, \ldots, E_\pi\}$ be a finite set of hyperplanes in R^n, P the set $\{P_1, \ldots, P_\rho\}$ of open polyhedral sets which have $R^n \backslash \cup E$ as their disjoint union, and $a \in R^n$. Then $\overline{P} = \{\overline{P}_1, \ldots, \overline{P}_\rho\}$. Also, $E(a)$ is the set of hyperplanes $E_i \in E$ with $a \in E_i$ considered empty when $a \notin \cup E$, $P(a)$ the set of polyhedral sets $P_j \in P$ such that $a \in \overline{P}_j$, and $\overline{P}(a) = \{\overline{P}_j \mid P_j \in P(a)\}$. Finally, for $P_i \in P$ let $C(a, P_i)$ be the cone determined by P_i at a, and $P^C(a) = \{C(a, P_i) \mid P_i \in P(a)\}$.

With every p.l. function F and every point $a \in R^n$ we associate the p.l. function F^a which in every cone $C(a, C_i) \in C^C(a)$ continues $F|_{C_i}$ linearly.

Definition 5: Let $a \in R^n$ and let F_j^L be nonsingular for all $C_j \in C(a)$. For each $C(a, C_i) \in C^C(a)$ let

$$F_i^a : \overline{C(a, C_i)} \to R^n \quad \text{be the linear continuation of } F_i \text{ to } \overline{C(a, C_i)} .$$

Then $F^a : R^n \to R^n$ is the unique p.l. function with

$$F^a|_{\overline{C(a, C_i)}} = F_i^a \quad \text{for all } C(a, C_i) \in C^C(a) .$$

The proof of the contentions implicitly contained in Definition 5 is straightforward.

3. HOW DOES THE NUMBER OF SOLUTIONS OF $F(x) = \gamma$ DEPEND ON γ ?

Fujisawa and Kuh [10] and Rheinboldt and Vandergraft [22] have shown that when all the determinants associated with a p.l. function F have the same sign, F is a surjection. We present an analogous local result which in some cases of widespread occurence permits us to decide whether an equation $F(x) = \gamma$ is solvable in a given bounded region.

Theorem 1: Let $F : R^n \to R^n$ be p.l. Let a be a point in R^n such that the determinants associated with F in $C(a)$ have the same sign. Define F^a as in Definition 5 , H^a as the set of hyperplanes in H which contain facets of polyhedral sets in $C(a)$, and Γ^a as the set of polyhedral sets generated by H^a through F^a (c.f. Def. 3). Then $\cup\Gamma^a[F(a)] \subset F[\cup\overline{C}(a)]$.

□

It turns out that under the conditions of Fujisawa and Kuh, which are identical with those of Theorem 2, the number of solutions of the p.l. equation $F(x) = \gamma$ is "almost" independent of γ .

Theorem 2: Let the p.l. function $F : R^n \to R^n$ be associated with the set of hyperplanes H and the set of polyhedral sets C . Also, let the determinants associated with F have the same sign. Then the following is true: (i) For $\gamma \in R^n \backslash F(H^*)$ the number m of solutions of the equation $F(x) = \gamma$ in R^n is independent of γ . (ii) When Γ is the set of polyhedral sets generated by H through F , the number $\mu(\Gamma_i)$ of polyhedral sets $C_j \in C$ with $\Gamma_i \subset F(C_j)$ is m for all $\Gamma_i \in \Gamma$. (iii) For $\gamma \in F(H^*)$ the equation $F(x) = \gamma$ has at most m solutions in R^n . (iv) The uniqueness of the solution of one equation $F(x) = \gamma$, where $\gamma \in R^n \backslash F(H^*)$ is arbitrary, is a necessary and sufficient condition for F to be a homeomorphism.

□

Theorem 3: Define F^a as in Definition 5. Then the following is true:
(i) Under the assumptions of Theorem 1, for
$\delta \in A = \cup\Gamma^a[F(a)]\backslash F^a([H(a)]^*)$ the number $\mu(a)$ of solutions of the equation $F(x) = \delta$ in $\cup\overline{C}(a)$ is independent of δ .
(ii) For each $\Gamma_\ell^a \in \Gamma[F(a)]$ exactly $\mu(a)$ polyhedral sets $C_j \in C(a)$ satisfy $\Gamma_\ell^a \subset F(C_j)$.
(iii) For
$\delta \in (\cup\overline{\Gamma}^a[F(a)])\cap F^a([H(a)]^*)$ the equation $F(x) = \delta$ has at most $\mu(a)$ solutions in $\cup\overline{C}(a)$.
(iv) The uniqueness of the solution of one equation $F(x) = \delta$, where $\delta \in R^n\backslash F(H^*)$ is arbitrary, is a necessary and sufficient condition for F to be a homeomorphism.

□

A completely analogous theorem exists for the number of solutions of $F(x) = \delta$ for the case of points δ which belong to *unbounded* sets $\Gamma_i \in \Gamma$, and which are situated at a sufficiently large distance from the origin.

4. THE UNIQUE SOLVABILITY OF PL EQUATIONS

The previous theorems directly lead to another necessary and sufficient condition for a p.l. function to be a homeomorphism, best stated in geometrical language. We have seen that when $a \in R^n$ is arbitrary, when $\Gamma_j \in \Gamma[F(a)]$, and when the determinants associated with F in $C(a)$ all have the same sign, then the number of polyhedral sets $C_i \in C$ with $\Gamma_j \in F(C_i)$ is positive and independent of Γ_j (Theorems 1 and 3). Under these conditions, therefore, the sum of the n-angles subtended at $F(a)$ by the polyhedral seta $F(C_i)$, with $C_i \in C(a)$, is a multiple of one complete n-angle. Also, equality of signs of the determinants associated with F in $C(a)$ is certainly *necessary* for $F|_{\cup\overline{C}(a)}$ to be a homeomorphism. Since a norm-coercive function $R^n \to R^n$ which is a local homeomorphism is also a homeomorphism, we obtain the following condition.

Theorem 4: Let the p.l. function $F : R^n \to R^n$ be associated with the finite set of polytopes C .

(i) Let a be a point in R^n . Then for $F|_{\cup\overline{C}(a)}$ to be a homeomorphism, it is necessary and sufficient that the following conditions hold:

1) The determinants associated with F in $C(a)$ have the same sign.
2) The sum of the n-dimensional angles subtended at $F(a)$ by the images $F(C_j)$ of the polytopes $C_j \in C(a)$ amounts to less than two complete n-angles.

(ii) Let Φ be the set of those closed facets of polytopes in C which have no proper subfaces, and $X \subset \cup\Phi$ a set of points representative of the faces in Φ . Then for F to be a homeomorphism it is necessary and sufficient that 2) holds for every $a \in X$ and that the following condition is satisfied:

1') The determinants associated with F have the same sign.

□

Remark 1: In principle condition 2) can be checked for every given p.l. function, since Φ is finite. In practice, however, expressions for the magnitude of general n-angles are not available. Therefore Theorem 4 will be applied directly only in exceptional cases, even for moderate values of n .

In order to illustrate Theorem 1, we shall formulate condition 2) analytically for $n = 2$.

Example 1: Let F be a p.l. function from R^2 to R^2 , C_1 any polygon in C , and v one of its vertices. Starting from C_1 , proceed clockwise about v , and consecutively denote the other polygons in $C(v)$ by $C_2, \ldots, C_J$. Starting from the side of C_1 which C_1 shares with C_2 and proceeding clockwise again, let $x_1, \ldots, x_J$ be points on the sides of $C_1, \ldots, C_J$, respectively, none of which coincides with v . The matrix of $F|^L_{C_i}$ is denoted by A_i , as usual. Finally, for $i = 1, \ldots, J$ let $\nu_i = x_i - v$ be column vectors and $\nu_{J+1} = \nu_1$. Then F transforms the angle subtended by ν_i and ν_{i+1} into

$$\Theta_i = \arc\cos \frac{\nu_i^T A_i^T A_i \nu_{i+1}}{|A_i \nu_i| \cdot |A_i \nu_{i+1}|}$$

Therefore local homeomorphism at v obtains when $\sum_{i=1}^{J} \Theta_i = 2\pi$ or $\sum_{i=1}^{J} \Theta_i < 4\pi$, assuming that $\det A_1, \ldots, \det A_J$ all have the same sign. Thus, when in a given case the geometry of C is sufficiently simple, homeomorphism of F can be ascertained by a direct application of Theorem 4 .

For $n = 3$ an analogous procedure can be formulated, but its application will obviously be much more cumbersome.

By Theorem 1 condition 1') of Theorem 4 is a sufficient condition for F to be surjective. It is therefore interesting that when a certain simple condition is satisfied by C then condition 1') is necessary and *sufficient* for F to be a homeomorphism. Since in many cases the geometric condition on C is easily seen to be satisfied, Theorem 5 possesses a wide range of applicability as well.

Theorem 5: Let the p.l. function $F : R^n \to R^n$ be associated with the sets of hyperplanes H and polyhedral sets C .

(i) Let $b \in \cup H$ be a point such that the hyperplanes $H_i \in H$ which contain b have linearly independent normals. Then for $F|_{\cup\overline{C}(b)}$ to be a homeomorphism it is necessary and sufficient that the determinants associated with F in $C(b)$ have the same sign.

(ii) Let Φ be the set of those closed facets of polyhedral sets in C which have no proper subfaces and $X \subset \cup\Phi$ a set of points representative of the faces in Φ .

Let (i) be satisfied by every $b \in X$.

Then for F to be a homeomorphism it is necessary and sufficient that the determinants associated with F have the same sign.

□

Following Theorem 5, condition 2) in Theorem 4 can be relaxed:

Theorem 6: The assertion of Theorem 4 (ii) is still valid when X is replaced by the set Y of points $x \in X$, such that the normals to the hyperplanes $H_i \in H$ which contain x are linearly independent.

□

The proof is clear.

Example 2: Let m be a positive integer. Consider the case of a p.l. planar function such that H consists of the lines $x = 1, \ldots, m$, $y = 1, \ldots, m$ and $y = x$. The cardinalities of X and Y are m^2 and m, respectively.

The practical value of Theorem 6 is thus evident.

5. ALGORITHMS TO DETERMINE THE HOMEOMORPHISM OF PL FUNCTIONS

We present schematically four algorithms to determine the homeomorphism of p.l. functions. Algorithm 1, based on Theorem 3 (iv), and Algorithm 2, based on Theorem 5 (i), are applicable to the determination of homeomorphism of $F|_{\cup\overline{C}(a)}$ for given $a \in R^n$.

Algorithm 3, based on Theorem 2 (iv), and Algorithm 4, based on Theorem 6 and 3 (iv) are applicable to the determination of homeomorphism in R^n as a whole.

Algorithm 1: Let $a \in R^n$ be given.

1: Determine whether all the determinants associated with F in $C(a)$ have the same sign. If this is not the case, $F|_{\cup\overline{C}(a)}$ is not a homeomorphism. If it is, go to 2.

2: Fix $c \in \cup\overline{C}(a) \backslash H^*$ arbitrarily.

3: For each $C_i \in C(a)$ determine whether the equation $F|_{\overline{C}_i(x)} = F(c) = \delta$ has a solution in $\overline{C}_i$.

We then may assert that $F|_{\cup\overline{C}(a)}$ is a homeomorphism if and only if c is the only δ-point of F.

Algorithm 2: Let $a \in R^n$ be given.

1: Step 1 is identical with step 1 of Algorithm 1.

2: Examine whether the normals to the hyperplanes $H_i \in H$ which contain a are linearly independent. If this happens to be the case, then $F|_{\cup\overline{C}(a)}$ is a homeomorphism, if not, the algorithm is inconclusive.

Algorithm 3:

1: Examine whether all the determinants associated with F have the same sign. If this is not the case, F is not a homeomorphism. If it is, go to 2.

2: Fix $c \in R^n \backslash H^*$ arbitrarily.

3: For each $C_i \in C$ determine whether the equation $F|_{\overline{C}_i(x)} = F(c) = \delta$ has a solution in $\overline{C}_i$.
We then may assert that F is a homeomorphism if and only if c is the only δ-point of F .

Algorithm 4:

1: is identical with 1 of Algorithm 3.

2: Determine a set X as defined in Theorem 5.

3: For each $a \in X$ apply Algorithm 2 or, when this algorithm happens to be inconclusive, apply Algorithm 1.
We then may assert that F is a homeomorphism if and only if $F|_{\cup\overline{C}(a)}$ is a homeomorphism for every $a \in X$.

Remark 2: The solution of each equation $F_i(x) = \delta$, regarded as a separate task, is associated with a computational complexity of order n^3 or, possibly

$$n^{\log_2 7} \sim n^{2.8} \quad .$$

However, when Algorithms 1 or 3 are used, a complexity of this magnitude appears in the solution of one of the equations $F|_{C_i(x)} = \delta$ only.
We shall explain this for Algorithm 3:
Assume that the matrix A_I^{-1} , associated with some $C_I \in C$, has already been computed and that $C_J \in C$ shares a facet with C_I . By [11] $\text{rank}(A_J - A_I) \leq 1$ and by the Householder formula ([16], page 79), an $O(n^2)$-computational complexity only is associated with the inversion of A_J . Therefore, all the matrices A_i , except for A_I , can be inverted with an $O(n^2)$-complexity. Thus the overall computational complexity is $O(n^{2.8} + qn^2)$.

For Algorithm 1 the situation is analogous.

Example 3: Let $n = 2$ and let H consist of the lines $y = 0$, $y = \pm 1$, $y = \pm x$. Assume that we have already learned that the determinants associated with the p.l. function F have the same sign. By Theorem 5 (i) F is a local homeomorphism at the points $(\pm 1, \pm 1)$ of H^*. For F to be a homeomorphism it is necessary and sufficient that $F|_{\cup \overline{C}((0, 0))}$ is a homeomorphism. In order to ascertain whether this homeomorphism obtains, choose c arbitrarily in the triangle $|x| \leq y \leq 1$, $c \neq (0, 0)$ and look for solutions of the equation $F(x) = F(c)$ in $\cup \overline{C}(0, 0)$ which are different from c. It is not difficult to show that when two polyhedral sets C_i and C_j share a facet and $\det A_i . \det A_j > 0$ then the restriction of F to $C_i \cup C_j$ is a homeomorphism. Hence the unbounded polygons in the strip $0 \leq y \leq 1$ need not be examined. Thus, although C has 12 polygons, only the three matrices corresponding to the strip $-1 \leq y < 0$ must be inverted. In principle, two of the inversions can be carried out by means of the Householder formula.

6. THE RIEMANN MANIFOLD OF A PIECEWISE LINEAR FUNCTION

When all the determinants associated with a piecewise linear function F have the same sign, F can be represented by a piecewise linear manifold M, best thought of as an analogy to the Riemann surface of a complex function. The condition that the determinants associated with F all have the same sign is <u>not</u> necessary for the construction of M. However, for the purposes of this paper it will be helpful to consider only functions of the type indicated. For these, as in the complex case, the inverse F^{-1} of F is uniquely determined on the manifold that represents F. We shall now describe the construction of M.

The polyhedral sets $F(\overline{C}_1), \ldots, F(\overline{C}_q)$ serve as the building blocks for the construction of M. As in the complex case, however, these polyhedral sets are not considered just as point sets in R^n. For each $x \in R^n$ we shall discriminate between the point $F(x) \in R^n$ and the point $F^*(x) \in M$ which represents $F(x)$ on M. For each $\overline{C}_k \in \overline{C}$ we may

visualize $F^*(\overline{C}_k) = F^*_k(\overline{C}_k)$ by $F(\overline{C}_k)$, but $F^*(C_k)$ and $F^*(C_\ell)$ are always considered disjoint when $1 \leq k < \ell \leq q$, even when $F(C_k)$ and $F(C_\ell)$ do meet. However, when φ, the intersection of $\overline{C}_k$ and $\overline{C}_\ell$, is not empty, then $F^*(\overline{C}_k)$ is to be "glued" to $F^*(\overline{C}_\ell)$ at $F^*(\varphi)$. Stated differently, $F^*_k(\varphi)$, $F^*_\ell(\varphi)$ and $F^*(\varphi)$ are identified with each other. The mapping
$F^* : R^n \to M$ is a homeomorphism. In order to see this, add a point at infinity to each of R^n and M. The continuous function F^* can be extended to a continuous function on the compact set $R^n \cup \{\infty\}$, since $\lim_{|x| \to \infty} |F(x)| = \infty$. Since M is Hausdorff and since F^* is a bijection, F^* is a homeomorphism.

The following example, due to Fujisawa and Kuh [10] is, in a certain sense, the simplest example of a piecewise linear function F, such that the manifold which represents F has more than one sheet: we learn from Theorem 5(ii) that when $n = 2$ and no point in H^* belongs to more than two lines $H_i \in H$, then F is a homeomorphism. Thus the manifold that represents F consists of one sheet. Hence, for the surface that represents F to have more than one sheet, some point on H^* must be common to at least three lines in H.

Example 4: ([10], p. 324). Let $R_1, \ldots, R_6$ denote the clockwise numbered closed polygons generated by the lines
$y = 0$, $y = \sqrt{3}\,x$, $y = -\sqrt{3}\,x$, letting
$R_1 = \{(x, y) \mid y \geq 0 , y \geq \sqrt{3}\,x\}$. Also, let $F_i = F|_{R_i}$,

$$A_1 = \begin{bmatrix} 1 & -2/\sqrt{3} \\ 0 & 1 \end{bmatrix}, \quad A_2 = \begin{bmatrix} 0 & -1/\sqrt{3} \\ \sqrt{3} & 0 \end{bmatrix}, \quad A_3 = \begin{bmatrix} -1 & -2/\sqrt{3} \\ 0 & -1 \end{bmatrix},$$

$$A_4 = -A_1, \quad A_5 = -A_2, \quad A_6 = -A_3,$$

$$z = \begin{bmatrix} x \\ \\ y \end{bmatrix} \quad , \text{ and } \quad F_i(z) = A_i \, z \quad .$$

Then $F(R_1) = R_1 \cup R_2$, $F(R_2) = R_3 \cup R_4$, $F(R_3) = R_5 \cup R_6$, $F(R_4) = R_1 \cup R_2$, $F(R_5) = R_3 \cup R_4$, $F(R_6) = R_5 \cup R_6$.

We immediately see that the surface which represents F is topologically equivalent to the Riemann surface of the complex function $w = \sqrt{z}$, the origin being the only branch point of both functions.

7. UPPER BOUNDS FOR THE NUMBER OF SOLUTIONS

The following theorem is easily proved:

Theorem 7: Let all the determinants associated with the piecewise linear function $F : R^n \to R^n$ have the same sign, let H contain a pair of nonparallel hyperplanes and let $\gamma \in R^n$. Then the number of sheets of the Riemann manifold that represents F , is $[q/2]$ at most. For q even this bound can be reduced to $(q/2) - 1$.

Proof: Let the equation $F(x) = \gamma$ have exactly m solutions $x \in R^n$ for $\gamma \notin F(H^*)$ (Theorem 2(i)). Then the Riemann manifold of F has m sheets, each of which extends over the whole of R^n . Since each $\overline{C}_i \in \overline{C}$ is a proper subset of some half space, and $F^L_{\overline{C}_i}$ is an affine function, $F(\overline{C}_i)$ is also a proper subset of some half space. Therefore between them the q sets $F(\overline{C}_i)$ fill up less than q halves of a sheet, i.e. $[q/2]$ full sheets at most for all q , and $(q/2) - 1$ for q even.

□

For the formulation of a sharper theorem the following definitions will be useful.

Definition 6: A *slab* is the closure of the set of points in R^n contained between two fixed hyperplanes.

Definition 7: A set in R^n is *n-dimensionally unbounded* when it is not a subset of the union of a finite number of slabs.

In the sequel we shall assume that the equation of the hyperplane $H_i \in H$ is $\langle a_i , x \rangle = d_i$.

We are now in a position to formulate a theorem which is sharper than Theorem 7.

Theorem 8: Let the determinants associated with the p.l. function $F : R^n \to R^n$ in the unbounded set of C have the same sign. Define r as the maximum of the numbers $|F(x)|$ for points x belonging to one of the bounded sets of $\overline{C}$, let $B = \{x \in R^n \mid |x| \leq r\}$, and let the number of n-dimensionally unbounded sets in C be q^* . Then q^* is even, and for $\gamma \in R^n \backslash B$ the equation $F(x) = \gamma$ has at most $(q^*/2) - 1$ solutions in R^n .

□

Example 5: Let $n = 2$ and let H consist of the lines $y = 0$, $y = 1$, $y = 1 - x$, and $y = 1 + x$. Here $q = 9$. However, q^* is 6 only. Therefore, when F satisfies the relevant assumptions, the equation $F(x) = \gamma$ has $(6/2) - 1 = 2$ solutions at most.

Example 6: The configuration of Example 4 is easily generalized. We first take up the case $n = 2$. Let p straight lines pass through the origin. thus generating $q^* = 2p$ rays and q^* angular domains. Assume that consecutive rays are at an angle of $2\pi/q^*$ to each other. A p.l. function F with positive determinants, analogous to the one of Example 4 , can now be defined such that each of the angular domains is itself mapped on an angular domain of $\pi - 2\pi/q^*$ radians. The image angles thus add up to $2\pi((1/2)q^* - 1)$ radians or $(1/2)q^* - 1$ complete angles. Thus, for $\gamma \neq 0$ the equation $F(x) = \gamma$ has *exactly* $(1/2)q^* - 1$ solutions.

Example 7: For a given even number q^* and for every $n > 2$ a p.l. function $F: R^n \to R^n$ with a Riemann manifold of $(1/2)q^* - 1$ sheets can now easily be constructed. We shall obtain F from the function of Example 6, as follows:

Let $x = (x_1 , . . . , x_n)$, $F(x) = (u_1 , . . . , u_n)$. For u_1 , u_2 employ the expressions which correspond to the function of Example 6, and for $i > 2$ let $u_i = x_i$. Obviously the number of sheets of the Riemann manifold of F is $(1/2)q^* - 1$.

The result of Theorem 8 is thus optimal. It would be of interest to investigate whether relevant multidimensional examples of p.l. functions F exist, such that the Riemann manifold of F has $(1/2)q^* - 1$ sheets and such that F is *not* reducible to a transformation of the plane.

The expressions for the number of solutions of p.l. equations given in Theorem 7 and 8 were obtained by solely taking account of the fact that each $F(C_i)$ is a proper subset of some half space. In a theorem not recorded here ([24], Theorem 4) we have obtained an improved upper bound for the number of solutions of $F(x) = \gamma$, by taking account of the matrices A_i as well. The idea is simple. When e_1 and e_2 are half spaces which contain $F(C_i)$ then $F(C_i) \subset e_1 \cap e_2$. When, in particular, ∂e_1 and ∂e_2 contain facets of $F(C_i)$, then ∂e_1 , ∂e_2 are the transformations by F_i of certain hyperplanes bordering on C_i . Employing, therefore, the matrix A_i , we can easily formulate an expression for the smallest n-angle subtended by pairs of hyperplanes bordering on $F(C_i)$. By summing over all C_i the values thus obtained, we obtain an improved upper bound for the number of solutions of $F(x) = \gamma$.

REFERENCES

1. Arrow, K.J.; Hahn, F.H., General Competitive Analysis, Holden Day, Oliver and Boyd, San Francisco and Edinburgh 1971.

2. Charnes, A.; Garcia, C.B.; Lemke, C.E., Constructive Proofs of Theorems Relating to $F(x) = y$, with Applications, Math. Programming 12, 328-343 (1977).

3. Chien, M.J.; Kuh, E.S., Solving Piecewise Linear Equations for Resistive Networks, Circuit Theory and Appl. 4, 3-24 (1976).

4. Eaves, B.C., An Odd Theorem, Proc. Amer. Math. Soc. 26, 509-513 (1970).

5. Eaves, B.C., Solving Piecewise Linear Convex Equations, Math. Programming Study 1, 96-119 (1974).

6. Eaves, B.C., A Short Course in Solving Equations with PL Homotopies, SIAM-AMS Proc. 9, 73-143 (1976).

7. Eaves, B.C.; Saigal, R., Homotopies for Computation of Fixed Points, Math. Programming 3, 1-22 (1972).

8. Eaves, B.C.; Saigal, R., Homotopies for Computation of Fixed Points On Unbounded Regions, Math. Programming 3, 225-237 (1972).

9. Eaves, B.C.; Scarf, H., The Solution of Piecewise Linear Equations, Math. Oper. Res. 1, 1-27 (1976).

10. Fujisawa, T.; Kuh, E.S., Piecewise Linear Theory of Nonlinear Networks, SIAM J. Appl. Math. 22, 307-328 (1972).

11. Fujisawa, T.; Kuh, E.S.; Ohtsuki, T., A Sparse Matrix Method for Analysis of Piecewise Linear Resistive Networks, IEEE Trans. Circuit Theory CT-19, 571-584 (1972).

12. Garcia, C.B., Computation of Solutions to Nonlinear Equations under Homotopy Invariance, Math. Oper. Res. 2, 25-29 (1977).

13. Garcia, C.B.; Zangwill, W.I., Determining All Solutions to Certain Systems of Nonlinear Equations, Math. Oper. Res. 4, 1-14 (1979).

14. Garcia, C.B.; Zangwill, W.I., Finding All Solutions to Polynomial Systems and Other Systems of Equations, Math. Programming 16, 159-176 (1979).

15. Harvey, C.; Stenger, F., A Two-Dimensional Analogue to the Method of Bisection for Solving Nonlinear Equations, Quarterly of Appl. Math. 351-368 (1976).

16. Householder, A.S., Principles of Numerical Analysis, McGraw Hill, New York 1953.

17. Katznelson, J., An Algorithm for Solving Nonlinear Resistor Networks, Bell. System Tech. J. 44, 1605-1620 (1965).

18. Kojima, M., Studies on Piecewise Linear Approximations of Piecewise C^1 Mappings in Fixed Points and Complementarity Theory, Math. Oper. Res. 3, 17-36 (1978).

19. Kojima, M.; Nishino, H.; Arima, N., A PL Homotopy for Finding All the Roots of a Polynomial, Math. Programming 16, 37-62 (1979).

20. Kuh, E.S.; Hajj, I.N., Nonlinear Circuit Theory: Resistive Networks, Proc. IEEE 59, 340-355 (1971).

21. Ohtsuki, T.; Fujisawa, T.; Kumagai, S., Existence Theorems and a Solution Algorithm for Piecewise Linear Resistive Networks, SIAM J. Math. Anal. 8, 69-99 (1977).

22. Rheinboldt, W.C.; Vandergraft, J.S., On Piecewise Affine Mappings in R^n , SIAM J. Appl. Math. 29, 680-689 (1975).

23. Schramm, R., On Piecewise Linear Functions and Piecewise Linear Equations (submitted for publication).

24. Schramm, R., Bounds for the Number of Solutions of Certain Piecewise Linear Equations (to appear in Proc. Conf. on Geometry and Differential Geometry, Haifa University, March 18-23, 1979).

25. Stynes, M., An N-Dimensional Bisection Method for Solving Systems of N Equations in N Unknowns (to appear).

Numerical Solution of Highly Nonlinear Problems
W. Forster (ed.)
© North-Holland Publishing Company, 1980

THEOREMS RELATING TO THE APPROXIMATION OF FIXED POINTS IN FUNCTION SPACES

Walter Forster

University of Southampton,
Southampton, England

1. INTRODUCTION

The computation of fixed points in function spaces (see e.g. [1]) relies on the fact that the properties required e.g. for the Schauder fixed point theorem [2] allow these spaces to be approximated by finite dimensional spaces. For practical purposes a problem given in a function space (e.g. an ordinary or partial differential equation, etc.) is approximated by a finite-dimensional problem by a process known as discretization. It is necessary to show that for a certain set of functions the discretized version of our original problem tends in the limit (e.g. mesh size goes to zero) to our originally given equation. This property of approximations is known as "consistency", i.e. the approximated equation is consistent with the original equation. Furthermore, it is important that the sequence of solutions of the approximated equation tends in the limit (e.g. mesh size goes to zero) to the solution of the originally given equation. This property is known as "convergence", i.e. the approximate solutions converge to the solution of our originally given problem. A further important property, which usually shows up first in computations, is "numerical stability". When we compute a solution function step by step we do not want the function to shoot off to infinity. We want the function to remain bounded. The relation between the concepts of consistency, convergence and stability has been clarified in various theorems (e.g. Kantorowitch [3], Dahlquist for ordinary differential equations [4], see also [5], [6], and Lax for certain partial differential equations [7]). The novel treatment we give in this paper relies on algebraic properties of the space of continuous functions, in particular on the Stone-Weierstrass theorem. With this approach it is possible to drop the usual assumption that the functions involved are "sufficiently often" differentiable. We only use the fact that continuous functions form an algebra.

2. THE ALGEBRA OF CONTINUOUS FUNCTIONS

We first consider the algebraic properties of the space of bounded continuous functions $C[0, 1]$ defined on the interval $[0, 1]$. The space $C[0, 1]$ is a ring, i.e. it satisfies the following axioms:

1) associativity of addition: $a + [b + c] = [a + b] + c$
   for all $a, b, c \in C$, where C is a set (here the continuous functions),
2) identity element for addition: $a + 0 = a = 0 + a$
   for each $a \in C$; $0 \in C$,
3) inverse element: $a + [-a] = [-a] + a = 0$
   for each $a \in C$, $[-a] \in C$, $0 \in C$,
4) commutativity of addition: $a + b = b + a$
   for all $a, b \in C$,
5) associativity of multiplication: $a.[b.c] = [a.b].c$
   for all $a, b, c \in C$,
6) identity element for multiplication: $a.1 = a = 1.a$
   for all $a \in C$; $1 \in C$,
7) distributivity of multiplication over addition: $a.[b + c] = a.b + a.c$
   and $[a + b].c = a.c + b.c$ for all $a, b, c \in C$,
8) commutativity of multiplication: $a.b = b.a$
   for all $a, b \in C$.

Furthermore, the space $C[0, 1]$ is a linear space, i.e. it satisfies the axioms: 1), 2), 3), 4),

9) $\lambda[a + b] = \lambda a + \lambda b$
   for all $a, b \in C$ and all $\lambda \in R$, where R is a field (usually the real numbers),

10) $[\lambda + \mu]a = \lambda a + \mu a$
    for all $a \in C$ and all $\lambda, \mu \in R$,
11) $[\lambda\mu]a = \lambda[\mu a]$
    for all $a \in C$ and all $\lambda, \mu \in R$,
12) $1a = a$
    for all $a \in C$ and $1 \in R$.

C[0, 1] satisfies the following additional axiom:

13) relation of scalar multiplication to multiplication:
    $\lambda[a.b] = [\lambda a].b = a.[\lambda b]$
    for all $a, b \in C$ and $\lambda \in R$.

An algebraic structure which satisfies axioms 1) to 13) is called a commutative algebra.

According to the Weierstrass approximation theorem (see e.g. [8]) every continuous function can be approximated as closely as we wish by polynomials.
The polynomials $p(t)$ can be thought of as being generated by the functions (generators of the algebra)

$g_0 : t \to 1$ i.e. the constant function (each element t is mapped to the constant 1),

$g_1 : t \to t$ i.e. the identity function (each element t is mapped to itself),

and the following three operations:
addition,
multiplication,
and multiplication by real numbers R .

3. AN EQUIVALENCE THEOREM

The relation between the original problem and approximations to the original problem is usually expressed as a theorem, called "equivalence theorem". The novel approach to equivalence theorems given in this paper utilizes the Stone-Weierstrass theorem and the above given algebraic properties of continuous functions. Together with the fixed point formulation of the problem this leads to a very short proof.

The solution of our original problem is assumed to be a continuous function. The result of a computation is a set of discrete points. We can interpolate by linear functions and we obtain a continuous function consisting of piecewise linear functions. If we consider continuous functions consisting of piecewise polynomial functions, then they form an algebra.
We define

$$C_N = \{f \in C : \; f \text{ is a piecewise polynomial function consisting of } N \text{ pieces of polynomials}\}.$$

Next we construct the following space consisting of piecewise polynomial functions

$$C_\infty^- = \bigcup_{\text{for all } N} C_N \quad .$$

One can easily see that C_∞^- is an algebra.
We complete the space C_∞^- by constructing Cauchy sequences of piecewise polynomial functions consisting of an increasing number of pieces of polynomials (the maximum diameter of the region over which the polynomials are taken is assumed to go to zero). We append the continuous function obtained by this limiting process to our space of continuous functions C_∞^- and obtain a complete metric space (we use the uniform norm) C_∞ .

We now consider the approximation process. We have the following diagram.

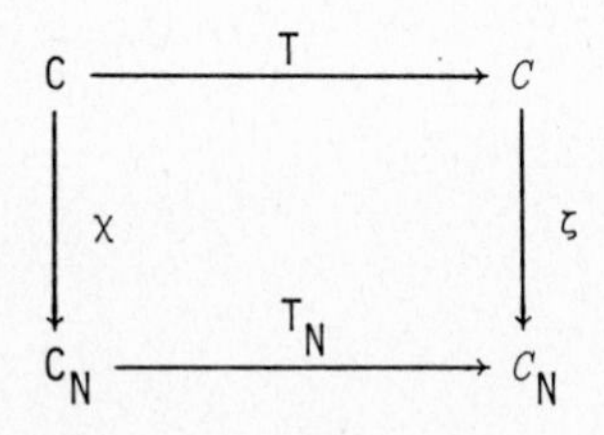

Diagram 1.

A mapping $T: C \to \mathcal{C}$ (in general nonlinear) is approximated by $T_N: C_N \to \mathcal{C}_N$. The mapping T maps continuous functions C into continuous functions $\mathcal{C}$ (It makes the discussion easier if we distinguish

between the two copies of $C[0, 1]$ and denote them C and $\mathcal{C}$). The mapping T_N maps C_N functions defined on a specified grid into $\mathcal{C}_N$ functions on the same grid. T_N approximates T in some sense. C_N (respectively $\mathcal{C}_N$) approximates C (respectively $\mathcal{C}$).
In the limit $N \to \infty$ we have the diagram:

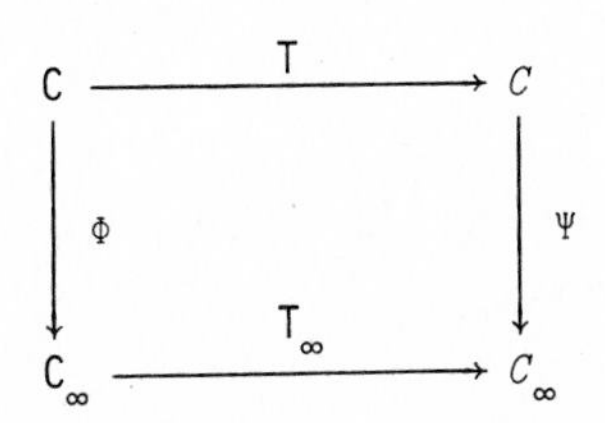

Diagram 2.

For an approximation to make sense we require $||y_N - y|| \to 0$ with $N \to \infty$, i.e. the solution of the approximate problem $y_n \in C_N$ should tend to the solution of the original problem $y \in C$. This is usually called convergence. Another way of looking at convergence is the following. For every $y_\infty \in C_\infty$ we have to have a $y \in C$ which is mapped to y_∞ , and vice versa. We therefore have the following algebraic definition of convergence.

Definition (convergence): A numerical method is convergent if $\Phi: C \to C_\infty$ is an isomorphism (of the algebra of continuous functions) with the same generators.
That means the spaces C and C_∞ have to be identical.

Furthermore, we require for a reasonable approximation $||T_N y_N - Ty|| \to 0$ with $N \to \infty$ for a set of functions $y_N \in C_N$ and $y \in C$ and a class of mappings T_N and T . This property can again be formulated as an isomporphism (of the algebra of continuous functions).
To consider morphisms of the algebra of continuous functions it is best to look at the various substructures, i.e. homomorphisms of the additive group, homomorphisms of the multiplicative monoid, and homomorphisms of linear spaces.
We split above isomorphism into several parts.

First we assume that Ψ acts on the scalars $\lambda \in R$ as identity, i.e. maps all scalars λ to itself. Next we split the remaining part into an isomorphism of the multiplicative monoid and into a monomorphism of the additive group.

Definition (consistency): A numerical method is consistent if $\Psi : C \to C_\infty$ is an isomorphism of the multiplicative monoid and both monoids have the same generators.

Definition (stability): A numerical method is stable if $\Psi : C \to C_\infty$ is a monomorphism of the additive group.

The multiplicative monoid contained in the algebra of continuous functions is generated by $g_1 : t \to t$, or in the context of polynomials $g_1 = x^1$, and we have the constant function $g_0 : t \to 1$, or in the context of polynomials $g_0 = x^0$. The functions generated can be denoted by
x^0 (constant function = identity of the multiplicative monoid),
x^1 (identity function = generator of the monoid),
x^2 ,
.. ,
.. ,
x^n ,
..
The set of all these functions is a free monoid (here with one generator x^1). Free monoids are unique to within isomorphisms.
If we have an isomorphism of the free multiplicative monoid (and both have the same generators, here e.g. $g_1 : t \to t$, or $g_1 = x^1$; we assume the identity $g_0 : t \to 1$, or $g_0 = x^0$, is mapped to the identity), then it follows that we have an isomorphism of the free additive group (because Ψ maps the generators of the additive group C into the generators of the additive group C_∞).
We have assumed that Ψ acts on the scalars like an identity. This, together with the group isomorphism, shows that we have an isomorphism of linear spaces.
Therefore, we have an isomorphism $\Psi : C \to C_\infty$ of the algebra of continuous functions.

Theorem (stability condition): Numerical stability is equivalent to $T_\infty(0) = 0$.

Proof: From the stability definition we have the requirement that $\Psi : C \to C_\infty$ is a monomorphism of the additive group. This is equivalent to the identity O (zero function: every element is mapped to zero) of the additive group C to be mapped to the identity O' (zero function) of the additive group C_∞ . From the fixed point formulation $T(0) = O$ and convergence it then follows that diagram 2 has to be commutative and this implies $T_\infty(0') = O'$.
On the other hand we have from the fixed point formulation $T(0) = O$ and if $T_\infty(0') = O'$, then by convergence diagram 2 has to be commutative and it follows that O is mapped to O' by Ψ (and this is the only element mapped to O'), i.e. we have a monomorphism.

□

We call $T_\infty(0) = 0$ stability condition.

Theorem (consistency condition): Consistency is equivalent to $T_\infty(1) = 1$ and $T_\infty(x) = x$, where $\{1, x\}$ are the generators of the algebra.

Proof: The requirement that $\Psi : C \to C_\infty$ is an isomorphism of the multiplicative monoid (both have the same generators) means that $\Psi : x \to x'$, where x is the monoid generator in C and x' is the monoid generator in C_∞ . Furthermore, the identity of the monoid is mapped to the identity, i.e. $\Psi : 1 \to 1'$, where $1 \in C$ and $1' \in C_\infty$. It then follows from the fixed point formulation and by convergence that diagram 2 is commutative, i.e. that $T_\infty(1') = 1'$ and $T_\infty(x') = x'$.
On the other hand if $T_\infty(1') = 1'$ and $T_\infty(x') = x'$, we can generate all functions in C_∞ (free monoid). From the fixed point formulation we have $T(1) = 1$ and $T(x) = x$. By convergence diagram 2 has to be commutative and this requires $\Psi : 1 \to 1'$ and $\Psi : x \to x'$, i.e. we have an isomorphism of free monoids.

□

We call $T_\infty(1) = 1$ and $T_\infty(x) = x$ consistency condition.

<u>Theorem</u> (Equivalence theorem): The necessary and sufficient condition for a numerical method to be convergent is that it is consistent and stable.

<u>Proof</u>: Consider diagram 2.
First we give the sufficiency part of the proof.
If we have a monomorphism of the additive group (stability) and an isomorphism of the multiplicative monoid (consistency) on the right hand side of diagram 2, i.e. Ψ is an isomorphism, then via the fixed point formulation of the problem (condition on T) and the stability and consistency condition (conditions on T_∞) all functions available in C and C_∞ are also available in C and C_∞. For each function diagram 2 commutes and Φ is an isomorphism of the algebra and both algebras have the same generators, i.e. we have convergence.
Necessity part of the proof.
If Ψ is only a monomorphism of the additive group (stability) but not an isomorphism of the multiplicative monoid (consistency), or vice versa, then the fixed point formulation again makes all the functions which are available in C available in C, but the consistency or stability condition (condition on T_∞) is not satisfied and the functions available in C_∞ are not the same as in C_∞. It follows, that diagram 2 does not commute for all functions and Φ is not an isomorphism of algebras, we do not have the same generators and in general we do not have convergence.

□

We summarize part of our results in form of a diagram.

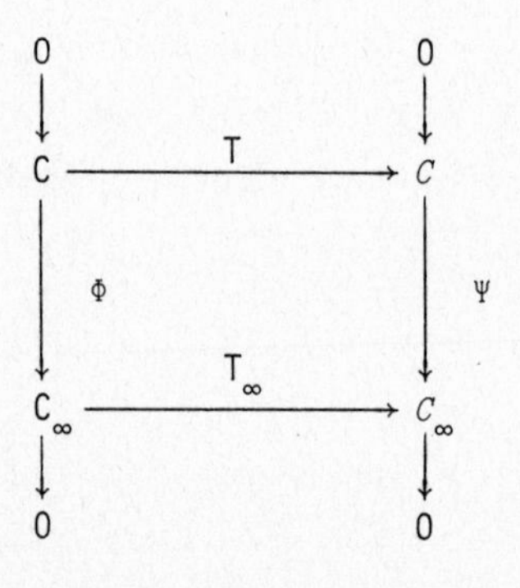

On the left hand side and on the right hand side of the diagram we have exact sequences, which is just another way to say that C and C_∞ are isomorphic groups and $\mathcal{C}$ and $\mathcal{C}_\infty$ are isomorphic groups.

4. EXAMPLES

4.1. Ordinary differential equations and linear multistep methods.

Given the following initial value problem

$$\frac{dy}{dt} = f(t, y) \quad , \quad y(0) = y_0$$

we assume that the solution satisfies the conditions of the existence and uniqueness theorem, see e.g. [8]. We rewrite the problem as a fixed point problem (or integral equation)

$$y(t) = y_0 + \int_0^t f(s, y(s))\, ds \quad .$$

This is a fixed point equation with

$$Ty = y_0 + \int_0^t f(s, y(s))\, ds$$

or $$T : C \to \mathcal{C} \quad .$$

The differential equation is solved by a linear multistep method, see e.g. [5], which we write as follows

$$h \sum_{i=0}^{k} \beta_i f_{n+i} = \sum_{i=0}^{k} \alpha_i y_{n+i}$$

(h is the step size, α_i and β_i are constants)
or

$$T_N : C_N \to \mathcal{C}_N \quad .$$

If we use the stabilty condition and the consistency condition, i.e. we use the test functions $\{0, 1, x\}$, and substitute into the linear multistep method, then we obtain the well known stability and consistency conditions for linear multistep methods as e.g. given in [5].

In the case of linear multistep methods one can easily see that after substitution of $\{1, x\}$ there is no dependance on h (for $h \to 0$) or N (for $N \to \infty$). Therefore, we have the following consistency condition for linear multistep methods:

$T_N(1) = 1$ and $T_N(x) = x$ for all N .

4.2. Ordinary differential equations and methods which are not linear multistep methods.

It can be checked that the consistency condition is satisfied e.g. by Runge-Kutta methods. Here we discuss an example of a method given in [9] which is not consistent. The method is

$$y_{n+1} - e^{-h} y_n = (1 - e^{-h}) f_n \quad .$$

We take $y = 1$ or $y' = 0$ with $y(0) = 1$. Substitution of this test function shows that for $h \to 0$ (or $N \to \infty$) the first part of the consistency condition is satisfied.

Now take $y = x$ or $y' = 1$ with $y(0) = 0$. Substitution gives

$$(n + 1)h - e^{-h} nh = (1 - e^{-h})$$

or

$$nh + h - (1 - h + h^2/2 - \dots)nh = (1 - 1 + h - h^2/2 + \dots) .$$

With nh remaining constant we obtain

$$nh = \text{arbitrary} = 0$$

i.e. the second part of the consistency condition is not satisfied.

4.3. Ordinary differential equations and periodic problems with known period.

For periodic problems with known period we could take periodic functions as generators, e.g. $\{\exp(inx), \exp(-inx)\}$, or $\{\cos nx, \sin nx\}$, where n is an integer. Methods of this kind are e.g. discussed in [10], [11]. No background theory is given there. Considering algebraic properties of continuous functions, in this case periodic functions, provides an explanation why such methods work.

4.4. Systems of ordinary differential equations.

The extension of above algebraic approach to systems of ordinary differential equations is straightforward. y and f are replaced by vectors

$$y = (y_1 , \ldots , y_m) \quad ,$$

$$f = (f_1 , \ldots , f_m) \quad .$$

The mapping is now $T : C^m \rightarrow C^m$,
where $C^m[0, 1] = C \times C \times \ldots \times C$ is the m-fold Cartesian product of $C[0, 1]$. In the stability condition 0 is replaced by a vector $(0, 0, \ldots , 0)$ and the generators of the algebra $C^m[0, 1]$ are the appropriate products of the generators of $C[0, 1]$.
For systems of ordinary differential equations the consistency condition then reads:
A method is consistent if $T_\infty(g) = g$ for all generators $\{g\} = \{g_1 , g_2 , \ldots , g_k\}$ of the algebra.

4.5. Partial differential equations.

The extension of above algebraic approach to partial differential equations is more complicated, because even simple partial differential equations (e.g. $u_{xx} = u_t$) lead to relations between the generators of the algebra. This restricts the possible solution functions to a subspace of the space of continuous functions (e.g. the function $u(x, t) = t$ is not a solution of above partial differential equation, but $u(x, t) = 2t + x^2$ or $u(x, t) = t^2 + tx^2 + (1/12)x^4$, etc. satisfy above partial differential equation). Furthermore, the simplicity achieved by utilizing the free additive group and the free multiplicative monoid is destroyed. Nevertheless, the stability condition given (with obvious modification) is still applicable (and coincides for certain generators with e.g. von Neumann stability).

5. CONCLUSIONS

Fixed point formulation and the utilization of the algebraic structure of continuous functions allow a considerable simplification in the proof of the equivalence theorem. In addition it leads to a practical criterion for testing consistency.

REFERENCES

1. Allgower, E.L.; Jeppson, M.M., The Approximation of Solutions of Mildly Nonlinear Elliptic Boundary Value Problems having Several Solutions, in Springer Lecture Notes 333, Berlin-Heidelberg-New York 1973.
2. Schauder, J., Der Fixpunktsatz in Funktionalräumen, Studia Math. 2, 171-180 (1930).
3. Kantorowitch, L.W., Functional Analysis and Applied Mathematics, Uspekhi Mat. Nauk USSR 3, 89-185 (1948).
4. Dahlquist, G. Convergence and Stability in the Numerical Integration of Ordinary Differential Equations, Math. Scand. 4, 33-53 (1956).
5. Henrici, P., Discrete Variable Methods in Ordinary Differential Equations, Wiley, New York 1962.
6. Henrici, P., Error Propagation for Difference Methods, Wiley, New York 1963.
7. Lax, P.D.; Richtmyer, R.D., Survey of the Stability of Linear Finite Difference Equations, Comm. Pure Appl. Math. 9, 267-293 (1956).
8. Simmons, G.F., Introduction to Topology and Modern Analysis, McGraw Hill, New York 1963.
9. Smith, J.M., Mathematical Modeling and Digital Simulation for Engineers and Scientists, Interscience, New York 1977.
10. Gautschi, W., Numerical Integration of Ordinary Differential Equations Based on Trigonometric Polynomials, Numer. Math. 3, 381-397 (1961).

11. Lambert, J.D., Computational Methods in Ordinary Differential Equations, Wiley, New York 1973.

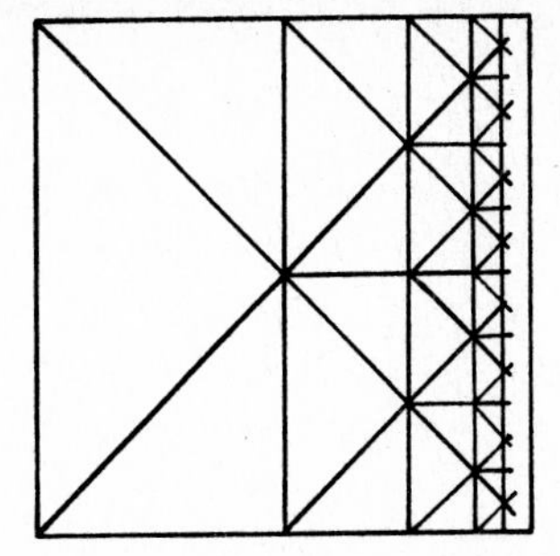

This section includes papers concerned with complementarity theory (including an application) and variational inequalities.

Numerical Solution of Highly Nonlinear Problems
W. Forster (ed.)
© North-Holland Publishing Company, 1980

A THEOREM ON THE PARTITIONING OF SIMPLOTOPES AND ITS APPLICATION TO A PROPERTY OF Q-MATRICES

Richard W. Cottle

Stanford University,
Stanford, California, U.S.A.

and

Rabe von Randow

Universität Bonn,
Bonn, West Germany

A <u>simplotope</u> is by definition, the cartesian product of two simplices, say Δ_r (of dimension r) and Δ_s (of dimension s). Accordingly, the simplotope $\Delta_r \times \Delta_s$ is a special sort of (r+s)-dimensional polytope.

Research on this report was partially supported by the National Science Foundation Grant MCS76-81259 A01 and the Office of Naval Research Contract N00014-75-C-0267.
Work on this paper was carried out at the Mathematisches Institut, Universität zu Köln and the Institut für Ökonometrie und Operations Research der Universität Bonn during the first author's visit there under the auspices of the Alexander von Humboldt Stiftung as a Senior U.S. Scientist Awardee.

Theorem 1: The simplotope $\Delta_r \times \Delta_s$ can be partitioned into $\binom{r+s}{r}$ simplices of dimension $r+s$ whose vertices are all vertices of $\Delta_r \times \Delta_s$. Furthermore, exactly two vertices of $\Delta_r \times \Delta_s$ belong to each of the simplices in the partition.

□

It is an interesting consequence of the last assertion of Theorem 1 that there is a connection between this partitioning and the so-called Northwest Corner Rule (which can be) used to determine a basic feasible solution of the Transportation Problem. (See Dantzig [3, p. 361].).

As shown by Sommerville [11], simplotopes are isomorphic to frusta of simplices. (The dimensions of the "factors" Δ_r and Δ_s depend on the number of vertices of the original simplex in each of the frusta.) This fact enables us to handle one of the three cases arising in the proof of Theorem 2 below. To state this result, we first need to introduce a bit of notation.

Let B^n denote the ball of radius 1 centered at the origin 0 in R^n, and let $S^{n-1} = \partial B^n$ be the corresponding unit sphere. If C is a nonsingular $n \times n$ matrix whose columns C_i have Euclidean length 1, the points $C_1, \ldots, C_n$ are the vertices of a <u>spherical</u> (n-1)-<u>simplex</u> $\sigma(C)$. In fact

$$\sigma(C) = S^{n-1} \cap \text{pos } C \quad .$$

The corresponding <u>spherical sector</u> is the set

$$\tau(C) = B^n \cap \text{pos } C = \text{conv } \{0, \sigma(C)\} \quad .$$

Regarding B^n as a convex body with homogeneous distribution of mass, we can express the <u>centroid</u> (or <u>center of mass</u>) of $\tau(C)$ as the point

$$\overline{x}(C) := \frac{1}{v(C)} \int_{\tau(C)} x \, dx$$

where $v(C)$ denotes the volume of the sector $\tau(C)$. Using these notations, we have

Theorem 2: $\langle C_i , \overline{x}(C) \rangle > 0 , \quad i = 1, \ldots , n .$

□

An immediate consequence of this is the

Corollary: $\langle d , \overline{x}(C) \rangle > 0$ for all $d \in \sigma(C)$.

□

The Corollary above is instrumental in proving a result on the linear complementarity problem which we pose in the form

Given real $n \times n$ matrices A and B and a real n-vector c , find n-vectors w and z satisfying

$$\begin{aligned} Aw + Bz &= c \\ w \geq 0 , \quad z &\geq 0 , \\ \langle w , z \rangle &= 0 . \end{aligned}$$

Denote this system by the triple (A, B, c).

We call $[A, B]$ a Q-matrix iff (A, B, c) has at least one solution for each c .

One of the outstanding problems in linear complementarity theory is to characterize Q-matrices. This is not exactly an unsolved problem as the papers of Aganagic-Cottle [1], Kelly-Watson [6], Mangasarian [7], and Pang [9], readily show. Despite this, it is still of interest to find a "useful" characterization of the class Q .

It is clear that when $c = 0$, the problem is trivial; thus, we may assume $c \neq 0$. From this and the definition of (A, B, c) , it is clear that positive scaling of c and the columns of $[A, B]$ result in an equivalent problem. Accordingly, it is not restrictive to assume

$$||A_i|| = ||B_i|| = ||c|| = 1 \qquad i = 1, \ldots , n .$$

In attempting to solve (A, B, c) one seeks a complementary "submatrix" C of $[A, B]$ such that

$$Cx = c , \quad x \geq 0 .$$

A complementary submatrix of $[A, B]$ is one for which

$$C_i \in \{A_i , B_i\} \qquad i = 1, \dots , n .$$

Let comp $[A , B]$ denote the set of all such C . If $\det C \neq 0$ for all $C \in$ comp $[A, B]$, we call $[A, B]$ nondegenerate. We assume (for simplicity) that $[A, B]$ is nondegenerate, though this is not really necessary if instead one concentrates on the nonsingular $C \in$ comp $[A, B]$.

It is now clear that $[A, B]$ is a Q-matrix iff

$$\cup\{\sigma(C) : C \in \text{comp } [A, B]\} = S^{n-1}$$

or, equivalently, iff

$$\cup\{\tau(C) : C \in \text{comp } [A, B]\} = B^n .$$

The formulation of the question in terms of covering the sphere S^{n-1} by sperical simplices is due to Kelly and Watson [6]. Obviously, $[A, B]$ can be a Q-matrix only if the $\tau(C)$ - for $C \in$ comp $[A, B]$ - and hence their centroids are "sufficiently well distributed" in B^n . With this in mind, define the set

$$X[A, B] = \text{conv } \{\overline{x}(C) : C \in \text{comp } [A, B]\} .$$

Proposition: If $n = 2$, $[A, B]$ is a Q-matrix if and only if $0 \in \text{int } X[A, B]$.

□

For higher dimensions only the necessity statement is valid.

Theorem 3: For all n , if $[A, B]$ is a Q-matrix, then $0 \in \text{int } X[A, B]$. For $n \geq 3$, the converse is false.

□

The results announced here will be submitted for publication elsewhere.

REFERENCES

1. Aganagic, M.; Cottle, R.W., On Q-Matrices, Technical Report SOL 78-9, Department of Operations Research, Stanford University, September 1978.

2. Bonnesen, T.; Fenchel, W., Theorie der konvexen Körper, Chelsea Publishing Company, New York 1948.

3. Dantzig, G.B., Application of the Simplex Method to a Transportation Problem, in Activity Analysis of Production and Allocation, T.C. Koopmans (Ed.), Wiley, New York 1951, pp. 359-373.

4. Grünbaum, B., Convex Polytopes, Interscience, New York 1967.

5. Kelly, L.M.; Watson, L.T., Erratum: Some Perturbation Theorems for Q-Matrices, SIAM Journal of Applied Mathematics 34, 320-321 (1978).

6. Kelly, L.M.; Watson, L.T., Q-Matrices and Spherical Geometry, to appear in Linear Algebra and its Applications.

7. Mangasarian, O.L., Characterization of Linear Complementarity Problems as Linear Programs, Mathematical Programming Study 7, 74-87 (1978).

8. Murty, K.G., On the Number of Solutions to the Complementarity Problem and Spanning Properties of Complementary Cones, Linear Algebra and its Applications 5, 65-108 (1972).

9. Pang, J.S., A Note on an Open Problem in Linear Complementarity, Mathematical Programming 13, 360-363 (1977).

10. Samelson, H.; Thrall, R.M.; Wesler, O., A Partition Theorem for Euclidean N-Space, Proceedings of the A.M.S. 9, 805-807 (1958).

11. Sommerville, D.M.Y., An Introduction to the Geometry of N Dimensions, Dover Publications, New York 1958.

12. Watson, L.T., A Variational Approach to the Linear Complementarity Problem, Ph.D. Thesis, University of Michigan, 1974.

13. Watson, L.T., Some Perturbation Theorems for Q-Matrices, SIAM Journal of Applied Mathematics 31, 379-384 (1976).

Numerical Solution of Highly Nonlinear Problems
W. Forster (ed.)
© North-Holland Publishing Company, 1980

AN EQUILIBRIUM MODEL FOR AN OPEN ECONOMY WITH INSTITUTIONAL CONSTRAINTS ON FACTOR PRICES

Lars Mathiesen and Terje Hansen

Norwegian School of Economics
and Business Administration,
Bergen, Norway

The purpose of this paper is to present (a general equilibrium model within) a computational framework which allows for comparative static analysis. The theory of general economic equilibrium is concerned with both the quantities and prices of all commodities. This theory will be applied to a single open economy by allowing activities of exports and imports and interpreting one commodity as foreign currency. The model is then closed through a constraint on the balance of payments.

There exist numerous different definitions of a general equilibrium. For our purpose the following definition posed as a complementarity problem is convenient:

A price vector π^* and a vector of activity levels y^* constitute a general competitive equilibrium if:

i) prices and activity levels are non-negative; $\pi^* \geq 0$, $y^* \geq 0$,

ii) excess supply is non-negative in all markets; i.e. $- By^* + \omega - \xi(\pi^*) \geq 0$,

iii) no activity earns a positive profit; $B'\pi^* \geq 0$,

iv) an activity that makes a negative profit is not used and an activity that is operated makes zero profit; $\pi^{*\prime}By^* = 0$,

v) a commodity in excess supply has zero price and a positive price implies market clearance; i.e. $(-By^* + \omega - \xi \; \pi^*))'\pi^* = 0$.

Equilibrium prices for the above economy could be calculated using a fixed point algorithm. With the loss of some generality, however, larger problems may be solved using linear complementarity in the following way. We make an initial guess, π^0 , on equilibrium values of π^* , and replace market demand by their first order Taylor expansions

$$\xi(.)_{\pi^0} = d - D\pi \; .$$

The solution to this linearized problem, $\bar{\pi}$ and $\bar{y}$ will then in a sense represent an approximation of a competitive equilibrium.

Within this general framework we introduce constraints on certain prices. The presence of such constraints implies that market prices and shadow prices will not necessarily coincide. Unless such constraints are introduced models cannot explain the simultaneous existence of excess supply of a factor of production and yet a positive market price. We assume that unemployed workers receive a compensation such that their disposable income is the same as those who are employed at the minimum wage. For the Scandinavian countries this is a reasonable description of the labour market. This assumption simplifies the characterization of market demand in our model.

The minimum wage requirements can within the linear model be reflected by both lower bounds on individual wages and ties between wages and a cost of living index.

Making transfer to the unemployed explicit in the model necessitates the inclusion of a public sector. Hence we introduce a public sector with taxes on commodities, tariffs on imports, an unemployment compensation program and a public expenditure plan. Tax- and tariff rates are exogenous, while the total public receipts and expenditures and the total transfer are endogenous.

A worked numerical example illustrates how the model may be used to study the consequences of a 10% alternatively 20% increase in the nominal minimum wage. In this specific example international trade increases considerably, one sector contracts significantly, whereas the others contract or expand moderately. Unemployment of unskilled labour increases, and public expenditure is reduced in order to finance the unemployment compensation program.

1. INTRODUCTION

Economic planning in an open economy has been approached through a variety of models. For multi-sector modelling the input-output technique and linear programming are often used. Excellent reviews are Manne (1974) and Blitzer et al. (1975). A great advantage of linear programming is that it provides means for efficient, systematic exploration of the economy's choice set as delimited by technological and other constraints [1]. The shadow prices associated with the optimal solution of linear programs may in principle be used for decentralized planning. This use of shadow prices, however, should not be taken literally. The shadow price system will often be too close to that of a competitive system to be regarded as appropriate for direct use. One notorious example is the zero shadow price imputed to surplus labour in models of developing countries. Another problem is that ad hoc restrictions necessary for primal stability pick up their own shadow prices, thus making interpretation more difficult. One main reason for such defects in the shadow price system is obviously the often rather rudimentary way in which the actual substitutabilities of the economy are modelled [2]. More often that not domestic supply of resources and final demand for goods is stipulated as fixed and thus exogenous to the model.

Public policies consistent with such a model will largely be limited to resource allocation. For countries with developed markets there is, however, a whole set of public policies related to the price side. These policies are concerned with taxes, subsidies, tariffs and the exchange rate. As such these policies relate to the dual of the resource planning model. A model to accommodate for both types of policies should thus contain behavioural relationships which constrain responses of individual economic agents to policy changes as well as the physical constraints on material balance and resource availability [3].

Using the complementarity format we present a multisector model of an economy engaged in international trade. The central feature of our model is that demand and supply are endogenous which implies that prices depend directly on demand and supply. In this way the model allows for greater substitution between factors of production and between commodities. The price system will thereby reflect a larger set of its alleged determinants. The format of complementarity treats quantities and prices in a symmetric manner. This allows one to put institutional restrictions on endogenous prices as well as quantities and even tying these two sets of variables together. Consequently, such phenomena as minimum wage laws may be incorporated in the analysis. Without such constraints models cannot explain the simultaneous existence of excess supply of a factor of production and yet a positive market price. Our model will in this respect resemble fixed price models; i.e. Barro and Grossman (1971) and (1976), Drèze (1975) and Malinvaud (1977).

The purpose of this paper is to present a computational framework which allows for comparative static analysis. In our model, there is a public sector with taxes on goods and factors of production, tariffs on imports, an unemployment compensation program and a program for expenditures. Like in other analyses, notably Shoven and Whalley (1973) and Shoven (1974), tax rates etc. are exogenous to the model.

In the next section we briefly review the concepts of equilibrium and complementarity. In section 3. we present the basic model and make economic interpretations. In section 4. we extend the model to include institutional constraints on prices and a public sector. Finally, we illustrate the applicability of the model by a numerical example in section 5.

2. GENERAL EQUILIBRIUM AND THE COMPLEMENTARITY FORMAT

In this section we review the notion of general equilibrium and the complementarity format. These two concepts constitute the building blocks of our model and are essential to the subsequent discussion of the model.

Consider an economy with production described by the technology matrix B, where $b_{kj} \geq 0$ represents an input of the k-th commodity in the j-th process and where $b_{kj} \leq 0$ represents an output of the k-th commodity in the j-th process. Further let ω represent initial holdings, let π denote prices, and let $\xi(\pi)$ denote market demand.

Because of the generality of the theory of economic equilibrium there exist numerous definitions of a general equilibrium though their contents very much coincide. We shall use the following definition of a general competitive equilibrium:

Definition: A price vector π^* and a vector of activity levels y^* constitute a general competitive equilibrium if

i) Excess supply is non-negative in all markets;
i.e. $-By^* + \omega \geq \xi(\pi^*)$.

ii) No activity earns a positive profit;
$B'\pi^* \geq 0$.

iii) Prices and activity levels are non-negative;
$\pi^* \geq 0 \ , \ y^* \geq 0$.

iv) An activity that earns a deficit is not run and operated processes run at a balance;
$\pi^{*\prime}By^* = 0$.

v) A commodity in excess supply has zero price and a positive price implies market clearance;
i.e. $(-By^* + \omega - \xi(\pi^*))'\pi^* = 0$.

□

The complementarity format is as follows: Find z such that

$$F(z) \geq 0 \ , \quad z \geq 0 \quad \text{and} \quad z'F(z) = 0 \ .$$

When the mapping F of R^n into itself is an affine transformation, say $F(z) = q + Mz$, the corresponding complementarity problem is said to be linear, otherwise it is nonlinear. The complementarity format has long been known to economists through the equilibrium conditions of a general equilibrium and the Kuhn-Tucker conditions of a local extremum of a mathematical programming problem. The former is seen by the association

$$z = \begin{bmatrix} \pi \\ y \end{bmatrix} \ , \quad F\begin{bmatrix} \pi \\ y \end{bmatrix} = \begin{bmatrix} -\xi(\pi) - By + \omega \\ B'\pi \end{bmatrix} \ .$$

From this it is also seen that our definition of a general equilibrium will be a linear complementarity problem when market demand is linear in prices, i.e.

$$\xi(\pi) = d + D\pi \quad .$$

The complementarity format has gained considerable attention recently. Much effort is put into exploring the conditions for existence of solutions and devising algorithms for computing such solutions. For an up to date review see Balinski and Cottle (1978). The similarity between complementarity and fixed point theory (as for example applied to the computation of general equilibrium) is also generally acknowledged, see Karamardian (1977) and Balinski and Cottle (1978).

3. A MULTI-SECTOR PLANNING MODEL FOR AN OPEN ECONOMY

The model presented below is a static general equilibrium model for an open economy. Goods in foreign trade are assumed to be exported and imported at exogenously stipulated prices, while prices for resources, goods and services traded domestically are endogenously determined. We shall use the following notation:

P . . . . . denotes internationally stipulated prices (expressed in a foreign currency), with subscripts x and m referring to export and import prices respectively,

π . . . . . denotes endogenously determined prices for goods and services,

ω . . . . . denotes endogenously determined prices of resources,

γ . . . . . denotes the endogenously determined price of the local currency in terms of the foreign currency,

A , B . . . denote technology-matrices,

$a_{ij} > 0$ (< 0) indicate that the i-th good is an input to (output of) the j-th process;

$a_{ij} = 0$ ($b_{kj} = 0$) indicate that the i-th good (k-th resource) is not involved in the j-th process;

$b_{kj} > 0$ indicate that the k-th resource is an input to the j-th process;

$D(\pi, \omega, \gamma)$ denotes final demand for goods and services,
$S(\pi, \omega, \gamma)$ denotes supply of resources,
y . . . . denotes activity levels of domestic production,
y_x , y_m denotes activity levels of export and import respectively.

An equilibrium for our model is characterized by activity levels y , y_x and y_m and prices π , ω and γ , such that the following conditions (1) to (7) are satisfied.

(1)
$$- D(\pi, \omega, \gamma) - Ay - y_x + y_m \geq 0 ,$$
$$S(\pi, \omega, \gamma) - By \geq 0 ,$$

i.e. imports plus domestic output is at least as big as exports, domestic input and final demand in markets for all goods and services and in the case of a resource total demand must not exceed supply. In sum, excess demand is non-positive in all markets.

(2)
$$P'_x y_x - P'_m y_m \geq E ,$$

which constrains the balance of payments (in foreign currency) to be not worse than E . The parameter E will depend on this specific economy's potential for obtaining credit ($E < 0$) or its obligations to repay loans ($E > 0$) in the planning period.

(3)
$$A'\pi + B'\omega \geq 0 ,$$

i.e. no process earns a positive profit.

(4)
$$P_x\gamma \leq \pi \leq P_m\gamma ,$$

i.e. domestic prices for goods and services are bounded by the internationally stipulated prices expressed in the local currency. Observe that (4) is equivalent to

(4')
$$\pi - P_x\gamma \geq 0 ,$$
$$-\pi + P_m\gamma \geq 0 .$$

Furthermore, we must have

$$\text{(5)} \qquad \pi \geq 0 \;,\; \omega \geq 0 \;,\; \gamma \geq 0 \;,$$

$$y \geq 0 \;,\; y_x \geq 0 \;,\; y_m \geq 0 \;,$$

i.e. non-negative prices and activity levels.

$$\text{(6)} \qquad \{- D(\pi , \omega , \gamma) - Ay - y_x + y_m\}'\pi = 0 \;,$$

$$\{S(\pi , \omega , \gamma) - By\}'\omega = 0 \;,$$

$$\{P_x'y_x - P_m'y_m - E\}\gamma = 0 \;,$$

i.e. a commodity or resource in excess supply has a zero price and a positive price implies a zero excess supply.

$$\text{(7)} \qquad \{A'\pi + B'\omega\}'y = 0 \;,$$

$$\{\pi - P_x\gamma\}'y_x = 0 \;,$$

$$\{-\pi + P_m\gamma\}'y_m = 0 \;,$$

i.e. an activity that earns a deficit is not used and an activity that is operated runs at a balance.

From what was said in the preceeding section there is nothing remarkable on this model. We have only extended the interpretation of activities and markets of an equilibrium model of a closed economy to take account of characteristics of an open economy. Before we incorporate additional features like institutional constraints on endogenous prices we make some observations.

Summing up the three conditions in (6) and using (7) we obtain

$$\text{(8)} \qquad \omega'S(.) = \pi'D(.) + \gamma E \;,$$

which is Walras' Law and the equivalent in national accounting terms to:

$$\begin{bmatrix}\text{Factor}\\ \text{income}\end{bmatrix} = \begin{bmatrix}\text{Domestic}\\ \text{demand}\end{bmatrix} + \begin{bmatrix}\text{Value of net exports}\\ \text{in local currency}\end{bmatrix} .$$

According to economic theory assume that $D(.)$ and $S(.)$ are homogeneous of degree zero in π , ω and γ . Then, if $(\pi^* , \omega^* , \gamma^*)$ are equilibrium prices, so are $(\lambda\pi^* , \lambda\omega^* , \lambda\gamma^*)$ for $\lambda > 0$. Hence only relative prices are determined by the model (1) - (3), (4'), (5) - (7) .

Equilibrium prices for the above economy could be calculated using the fixed point algorithm of Hansen (1968), Eaves (1972) and Scarf and Hansen (1973). However, the problem with these algorithms is that if the number of markets is large, then they are computationally not feasible. However, with loss of some generality large problems may be solved using linear complementarity in the following way. We guess initially equilibrium values of π , ω and γ and replace $D(.)$ and $S(.)$ by their first order Taylor expansions. (1) - (3), (4'), (5) - (7) then correspond to a linear complementarity problem. The solution to this problem $\bar{\pi}$, $\bar{\omega}$, $\bar{\gamma}$, $\bar{y}$, $\bar{y}_x$ and $\bar{y}_m$ will then in a sense represent an approximation of a competitive equilibrium. The quality of the approximation will obviously depend on the price sensitivity of the demand and supply functions and the technology as well as on the quality of the initial guess.

Let

$$(9) \qquad D(\pi , \omega , \gamma) = d + D_1\pi + D_2\omega + D_3\gamma ,$$

$$S(\pi , \omega , \gamma) = s + S_1\pi + S_2\omega + S_3\gamma .$$

Substituting (9) into (1) we get the following linear complementarity problem for an open economy where there are no institutional constraints on factor prices.

Find non-negative vectors π , ω , γ , y , y_m and y_x such that

$$\begin{array}{lcl}
- D_1\pi - D_2\omega - D_3\gamma - Ay - y_x + y_m & \geq & d \\
S_1\pi + S_2\omega + S_3\gamma - By & \geq & - s \\
P_x'y_x - P_m'y_m & \geq & E \\
A'\pi + B'\omega & \geq & 0 \\
\pi \quad - P_x\gamma & \geq & 0 \\
- \pi \quad + P_m\gamma & \geq & 0
\end{array}$$

and

$$\{- D_1\pi - D_2\omega - D_3\gamma - Ay - y_x + y_m - d\}'\pi = 0 ,$$

$$\{S_1\pi + S_2\omega + S_3\gamma - By + s\}'\omega = 0 ,$$

$$\{P_x'y_x - P_m'y_m - E\}\gamma = 0 ,$$

$$\{A'\pi + B'\omega\}'y = 0 ,$$

$$\{\pi - P_x\gamma\}'y_x = 0 ,$$

$$\{- \pi + P_m\gamma\}'y_m = 0 .$$

4. CONSTRAINTS ON FACTOR PRICES AND A PUBLIC SECTOR

In the analysis so far we have assumed that there are no constraints on resource prices. The presence of such constraints implies, for example, that the market prices and the shadow prices of the factors of production will not necessarily coincide. Unless such constraints are introduced, models cannot explain the simultaneous existence of excess supply of a factor of production and yet a positive market price. Hansen and

Manne (1977) demonstrated that constraints on market prices could be easily incorporated in an equilibrium model through linear complementarity. Though the focus of their paper was somewhat different from ours, the following discussion is closely related to their work.

In order to illustrate how bounds on factor prices may be incorporated in our model, we shall consider the case with lower bounds on wages. We shall in that connection make the following assumption.

Assumption: Unemployed labour receives unemployment compensation such that disposable income is the same for those that are unemployed and those that are employed at the minimum wage.

□

We make this assumption in order to simplify the subsequent discussion since it implies that we do not have to distinguish in the demand function between those who are employed and those who receive unemployment compensation if there is unemployment and the minimum wage applies.

The minimum wage requirements are reflected by the constraints

$$\omega \geq \overline{\omega} \quad .$$

For factors of production other than labour the corresponding component of $\overline{\omega}$ is obviously 0 . If necessary, the model can accommodate more complicated minimum wage requirements, for example one that ties wages to a cost of living index.

In the subsequent discussion we shall have to distinguish between shadow prices and market prices for factors of production. We shall let λ denote shadow prices and let ω denote market prices. Let v denote the non-negative wedges between the market prices and the shadow prices, i.e. $\omega = v + \lambda$, and substitute $v + \lambda$ for ω throughout the model. The complementarity constraint associated with $v + \lambda \geq \overline{\omega}$ will then be $(v + \lambda - \overline{\omega})'v = 0$. Thus if a component of v is strictly positive, i.e. there is a positive wedge between the market and the shadow price, then the corresponding minimum wage constraint is effective. On the other hand, if the minimum wage constraint is not effective, then the wedge is zero.

Our market demand functions presuppose that unemployed factors of production are paid the market wage. The necessary transfer of income behind this assumption could be left outside the model and there would be no need for a public sector in the model. We will, however, make these transfers explicit. Hence we introduce a public sector with its fiscal means, taxes and tariffs, and its ends, an unemployment compensation program and a public consumption plan.

Let t be the rate of value added tax and let r be the tax which applies to factors of production. Furthermore, let c be tariff rates which apply to imports. Obviously a more complicated tax system could be incorporated in the model. Let

$\hat{\pi} = (1 + t)\pi$ denote the price paid for the end use of goods,

$\hat{\omega} = (1 - r)\omega$ denote the after tax remuneration to the factors of production,

$\hat{P}_m = (1 + c)P_m$ denote the prices paid in foreign currency for imported goods,

$\overline{\omega}$ denote the exogenously stipulated minimum market price for factors of production,

$D(\hat{\pi}, \hat{\omega}, \gamma)$ denote private demand for goods and services,

$\overline{G} + g\alpha - g\beta$ denote the public expenditure plan where $\overline{G}$ and g are vectors and α and β (≥ 0) are variables representing an increase and a reduction of the marginal part of the plan.

With these symbols, the income part of the public budget can be described as:

$$T = t.D(\hat{\pi}, \hat{\omega}, \gamma)'\pi + r.S(\hat{\pi}, \hat{\omega}, \gamma)'\omega + c.\gamma.P_m'y_m ,$$

and the expenditure part as:

$$G + Y = (\overline{G} + g\alpha - g\beta)'\begin{bmatrix}\pi \\ \omega\end{bmatrix} + \overline{\omega}'u .$$

The vector u denotes the excess supply of resources. $\overline{\omega}'u$ thus represents the cost of the unemployment compensation program. The public budget constraint is given by

$$T - G - Y = F \quad ,$$

where $F > 0$ denotes the surplus income in an overbalanced budget, and $F < 0$ is interpreted oppositely. The complementarity condition

$$\alpha . \beta = 0$$

completes the model. Observe the special representation of the complementarity condition corresponding to the public budget constraint.

The public budget constraint is nonlinear because of products of endogenous variables. We shall therefore have to linearize the constraint by taking its first order Taylor expansion. Furthermore, we substitute for the excess supply of factors of production (u) and obtain

$$H_1\pi + H_2\lambda + H_2 v + H_3\gamma + H_4 y + H_5 y_m + H_6\alpha - H_6\beta = \overline{F} \quad .$$

The complete model may now be stated as the following linear complementarity problem.

Find vectors π , λ , v , γ , y , y_x , y_m , α and β such that [4] :

(10)
$$
\begin{array}{rcl}
- D_1\pi - D_2\lambda - D_2 v - D_3\gamma - Ay - y_x + y_m - g_1\alpha + g_1\beta & \geq & d + \overline{G}_1 \ , \\
S_1\pi + S_2\lambda + S_2 v + S_3\gamma - By \qquad - g_2\alpha + g_2\beta & \geq & - s + \overline{G}_2 \ , \\
P'_x y_x - P'_m y_m & \geq & E \ , \\
A'\pi + B'\lambda + B'v & \geq & 0 \ , \\
\pi \qquad - P_x\gamma & \geq & 0 \ , \\
- \pi \qquad P_m\gamma & \geq & 0 \ , \\
\lambda + v & \geq & \overline{\omega} \ , \\
H_1\pi + H_2\lambda + H_2 v + H_3\gamma + H_4 y \qquad + H_5 y_m + H_6\alpha - H_6\beta & = & \overline{F} \ .
\end{array}
$$

(11)
$$
\begin{array}{l}
\pi \geq 0 \ , \quad \lambda \geq 0 \ , \quad v \geq 0 \ , \quad \gamma \geq 0 \ , \\
y \geq 0 \ , \quad y_x \geq 0 \ , \quad y_m \geq 0 \ , \quad \alpha \geq 0 \ , \quad \beta \geq 0 \ .
\end{array}
$$

(12)
$$
\begin{array}{l}
\{- D_1\pi - D_2\lambda - D_2 v - D_3\gamma - Ay - y_x + y_m - g_1\alpha + g_1\beta - \overline{G}_1 - d\}'\pi = 0 \ , \\
\{S_1\pi + S_2\lambda + S_2 v + S_3\gamma - By - g_2\alpha + g_2\beta - \overline{G}_2 + s\}'\lambda = 0 \ , \\
\{P_x y_x - P_m y_m - E\}\gamma = 0 \ , \\
\{A'\pi + B'v + B'\lambda\}'y = 0 \ , \\
\{\pi - P_x\gamma\}'y_x = 0 \ , \\
\{P_m\gamma - \pi\}'y_m = 0 \ , \\
\{\lambda + v - \overline{\omega}\}'v = 0 \ , \\
\alpha . \beta = 0 \ .
\end{array}
$$

5. NUMERICAL EXAMPLE

In order to illustrate the applicability of the model we shall consider an economy with 3 resources (capital, skilled and unskilled labour). Technology is described by the coeffcients in Table 1.

	Activity						
	1	2	3	4	5	6	7
Good 1	-1	-1	-1				
Good 2				-1	-1		
Good 3						-1	-1
Capital	.7	.2	.1	.6	.2	1.0	1.4
Skilled labour	.2	.2	.1	.2	.3	.4	.1
Unskilled labour	.3	.6	.7	.1	.2	.2	.6

Table 1. Technological coefficients

Private investment demand and public demand for goods and resources are as follows:

	Private investment	Public demand	
	(Fixed)	Fixed	Variable
Good 1	.5	3.7	.3 $(\alpha - \beta)$
Good 2	.5	2.7	.3 $(\alpha - \beta)$
Good 3	1.	.3	.2 $(\alpha - \beta)$
Capital	2.	2.9	.1 $(\alpha - \beta)$
Skilled labour	1.	.9	.1 $(\alpha - \beta)$
Unskilled labour	2.	.6	.4 $(\alpha - \beta)$

Tax rates t on goods and r on resources are .2 and .4 respectively. Private non-investment demand for goods in terms of producer prices are as follows:

$$D(\pi, \omega, \gamma) = \begin{bmatrix} 5.3 - 4.8\pi_1 + 1.2\pi_2 + .6\pi_3 + .18\omega_2 + .24\omega_3 \\ 2.8 + 1.2\pi_1 - 3.6\pi_2 + .6\pi_3 + .30\omega_2 + .30\omega_3 \\ 8.6 + .6\pi_1 + .6\pi_2 - 4.8\pi_3 + .12\omega_2 + .06\omega_3 \end{bmatrix} .$$

Supply of capital, skilled and unskilled labour are given by:

$$S(\pi, \omega, \gamma) = \begin{bmatrix} 10.2 \\ 5 - 1.2\pi_1 - 1.2\pi_2 - .6\pi_3 + 1.8\omega_2 \\ 11.6 - 1.2\pi_1 - 1.2\pi_2 - .6\pi_3 + 1.8\omega_3 \end{bmatrix}$$

Finally, we have a trade balance ($E = 0$) , a balanced public budget ($F = 0$) , and import and export prices are given by:

$$P_m = \begin{bmatrix} 1.2 \\ 1.0 \\ 2.2 \end{bmatrix}, \qquad P_x = \begin{bmatrix} 0.8 \\ 0.9 \\ 2.0 \end{bmatrix}.$$

In the base case there is no constraint on factor prices. In this case unskilled labour receives a compensation of 1 whereas skilled labour is paid 2 . We shall consider two alternatives to the base case.

Case 1: A minimum wage of 1.1 is introduced. Unemployed labour receives an unemployment compensation of 1.1 .

Case 2: A minimum wage of 1.2 is introduced. Unemployed labour receives an unemployment compensation of 1.2 .

The numerical model thus illustrates how the model may be used to study the consequences of a 10% alternatively 20% increase in the nominal minimum wage.

The Taylor expansion of the public budget constraint was taken at the solution of the model for the base case.

Solutions

		Base case	Case 1	Case 2
Domestic activities in use	3	8	7.56	7.07
	5	2	1.33	.64
	6	4	4.23	4.48
Import good 2		4	4.61	5.20
Export good 3		2	2.30	2.60
π_1		1	1.06	1.13
π_2		1	1	1
π_3		2	2	2
ω_1		1	1.01	1.01
ω_2		2	1.93	1.84
ω_3		1	1.1	1.2
γ		1	1	1
$\alpha - \beta$		1	.51	- .13
Unemployment of unskilled labour		0	.69	1.50

There are several interesting aspects of the solution. International trade increases considerably. There is a drastic change in the structure of production. The sectors which produce goods 1 and 2 contract considerably, whereas the sector producing good 3 (the exported good) expands significantly. As expected unemployment of unskilled labour increases and public demand is reduced in order to finance the unemployment compensation program. The numerical example thus illustrates the kind of question the model can answer.

Let us finally point out that the error due to linearization of the public budget constraint was insignificant.

NOTES

[1] L. Taylor (1975, p. 59) mentions three types of constraints used in such models:

i) real limitations such as commodity balances, bounds on total factor use and the balance of payments;

ii) "political" constraints such as lower bounds on the level of employment and upper bounds on imports, which would otherwise threaten already established but inefficient industries;

iii) "ad hoc" constraints for technical reasons to avoid overspecialization in trade, "flip-flop" consumption patterns and other forms of extreme behaviour to which linear systems are prone.

[2] Substitution could be introduced through international trade activities, through price sensitive demand or a nonlinear utility function, and through alternative domestic production activities.

[3] L. Westphal in a comment to A. Manne (1974) stresses the need for the development of models with greater emphasis on its dual aspects. He thus advocates "policy planning" models rather than "resource allocation" models.
"What we require is a methodology of policy planning' which aims at achieving a consistent set of policies within the framework of individual decision making in response to price incentives ...
Since the policies in question are primarily price policies concerned with taxes, tariffs, interest subsidies, the exchange rate, and the like, policy planning is really the dual of resource planning", op. cit. p. 496.

[4] The stated model is in terms of producer prices. Hence D(.) and S(.) are also functions of π and ω and not $\hat{\pi}$ and $\hat{\omega}$. This is accounted for by correcting the coefficients of the matrices D_1 , D_2 , . . . , S_3 accordingly.

Formally

$$
\begin{aligned}
D(\hat{\pi}, \hat{\omega}, \gamma) &= D((1+t)\pi, (1-r)\omega, \gamma) = \\
&= \hat{D}_1\{(1+t)\pi\} + \hat{D}_2\{(1-r)\omega\} + \hat{D}_3\gamma + d = \\
&= (1+t)\hat{D}_1\pi + (1-r)\hat{D}_2\omega + \hat{D}_3\gamma + d = \\
&= D_1\pi + D_2\omega + D_3\gamma + d \quad ,
\end{aligned}
$$

and analogously for $S(.)$.

REFERENCES

1. Balinski, M.L.; Cottle, R.W. (Eds.), Complementarity and Fixed Point Problems, Mathematical Programming Study 7, North Holland, Amsterdam 1978.

2. Barro, R.J.; Grossman, H.I., A General Disequilibrium Model of Income and Employment, American Economic Review, 82-93 (1971).

3. Barro, R.J.; Grossman, H.I., Money, Employment and Inflation, Cambridge University Press, Cambridge 1976.

4. Blitzer, C.R.; Clark, P.B.; Taylor, L. (Eds), Economy-Wide Models and Development Planning, Oxford University Press, London 1975.

5. Drèze, J., Existence of an Exchange Equilibrium under Price Rigidities, International Economic Review, 301-320 (1975).

6. Eaves, B.C., Homotopies for Computation of Fixed Points, Mathematical Programming 3, 1-22 (1972).

7. Hansen, T., On the Approximation of a Competitive Equilibrium, Ph.D. Thesis, Yale University, 1968.

8. Hansen, T.; Manne, A.S., Equilibrium and Linear Complementarity: an Economy with Institutional Constraints on Prices, in Equilibrium and Disequilibrium in Economic Theory, G. Schwodiauer (Ed.), D. Reidel, Dordrecht 1977, pp. 227-237.

9. Karamardian, S. (Ed.), Fixed Points, Algorithms and Applications, Academic Press, New York 1977.

10. Malinvaud, E., The Theory of Unemployment Reconsidered, Basil Blackwell, Oxford 1977.

11. Manne, A.S., Multi-Sector Models for Development Planning: a Survey, in Frontiers of Quantitative Economics, Vol. II., M.D. Intriligator; D.A. Kendrick (Eds.), North Holland, Amsterdam 1974.

12. Scarf, H.; Hansen, T., The Computation of Economic Equilibria, Yale University Press, New Haven 1973.

13. Shoven, J.B., A Proof of the Existence of a General Equilibrium with Ad Valorem Commodity Taxes, Journal of Economic Theory 8, 1-25 (1974).

14. Shoven, J.B.; Whalley, J., General Equilibrium with Taxes: a Computational Procedure and an Existence Proof, Review of Economic Studies, 475-489 (1973).

15. Taylor, L., Theoretical Foundations and Technical Implications, Ch. III. in Economy-Wide Models and Development Planning, C.R. Blitzer; et al. (Eds.), Oxford University Press, London 1975.

Numerical Solution of Highly Nonlinear Problems
W. Forster (ed.)
© North-Holland Publishing Company, 1980

A SIMPLICIAL DEFORMATION ALGORITHM WITH APPLICATIONS TO OPTIMIZATION, VARIATIONAL INEQUALITIES AND BOUNDARY VALUE PROBLEMS

Kurt Georg

Universität Bonn,
Bonn, West Germany

1. INTRODUCTION

In this paper we sketch a version of a complementary pivoting algorithm based on deformation in the sense of B.C. Eaves [13], and B.C. Eaves and R. Saigal [14], and give some applications to optimization, variational inequalities and boundary value problems.

The algorithm sketched here is presently being tested and shall be developed into a library program which can be handled also by the inexperienced user. The triangulation is based on M.J. Todd's J_3 [29], and since the labelling will not depend on the homotopy parameter, nodes with the same parameter are identified. Thus a modified triangulation with less simplices is obtained. In the same way, it would be possible to modify the more sophisticated triangulation of G. van der Laan and A.J.J. Talman [20] which allows faster shrinking. The pivoting steps are made by updating a Givens transformation of the labelling matrix and hence are numerically stable [7], [18], [31].

Supported by Deutsche Forschungsgemeinschaft, SFB 72 at the University of Bonn.

2. TRIANGULATION, LABELLING AND COMPLEMENTARY PIVOTING

2.1. Let us consider a triangulation in an abstract setting, i.e. a triangulation T (of dimension $N+1$) is a family $\{\sigma\}_{\sigma \in T}$ of sets (called simplices) such that

(a) $\# \sigma = N + 2$, $\sigma \in T$ ($\#$ = number of elements) ,

(b) $\# \sigma_1 \cap \sigma_2 \cap \sigma_3 \leq N$ whenever the three simplices $\sigma_1 , \sigma_2 , \sigma_3 \in T$ are all different.

We define the family of all k-faces, $k = 1 , \ldots , N$, of T by

(c) $T^k = \{\tau : \tau \subset \sigma , \sigma \in T , \# \tau = k+1\}$,

and the nodes (vertices) by

(d) $T^0 = \{p : p \in \sigma , \sigma \in T\}$.

For our purposes it will be sufficient to consider triangulations without boundary, i.e.

(e) each N-face $\tau \in T^N$ belongs to exactly two different simplices $\sigma_1 , \sigma_2 \in T$.

2.2. Let λ be a labelling on T , i.e. $\lambda : T^0 \to R^N$ is a map. An N-face $\tau \in T^N$ is completely labelled if $(\varepsilon , \ldots , \varepsilon^N)^T$ belongs to the convex hull $co(\lambda(\tau))$ for sufficiently small $\varepsilon > 0$. The standard technique of complementary pivoting consists of the following fact:

<u>Lemma</u>: A simplex $\sigma \in T$ either contains none or exactly two different completely labelled N-faces.

□

2.3. Let us briefly describe the algorithm.

<u>Algorithm</u>:

(a) Start: Choose $\tau_1 \in T^N$ completely labelled and $\tau_1 \subset \sigma_1 \in T$.
Set $i = 2$.

(b) Complementarity step: Find the completely labelled N-face $\tau_i \in T^N$ such that $\tau_i \subset \sigma_{i-1}$ and $\tau_i \neq \tau_{i-1}$.

(c) Pivoting step: Find the simplex $\sigma_i \in T$ such that $\sigma_i \neq \sigma_{i-1}$ and $\tau_i \subset \sigma_i$.
Increase i by one and go to (a) .

Using a graph theoretic argument, by the lemma of 2.2. it is readily seen that the algorithm has only one of the following two possibilities:

(d) Straight line: The N-faces of the sequence $\tau_1 , \tau_2 , \ldots$ are all different.

(e) Cycle: There exists an $n \in N$ such that the N-faces $\tau_1 , \tau_2 , \ldots , \tau_n$ are all different and $\tau_{n+i} = \tau_i \; , \; i \in \mathbb{N}$.

3. A MODIFICATION OF M.J. TODD'S TRIANGULATION J_3

3.1. M.J. Todd's triangulation J_3 [29] of dimension N+1 has the nodes

(a) $J_3^0 = \{(k , v) : k \in Z , v \in R^N , 2^k v \in Z^N\}$.

We refer to the integer k as the level of the node. The somewhat technical description of the simplices of J_3 and its pivoting rules can be found in [29]. Let us just mention here the refining property, i.e. any simplex $\sigma \in J_3$ has a vertex (k , v) such that $\sigma \subset [k-1 , k] \times B$, where B is the ball in R^N with center v and radius 2^{-k} (with respect to the maximum norm) .

3.2. Since we will use a labelling which does not depend on a homotopy parameter, we modify J_3 to a triangulation T (of dimension N+1) in the following way:

Denote by $P : R \times R^N \to R^N$, $P(t , x) = x$ for $t \in R$, $x \in R^N$, the canonical projection. Then

(a) $T^0 := P(J_3^0)$,

(b) $T := \{P(\sigma) : \sigma \in J_3 , \# P(\sigma) = N+2 \}$.

It is not hard to see that T is indeed a triangulation, and the pivoting steps 2.3.(c) can be programmed in an efficient way. We emphasize that the nodes of T are points in R^N , and an N-face of T forms a set of N+1 affinely independent points in R^N .

4. A ZERO PROBLEM AND ITS ALGORITHMIC SOLUTION

4.1. By R^{N*} we denote the family of compact convex nonempty subsets of R^N . Let X be a metric space. We call a multivalued map $G : X \to R^{N*}$ upper-semicontinuous (u.s.c.) if for every open subset

$V \subset R^N$ the "inverse image" $\{x \in X : G(x) \subset V\}$ is open.

The following lemma is easily shown and serves here only as a simple way for writing down u.s.c. maps:

Lemma: Let $X_1, \ldots, X_n$ be open mutually disjoint subsets of X such that $X_1 \cup \ldots \cup X_n$ is dense in X, and let u.s.c. maps

$G_i : cl(X_i) \to R^{N*}$ be given. Then there is a smallest u.s.c. map

$G : X \to R^{N*}$ which contains all the G_i's, namely

(a) $G(x) = co(\bigcup_{i \in I(x)} G_i(x)) , \quad x \in R^N ,$

where $I(x) = \{i : x \in cl(X_i)\}$ and "co" and "cl" denote the convex and closed hull, respectively.

□

In the sequel, we will briefly denote the situation of the above lemma in the following way:
The u.s.c. map $G : X \to R^{N*}$ is defined by

$$G(x) = \begin{cases} G_1(x) & \text{for} \quad x \in X_1 , \\ \ldots & \ldots \quad \ldots \\ G_n(x) & \text{for} \quad x \in X_n . \end{cases}$$

4.2. We now describe the type of zero-problem we want to solve with a simplicial algorithm. Let $G : R^N \to R^{N*}$ be an u.s.c. map with the following property:

(a) There exists a nonsingular $N{\times}N$-matrix M with real entries, a "starting point" $\bar{x} \in R^N$ and a radius $r > 0$ such that $G(x) = \{M(x - \bar{x})\}$ for all $x \in R^N$ with $||x - \bar{x}|| > r$.

The zero problem we want to solve is:

(b) Find $x^* \in R^N$ such that $0 \in G(x^*)$.

We immediately get the following necessary condition:

(c) $||x^* - \bar{x}|| \leq r$.

For reasons of simplicity, here and in the following we choose the Euclidean norm $|| \ldots || = || \ldots ||_2$ on R^N .

4.3. We define the labelling and the starting simplices to solve problem 4.2.(b) by the simplicial algorithm 2.3.
Choose a starting level

(a) $\bar{k} = \text{Max}\,\{k \in Z : 2^{-k} > \frac{8\sqrt{N}\,r}{3N+1}\}$

and define a "centre" $b = (\beta_1, \ldots, \beta_N)^T \in R^N$ by

(b) $\beta_i = (\frac{1}{2} + \frac{N+1-2i}{8N})\,2^{-\bar{k}}$, $i = 1, \ldots, N$.

It is easily checked that the distance from b (with respect to the Euclidean norm $\|\ldots\|$) to any node of T^0 with level $\leq \bar{k}$ is strictly greater than r . Furthermore, b is interior to the convex hull of the starting N-face τ_1 :

(c) $\tau_1 = \{v_0, \ldots, v_N\}$ where $(v_0, \ldots, v_N)$ denote the respective columns of the matrix

$$2^{-\bar{k}} \begin{bmatrix} 0 & 1 & . & . & . & 1 \\ . & . & . & & & . \\ . & & . & . & & . \\ . & & & . & . & . \\ 0 & . & . & . & 0 & 1 \end{bmatrix}$$

If we use the triangulation T of section 3. and define the labelling $\lambda : T^0 \to R^N$ by

(d) $\lambda(v) \in G(v + \bar{x} - b)$, $v \in T^0 \subset R^N$,
it follows that τ_1 is the only completely labelled N-face with all vertices of level $\bar{k}$. If we choose the starting simplex $\sigma_1 = \tau_1 \cup \{v_{N+1}\}$ in the "refining direction" by adding the vertex

(e) $v_{N+1} = 2^{-(\bar{k}+1)}\,(1, \ldots, 1)^T \in R^N$,
it is immediately seen that the simplicial algorithm 2.3., producing the sequence
(f) τ_1 , σ_1 , τ_2 , σ_2 , . . .
of completely labelled N-faces and corresponding simplices, cannot cycle and has at least one limit point, i.e. a point $v^* \in R^N$ such that

any neighbourhood of v^* contains at least one (and hence infinitely many) N-faces of (f) . Any limit point v^* yields a zero-point

(g) $x^* = v^* + \bar{x} - b$

of G and thus the simplicial algorithm solves problem 4.2.(b).

The rest of the paper is dedicated to some special nonlinear problems which may be solved in the above way by specifying the u.s.c. map 4.2. and applying the simplicial algorithm.

Normally a region Ω is fixed within which a zero of a map is sought. The user has the freedom of choice for the linear mapping $M(x - \bar{x})$ on the exterior of Ω . If a limit point $x^* \in \partial\Omega$ is obtained by the simplicial algorithm, this may be regarded as a "failure" . However, as our examples 5. and 6. will show, even this case gives information of considerable interest, provided that the u.s.c. map 4.2. is defined in a suitable way. In 7. an example is given where $x^* \in \partial\Omega$ is actually the interesting case.

4.5. Let us illustrate the above idea by showing how the fixed point theorem of Leray-Schauder may be realized [23]. Suppose that $\Omega \subset R^N$ is a bounded open neighbourhood of $\bar{x} = 0$, and take $r \geq \max \{ ||x|| : x \in \partial\Omega \}$. Let $F : cl(\Omega) \to R^N$ be a given continuous map. We define the u.s.c. map $G : R^N \to R^{N*}$ by (cf. 4.1.)

(a)
$$G(x) = \begin{cases} x & \text{for } x \notin cl(\Omega) , \\ x - F(x) & \text{for } x \in \Omega . \end{cases}$$

The above simplicial algorithm provides a zero point x^* of G .
Two cases are possible:

(I) $x^* \in \Omega$: Then x^* is a fixed point of F .

(II) $x^* \in \partial\Omega$: Then by construction 4.1. of G , there is a real $t \geq 0$ such that $x^* - F(x^*) + tx^* = 0$ or

(b) $F(x^*) = (1 + t)x^*$,

i.e. x^* is a point on the boundary $\partial\Omega$ where the condition of Leray-Schauder does not hold.

Both cases are interesting. If one knows for a special problem that (I) cannot occur, then a solution to a nonlinear eigenvalue problem (b) is approximated. On the other hand, if one can guarantee the Leray-Schauder condition, i.e. that (II) cannot occur, then a fixed point of F is approximated.

5. NONLINEAR PROGRAMMING

5.1. Let us begin with an application to nonlinear programming which is fairly well known, cf. [2], [22], [23], [27], [29], [30].
Suppose that Θ and $\phi : R^N \to R$ are convex. We consider the problem:

(a) Minimize $\Theta(x)$ subject to $\phi(x) \le 0$.

Given a starting point $\bar{x} \in R^N$ and a radius $r > 0$, we define the u.s.c. map $G : R^N \to R^{N*}$

(b)
$$G(x) = \begin{cases} \{x - \bar{x}\} & \text{for } ||x - \bar{x}|| > r , \\ \partial\phi(x) & \text{for } ||x - \bar{x}|| < r \text{ and } \phi(x) > 0 , \\ \partial\Theta(x) & \text{for } ||x - \bar{x}|| < r \text{ and } \phi \quad < 0 , \end{cases}$$

where "∂" denotes the subgradient, see for example [26], [28].
Let us now make the reasonable assumption that

(c) $\{x \in R^N : \phi(x) < 0 , ||x - \bar{x}|| < r\} \neq \emptyset$.

5.2. Thus as described in 4., the simplicial algorithm calculates a zero-point of G , i.e. a point $x^* \in R^N$ such that $0 \in G(x^*)$.
Let us briefly discuss the two cases that may occur:

(I) $||x^* - x|| < r$.
In this case it follows easily that $\phi(x^*) \le 0$ and $0 \in \partial\Theta(x^*) + t\, \partial\phi(x^*)$ for some $t \ge 0$, thus x^* is a minimal point of $\Theta + t\,\phi$ and hence solves 5.1.(a) .

(II) $||x^* - x|| = r$.

Since $x - \bar{x}$ is the gradient of $\frac{1}{2}||x - \bar{x}||^2 - \frac{1}{2}r^2$ at x , by arguments analogous to the ones above, it follows that x^* solves the more restricted problem:

(a) Minimize $\Theta(x)$ subject to $\phi(x) \le 0$ and $||x - \bar{x}|| \le r$.

Furthermore, if x^* solves 5.2.(a) , but not 5.1.(a) , there exists a $t \ge 0$ and an $s > 0$ such that $-s(x^* - x) \in \partial\Theta(x^*) + t\,\partial\phi(x^*)$, i.e. $x^* - \bar{x}$ is a direction of descent in x^* for the function $\Theta + t\,\phi$.

5.3. We summarize the above discussion in the following proposition.

Proposition: Let x^* be a zero-point of G . If 5.1.(a) possesses a solution in $\{x \in R^N : ||x - \bar{x}|| \le r\}$, then x^* is such a solution. If x^* does not solve 5.1.(a) , then $||x^* - \bar{x}|| = r$, and x^* solves the restricted problem 5.2.(a) . Furthermore, $x^* - \bar{x}$ is a direction of descent in x^* for the function $\Theta + t\,\phi$ with some $t \ge 0$.

□

5.4. In the non-convex case, a similar discussion can be given via Lagrange's equation:

Proposition: Let Θ and ϕ be convex or sufficiently smooth, and let x^* be a zero point of G as before. Then one of the following two statements is true:

(I) x^* is a critical point of a convex combination of Θ and ϕ .

(II) $||x^* - \bar{x}|| = r$, and $x^* - \bar{x}$ is a direction of descent in x^* for some convex combination of Θ and ϕ .

□

We emphasize that the simplicial algorithm also gives interesting information in case it fails to solve the problem 5.1.(a) .

6. QUASI-VARIATIONAL INEQUALITIES

6.1. We now tackle in a similar way an N-dimensional version of quasi-variational inequalities, see e.g. [8], [9], [10], [11], [21], [24] and the literature cited there. Since it is possible to solve such problems by KKM-maps [12], the approach by simplicial algorithms seems quite natural [11], [16].

Let $T : R^N \to R^N$ be a continuous map and $\phi : R^N \times R^N \to R$ a continuous function, convex for each first argument. We consider the problem:

(a) Find $x \in R^N$ such that $\phi(x , x) \leq 0$ and $(h - x)^T T(x) \geq 0$ for all $h \in R^N$ with $\phi(x , h) \leq 0$.

It is well known that (a) is equivalent to the fixed point problem:

(b) Find $x \in R^N$ such that $x = P_x(x - T(x))$.

Here P_x denotes the metric projection $R^N \to \{h \in R^N : \phi(x , h) \leq 0\}$, i.e. $P_x v = w$ if and only if $\phi(x , w) \leq 0$ and $||v - w|| \leq ||v - h||$ for all $h \in R^N$ with $\phi(x , h) \leq 0$. Hence, fixed point methods, e.g. monotone iteration [8], [11], [21], [24], may be used to solve (a) .

Denote by K the positive cone $[0 , \infty)^N$ of R^N , and define on R^N the usual partial ordering by

(c) $x \leq y \iff y \geq x \iff y - x \in K$.

In the special case that $\phi(x , y) \leq 0$ is equivalent to $y \leq f(x)$, $x , y \in R^N$, where $f : R^N \to R^N$ is some continuous map, the metric projection P_x is easily calculated, and problem (b) reduces to the nonlinear complementarity problem [9], [10], [11], [24] :

(d) Find $x \in R^N$ such that $T(x) \vee (x - f(x)) = 0$, where "$\vee$" denotes the supremum in the partial ordering (c) .

Problems such as those above can be solved by simplicial algorithms in many ways, see e.g. the survey [22]. In general however, applying a simplicial algorithm to the fixed point problem (b) means that for every pivoting step one has to calculate a metric projection. To avoid this, we will now give a different approach [16] .

6.2. As in 5., we choose a starting point $\bar{x} \in R^N$ and a radius $r > 0$ and define the u.s.c. map $G : R^N \to R^{N*}$ by (cf. 4.1.)

(a)

$$G(x) = \begin{cases} \{x - \bar{x}\} & \text{for } ||x - \bar{x}|| > r , \\ \partial_2 \phi(x , x) & \text{for } ||x - \bar{x}|| < r \text{ and } \phi(x , x) > 0 , \\ \{T(x)\} & \text{for } ||x - \bar{x}|| < r \text{ and } \phi(x , x) < 0 , \end{cases}$$

where "∂_2" denotes the subgradient with respect to the second argument. We make the following assumption:

(b) For each $x \in R^N$, the set $\{h \in R^N : ||h - \bar{x}|| < r , \phi(x , h) < 0\}$ is non-empty.

6.3. Again, the simplicial algorithm of section 4. produces a zero point x^* of G , and we discuss the two cases:

(I) $||x^* - \bar{x}|| < r$.
Then $\phi(x^* , x^*) \le 0$ and $0 \in \{T(x^*)\} + t\, \partial_2 \phi(x^* , x^*)$ for some $t \ge 0$. Since x^* is a minimal point of the convex function $h \to (h - x^*)^T T(x^*) + t\, \phi(x^* , h)$, it follows immediately that x^* solves 6.1.(a) .

(II) $||x^* - \bar{x}|| = r$.
By a similar method to that sketched above, one sees that in this case x^* solves the restricted quasi-variational inequality:

(a) Find $x \in R^N$ such that $\phi(x , x) \le 0$, $||x - \bar{x}|| \le r$ and $(h - x)^T T(x) \ge 0$ for all $h \in R^N$ with $\phi(x , h) \le 0$, $||h - \bar{x}|| \le r$.

6.4. We summarize the above approach in the following proposition.

Proposition: Let x^* be a zero point of G. If 6.1.(a) has in $\{h \in R^N : ||h - \bar{x}|| \leq r\}$ a unique solution, then this solution coincides with x^*. If x^* does not solve 6.1.(a), then $||x^* - \bar{x}|| = r$, and x^* solves the restricted problem 6.3.(a).

□

7. POSITIVE SOLUTIONS OF NONLINEAR EIGENVALUE PROBLEMS

7.1. Let us now apply the simplicial algorithm of section 4. to find a positive solution to a nonlinear eigenvalue problem. Consider the problem:

(a) Find $x \in \partial\Omega$, $x \geq 0$, $t > 0$ such that $Mx = tF(x)$.

Here $\Omega \subset R^N$ is a bounded open neighbourhood of $\bar{x} = 0$, M is a non-singular $N \times N$-matrix such that M^{-1} has non-negative entries, and $F : cl(\Omega) \to R^N$ is a continuous map such that

$$x \in cl(\Omega) \quad , \quad x \underset{\neq}{\geq} 0 \quad \text{implies} \quad F(x) \underset{\neq}{\geq} 0 \quad , \text{ cf. 6.1.(c)} .$$

We may think of (a) as a suitably discretized nonlinear elliptic eigenvalue problem in an ordered Banach space [5], [6], see e.g. [1], [2], [3], [4], [15], [17], [19], [25] for different numerical approaches.

7.2. Here we solve 7.1.(a) by defining the u.s.c. map $G : R^N \to R^{N*}$ as follows, cf. 4.1. :

(a)
$$G(x) = \begin{cases} \{-Mx\} & \text{for } x \notin cl(\Omega) , \\ \{F|x|\} & \text{for } x \in \Omega , \ ||x|| > \varepsilon , \\ \{d\} & \text{for } ||x|| < \varepsilon , \end{cases}$$

where $|x|$ denotes the supremum of x and $-x$ in the partial ordering 6.1.(c), $\varepsilon > 0$ is some number with

$\varepsilon < \mathrm{Min}\,\{\, ||x|| \;:\; x \in \partial\Omega \,\}$, and $d \underset{\neq}{\geq} 0$ is some positive vector. The starting point $\bar{x}$ in 4.2.(a) is zero, and the radius r may be chosen by $r \geq \mathrm{Max}\,\{\, ||x|| \;:\; x \in \partial\Omega \,\}$.

Clearly, the simplicial algorithm of section 4. finds a zero point x^* of G .

Proposition: Let x^* be a zero point of G . Then x^* yields a solution to the nonlinear eigenvalue problem 7.1.(a) .

Proof: We consider the following different cases:

(I) $||x^*|| < \varepsilon$: impossible .

(II) $||x^*|| \geq \varepsilon$, $x^* \in \Omega$: impossible, since

$F|x^*| \underset{\neq}{\geq} 0$, $d \underset{\neq}{\geq} 0$, and hence no convex combination of the two can vanish.

(III) $x^* \notin cl(\Omega)$: impossible since M is non-singular.

(IV) $x^* \in \partial\Omega$: This is the only possible case. It follows that $sMx^* = (1 - s)F|x^*|$ for some $s \in [0\,,\,1]$, and by the assumptions in 7.1. it is seen that $0 < s < 1$, hence for $t = s^{-1}(1 - s)$:

(b) $Mx^* = tf|x^*|$.

Since M^{-1} and F are positive, it follows that $x^* \geq 0$, i.e. $x^* = |x^*|$, and (b) solves 7.1.(a) .

□

REFERENCES

1. Allgower, E.L., Application of a Fixed Point Search Algorithm, in Fixed Points: Algorithms and Applications, S. Karamardian (Ed.), Academic Press, New York 1977, pp. 87-111.

2. Allgower, E.L.; Georg, K., Simplicial and Continuation Methods for Approximating Fixed Points and Solutions to Systems of Equations, to appear in SIAM Review, 1979.

3. Allgower, E.L.; Georg, K., Homotopy Methods for Approximating Several Solutions to Nonlinear Systems of Equations, in this volume.

4. Allgower, E.L.; Jeppson, M.M., The Approximation of Solutions of Nonlinear Elliptic Boundary Value Problems with Several Solutions, Springer Lecture Notes in Mathematics, 333, 1973, pp. 1-20.

5. Amann, H., Fixed Point Equations and Nonlinear Eigenvalue Problems in Ordered Banach Spaces, SIAM Review 18, 620-709 (1976).

6. Ambrosetti, A.; Hess, P., Positive Solutions of Asymptotically Linear Elliptic Eigenvalue Problems, Preprint, 1979.

7. Bartels, R.H.; Golub, G.H., The Simplex Method of Linear Programming Using LU Decomposition, Comm. ACM 12, 266-268 (1969).

8. Bensoussan, A.; Lions, J.L., Propriétés des inéquations quasi variationelles décroissantes, Lecture Notes in Economics and Mathematical Systems, 102, Springer, 1974.

9. Chandrasekaran, R., A Special Case of the Complementary Pivot Theory, Opsearch, 1970.

10. Dolcetta, I.C., Sistemi di complementarietà e disequazioni variazionali, tesi, Università di Roma, 1972.

11. Dolcetta, I.C.; Mosco, U., Implicit Complementarity Problems and Quasi Variational Inequalities, to appear in Proceedings of a Conference on Variational Inequalities, R.W. Cottle; F. Giannessi; J.L. Lions (Eds.), 1979.

12. Dugundji, J.; Granas, A., KKM Maps and Variational Inequalities, Annali della Scuola Normale Superiore di Pisa, Serie IV, Vol. V, 679-682 (1978).

13. Eaves, B.C., Homotopies for Computation of Fixed Points, Mathematical Programming 3, 1-22 (1972).

14. Eaves, B.C.; Saigal, R., Homotopies for Computation of Fixed Points on Unbounded Regions, Mathematical Programming 3, 225-237 (1972).

15. Georg, K., On the Convergence of an Inverse Iteration Method for Nonlinear Elliptic Eigenvalue Problems, Numer. Math. 32, 69-74 (1979).

16. Georg, K., An Application of Simplicial Algorithms to Variational Inequalities, in Approximation of Fixed Points and Functional Differential Equations, H.O. Peitgen; H.O. Walther (Eds.), Springer Lecture Notes, 1979, pp. 126-135.

17. Georg, K., An Iteration Method for Solving Nonlinear Eigenvalue Problems, to appear in the Proceedings of "Konstruktive Methoden bei nichtlinearen Randwertaufgaben und nichtlinearen Schwingungen", held at Oberwolfach, November 1978, ISNM, Birkhäuser Verlag, Basel 1979.

18. Gill, P.E.; Murray, W., A Numerically Stable Form of the Simplex Algorithm, Linear Algebra and its Applications 7, 99-138 (1973).

19. Jürgens, H.; Saupe, D., Methoden der simplizialen Topologie zur numerischen Behandlung von nichtlinearen Eigenwert- und Verzweigungsproblemen, Diplomarbeit, University of Bremen, 1979.

20. Laan, G. van der; Talman, A.J.J., A New Subdivision for Computing Fixed Points with a Homotopy Algorithm, Preprint, 1979.

21. Lions, P.L., Une methode iterative de resolution d'une inequation variationelle, Israel Journal of Math. 31, 204-208 (1978).

22. Lüthi, H.J., Komplementaritäts- und Fixpunktalgorithmen in der mathematischen Programmierung, Spieltheorie und Ökonomie, Lecture Notes in Economics and Mathematical Systems, 129, Springer, 1976.

23. Merrill, O.H., Applications and Extensions of an Algorithm that Computes Fixed Points of Certain Upper Semi-Continuous Point to Set Mappings, Ph.D. Thesis, University of Michigan, 1972.

24. Mosco, U., An Introduction to the Approximate Solution of Variational Inequalities, in Proceedings of a Conference on "Constructive Methods in Functional Analysis", CIME (1971), G. Geymonat (Ed.), 1973.

25. Peitgen, H.O.; Prüfer, M., The Leray-Schauder Continuation Method is a Constructive Element in the Numerical Study of Nonlinear Eigenvalue and Bifurcation Problems, in Approximation of Fixed Points and Functional Differential Equations, H.O. Peitgen; H.O. Walther (Eds.), Springer Lecture Notes in Mathematics, 1979.

26. Rockafellar, R.T., Convex Analysis, Princeton University Press, Princeton, 1970.

27. Saigal, R., The Fixed Point Approach to Nonlinear Programming, Math. Prog. Study 10, 142-157 (1979).

28. Stoer, J.; Witzgall, C., Convexity and Optimization in Finite Dimensions I, Springer, 1970.

29. Todd, M.J., The Computation of Fixed Points and Applications, Lecture Notes in Economics and Mathematical Systems, 124, Springer, 1976.

30. Todd, M.J., New Fixed Point Algorithms for Economic Equilibria and Constrained Optimization, Preprint, 1977.

31. Todd, M.J., Numerical Stability and Sparsity in Piecewise-Linear Algorithms, Preprint, 1979.

Numerical Solution of Highly Nonlinear Problems
W. Forster (ed.)
© North-Holland Publishing Company, 1980

SOME REMARKS ON QUASI-VARIATIONAL INEQUALITIES AND FIXED POINTS

I. Capuzzo Dolcetta and M. Matzeu

Universita di Roma,
Italy

Many problems arising in such different fields as stochastic control, economic equilibrium, filtration through porous media, plasma physics, etc. have been studied recently in the framework of the quasi-variational inequalities theory (see Bensoussan and Lions [3] and Mosco [12] for general references on the subject, Joly and Mosco [11], Baiocchi [2], Mossino [15] for specific applications in the above mentioned directions).
The aim of this paper is to point out some connections between quasi-variational inequalities and fixed point theory which seem to be interesting from both theoretical and numerical points of view.

The present work has been done in the program MONIF-ECO of the Façolta di Scienze, Universita di Roma.

1. A quasi-variational inequality (in short, QVI) involves, as a typical feature, some unilateral constraints which depend on the solution itself. They can be expressed in most applications through an inequality of the type

$$u \leq M(u) \quad .$$

To state the problem in a sufficiently general setting let us introduce a (real) Hilbert space H equipped with a partial order $\leq$ induced by a closed positive cone P. Let us suppose also that H is a vector lattice with respect to this order (that is, any two vectors u, v of H have a least upper bound and a greatest lower bound denoted respectively by $u \vee v$ and $u \wedge v$. As a consequence any $v \in H$ can be decomposed as

$$v = v^+ - v^-$$

with v^+ and v^- in P. We assume, moreover, that v^+ and v^- can be chosen to be mutually orthogonal.

In this context the QVI problem can be stated as follows:

QVI: find $u \in V \subseteq H$ such that

$$u \leq M(u)$$

$$\langle A(u) - f, v - u\rangle \geq 0$$

for all $v \in V$, $v \leq M(u)$,

where

V is a closed subspace and a sublattice of H,

$M: H \to H$ is a mapping (often called the obstacle map) satisfying the feasibility condition:

for all $w \in H$ there exists $v \in V$ such that $v \leq M(w)$,

A is an operator acting from V to its dual space V' ,

f is a given linear functional in V' ,

$\langle ..,..\rangle$ denotes the duality pairing between V and V' .

Let us observe that if $\bar{u}$ is a solution of QVI such that

$$A(\bar{u}) - f \in H \quad , \tag{1}$$

then $\bar{u}$ satisfies also the following implicit complementarity problem (in short, ICP):

ICP: find $\bar{u} \in V \subseteq H$ such that

$$\bar{u} \leq M(\bar{u})$$

$$A(\bar{u}) \leq f$$

$$(A(\bar{u}) - f, \bar{u} - M(\bar{u})) = 0 \quad ,$$

where $(..,..)$ is the inner product of H.

The proof of this fact can be easily performed by suitably choosing the "test" vector v in QVI .

This fact corresponds, in the present more general situation, to the well known relationship existing between complementarity problems and variational inequalities.

We want to stress here the fact that the possibility of interpreting $\bar{u}$ in the way described above, which is important for the numerical treatment of quasi-variational inequalities, heavily relies on the <u>regularity</u> information (1) . The validity of (1) can be deduced from the following estimates in the order of V' ([1]):

$$f \geq A(\bar{u}) \geq f \wedge A(M(\bar{u})) \quad , \tag{2}$$

provided one knows that both f and $A(M(\bar{u}))$ belong to $H \subseteq V'$.

We refer to Mosco [12] for the estimates (2) and to Caffarelli and Friedman [7] for the study of the regularity of the solution of some QVI's arising in the stochastic impulse control theory.

2. A main feature of the QVI problem is a sort of <u>circularity</u> between solutions and constraints, in the sense that the constraint $u \leq M(u)$ is not given in advance but depends on the solution itself. This suggests in a quite natural way a <u>fixed point</u> approach.

2.1. Let us associate to any w belonging to H the solution z of the <u>variational inequality</u>:

VI:
$$\begin{aligned} & z \in V \_ H \quad , \quad z \leq M(w) \\ & \langle A(z) - f, v - z \rangle \geq 0 \\ & \text{for all} \quad v \in V \quad , \quad v \leq M(w) \end{aligned}$$

(the solution of VI exists and is unique if, for example, A is monotone hemicontinuous and coercive, see Stampacchia [16]).

In this way a mapping is well defined

$$w \to z = S(w) \quad ,$$

called the <u>selection mapping</u> of the QVI problem, whose fixed points (if any) are, by its very definition, the solutions of QVI.

([1]) we say that $v' \in V'$ is non-negative and we denote this by $v' \geq 0$ if $\langle v', v \rangle \geq 0$ for all $v \in V$, $v \geq 0$.

An iterative procedure for the solution of the fixed-point equation

$$u \in V: \quad u = S(u)$$

has been proposed by Bensoussan and Lions [3]. They show that if A is a strictly T-monotone operator, i.e.

$$\langle A(u) - A(v), (u - v)^+\rangle > 0$$

for all $u, v \in V$ such that $(u - v)^+ \neq 0$, M is isotone and satisfies some continuity requirements, then the sequence

$$u_n = S(u_{n-1}) = S^n(u_0) \qquad n = 1, 2, \ldots$$

starting from the solution u_0 of the equation

$$u_0 \in V: \langle A(u_0), v\rangle = \langle f , v\rangle \quad \text{for all} \quad v \in V$$

converges weakly from below to a fixed point of S .

Non-constructive existence theorems for QVI have been proved under different assumptions on A and M by Boccardo and Capuzzo Dolcetta [4].

2.2. Different ways of treating a QVI as a fixed point problem are conceivable under the further assumption that

$$M(w) \in V \quad \text{for all} \quad w \in H \;;$$

let us describe them briefly.

Referring to a construction due to Georg [10] in the case $V = H = V' = R^N$, one can define a map $G: V \to V$ by

$$G(v) = v - (J^{-1}(A(v) - f)) \vee (v - M(v))$$

where J is the canonical isomorphism of V onto V' . It is not difficult to show, using the relations

$$u \vee v = u + (v - u)^+ = v + (u - v)^+ \quad ,$$

that u is a fixed point of G if and only if u solves QVI.

Another approach relates to duality theory: let us consider the following problem (the $\geq$ denotes here the order of V'):

(QVI)': find $u^* \in V'$ such that

$$u^* \geq 0$$

$$\langle v^* - u^* , A'(u^*) + M'(f - u^*) - A'(f)\rangle \geq 0$$

$$\text{for all} \quad v^* \in V' \quad , \quad v^* \geq 0 \quad ,$$

where A' and M' are defined for $v^* \in V'$ by

$$A'(v^*) = -A^{-1}(-v^*)$$

$$M'(w^*) = M(A'(v^*)) \quad .$$

(QVI)' can be seen as a dual problem of QVI (see Capuzzo Dolcetta and Mosco [8], Fusciardi [9]). It should be noticed that (QVI)' has quite a different nature from the 'primal' QVI, since there the constraints do not depend on the solution: (QVI)' is in fact a variational inequality and it is easy to show that u is a solution of QVI if and only if $u^* = -A(u) + f$ solves (QVI)' .

The connections between variational inequalities and fixed points have been observed by various authors (see Stampacchia [16], Mosco [14], Brezis and Sibony [6]); in the particular case under consideration u^* is characterized by the fixed point equation:

$$u^* = (u^* - J(A'(u^*) - A'(f) + M'(f - u^*)))^+ \quad . \tag{3}$$

3. In this final section we are concerned with a 1-dimensional QVI arising in the dynamic programming approach recently proposed by Bensoussan and Lions [3] for the study of some stochastic impulse control problems.

We shall discuss briefly a convergence result for finite element approximations and the use of a fixed point algorithm for the numerical solution of the associated discrete implicit complementarity system (for more details see Capuzzo Dolcetta and Mosco [8]).

The problem is to find a function u belonging to the Sobolev space $H^1(I)$ $(I = (a, b)$ is a given interval on the real axis) such that

$$u(x) \le 1 + \inf_{\substack{\xi \ge 0 \\ x+\xi \in \bar{I}}} u(x+\xi) = M\,u(x) \quad \text{in } I \tag{4}$$

$$\int_a^b u'\,(v' - u')\,dx + \int_a^b u\,(v - u)\,dx \ \ge \int_a^b f\,(v - u)\,dx$$

$$v \in H^1(I)\ , \quad v \le Mu \qquad \text{in } I \quad ,$$

where f is a given non-negative bounded measurable function on (a, b).

It is well known that (4) has a unique solution (see Bensoussan and Lions [3]). For a fixed integer $N \ge 2$ let us consider a grid of I having its nodes at the points

$$x_i = a + ih \qquad h = (b - a)/N \ , \quad 0 \le i \le N \ ;$$

denoting by V_h the finite dimensional subspace of $H^1(I)$ generated by the piecewise linear splines $\phi_i(x)$ $(i = 0 , \dots , N)$ defined as follows:

$$\phi_0(x) = \begin{cases} -1/h\ (x - x_1) & \text{if } a \le x \le x_1 \\ 0 & \text{otherwise} \end{cases} ,$$

$$\phi_i(x) = \begin{cases} 1/h\ (x - x_{i-1}) & \text{if } x_{i-1} \le x \le x_i \\ -1/h\ (x - x_{i+1}) & \text{if } x_i \le x \le x_{i+1} \\ 0 & \text{otherwise} \end{cases} , \qquad (i=1, \dots , N)$$

$$\phi_N(x) = \begin{cases} 1/h\ (x - x_{n-1}) & \text{if } x_{N-1} \le x \le b \\ 0 & \text{otherwise} \end{cases} ,$$

we are led to consider the approximate problem:

$$(4)_h \qquad u_h \in V_h : u_h(x) \le M_h\, u_h(x) \quad \text{in } I$$

$$\int_a^b u_h' \,(v' - u_h')\, dx + \int_a^b u_h\, (v - u_h)\, dx \ \ge \ \int_a^b f\, (v - u_h)\, dx$$

$$\text{for all } v \in V_h \ , \ v \le M_h\, u_h \quad \text{in } I \ ,$$

where $M_h : V_h \to V_h$ is defined as

$$M_h\, v(x) = 1 + \sum_{j=0}^{N} (\min c_j \ , c_{j+1} \ , \dots , c_N)\, \phi_j(x)$$

$$\text{for all } x \in I \quad \text{if} \quad v(x) = \sum_{j=0}^{N} c_j\, \phi_j(x) \ .$$

Now we are in a position to give the following

Theorem: Problem $(4)_h$ has a unique solution for every h . The sequence $\{u_h\}$ converges, as $h \to 0$, to the solution u of (4) uniformly on $[a , b]$.

The proof is obtained by considering the implicit system in R^{n+1} associated to $(4)_h$ and its dual problem in the sense of section 2. The corresponding fixed point equation is then solved by a suitable algorithm. The proof of convergence relies on some results about the continuous dependence of the solution on the data in variational inequalities (see Mosco [14], Boccardo and Capuzzo Dolcetta [2]).

At the moment we are carrying out some numerical experiments using the algorithms described by Allgower and Georg [1].

REFERENCES

1. Allgower; Georg; Simplicial and Continuation Methods for Approximating Fixed Points and Solutions to Systems of Equations, to appear in SIAM Review.

2. Baiocchi; Inèquations quasi-variationelles dans les problèmes a frontière libre en hydraulique, Lect. Notes Math., 503, Springer, Berlin 1976.

3. Bensoussan; Lions; Applications des inèquations variationelles en contrôle stochastique, Dunod, Paris 1978.

4. Boccardo; Capuzzo Dolcetta; Disequazioni quasi-variazionali con funzione d'ostacolo quasi limitata: esistenza di soluzioni e G-convergenza, Boll. UMI. vol. 5, 15-B (1978).

5. Bocardo; Dolcetta; Stabilità delle soluzioni di disequazioni variazionali ellittiche e paraboliche quasi lineari, Ann. Univ. Ferrara, vol. 24 (1978).

6. Brezis; Sibony; Méthodes d'approximation et d'itèration pour les operateurs monotones, Arch. Rat. Mech. Anal. 28, 59-82 (1969).

7. Caffarelli; Friedman; Regularity of the Solution of the Quasi-Variational Inequality for the Impulse Control Problem, Comm. P.D.E. 3 (8), 745-753 (1978).

8. Capuzzo Dolcetta; Mosco; Implicit Complementarity Problems and Quasi-Variational Inequalities, to appear in Variational Inequalities, R.W. Cottle; F. Giannessi; J.L. Lions (Eds.), Wiley.

9. Fusciardi; An Under Relaxed Method for the Numerical Solution of Quasi-Variational Inequalities, to appear.

10. Georg; An Application of Simplicial Algorithms to Variational Inequalities, to appear in Approximation of Fixed Points and Functional Differential Equations, H.O. Peitgen (Ed.), Springer Lecture Notes, 1979.

11. Joly; Mosco; A propos de l'existence et de la regularité des solutions des certaines inèquations quasi-variationelles, to appear.

12. Mosco; Implicit Variational Problems and Quasi-Variational Inequalities, Lecture Notes Math., 543, Springer, Berlin 1976.

13. Mosco; On Some Nonlinear Quasi-Variational Inequalities and Implicit Complementarity Problems in Stochastic Control Theory, to appear in Variational Inequalities, R.W. Cottle; F. Giannessi; J.L. Lions (Eds.), Wiley.

14. Mosco; An Introduction to the Approximate Solution of Variational Inequalities, in Constructive Aspects of Functional Analysis, CIME-Erice 1971, Ed. Cremonese, Roma 1973.

15. Mossino; Sur certaines inequations quasi-variationelles apparaissant en physique, C.R. Acad. Sc. Paris 282, 187-190 (1976).

16. Stampacchia; Variational Inequalities, Proc. NATO Adv. Study Inst., Venezia 1968, Ed. Oderisi 1969.

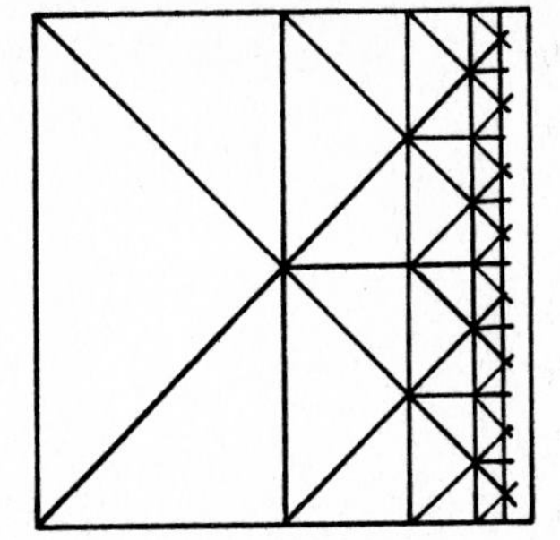

The following paper deals with the historical development of the Brouwer degree.

Numerical Solution of Highly Nonlinear Problems
W. Forster (ed.)
© North-Holland Publishing Company, 1980

BROUWER DEGREE: HISTORY AND NUMERICAL COMPUTATION

Hans W. Siegberg

Universität Bonn,
Bonn, West Germany

1. INTRODUCTION

One of the main tools for existence proofs in nonlinear analysis is the Brouwer degree: if a map is not locally surjective, then its degree vanishes.

Today, usually the Brouwer degree is defined at first for smooth maps F (via Sard's theorem [36]) as the algebraic number of preimages of a regular value of F. Then, using several approximation steps, the concept is extended to continuous mappings (cf. e.g. [4], [11], [26], [27], [33], [37]). Of course, this definition does not give much basis and insight into numerical access of the degree by a computer. It turns out that all previous computational formulas of the Brouwer degree are based essentially on older definitions of degree.

This work was supported by the "Deutsche Forschungsgemeinschaft, Sonderforschungsbereich 72, Universität Bonn".

2. THE ROLE OF GAUSS

Degree arguments, resp. degree-like arguments, can be traced back at least to the work of Gauss. In his dissertation "Demonstratio nova theorematis omnem functionem algebraicam rationalem integram unius variabilis in factores reales primi vel secundi gradus resolvi posse" (1799) [16.a], in which he proved the fundamental theorem of algebra, Gauss opened the door to degree theory.

Gauss published four proofs of the fundamental theorem [16.a], [16.b], [16.c], [16.d]. The common core of his first [16.a] and his last [16.d] proof may be roughly described as follows:

Let $P : C \to C$ be a polynomial, $P(z) = z^n + a_1 z^{n-1} + \dots + a_{n-1} z + a_n$ ($a_n \neq 0$) ; for simplicity we assume real coefficients. Split P into its real and imaginary parts, $P(z) = U(z) + iT(z)$. If B is a ball (centred at the origin) with a sufficiently large radius, then the sets $U^{-1}(0)$, $T^{-1}(0)$ intersect $S := \partial B$ transversally, and the intersection points $U^{-1}(0) \cap S =: \{q_1 , \dots , q_{2n}\}$ and $T^{-1}(0) \cap S =: \{p_1 , \dots , p_{2n}\}$ alternate on S ; see Fig. 1.

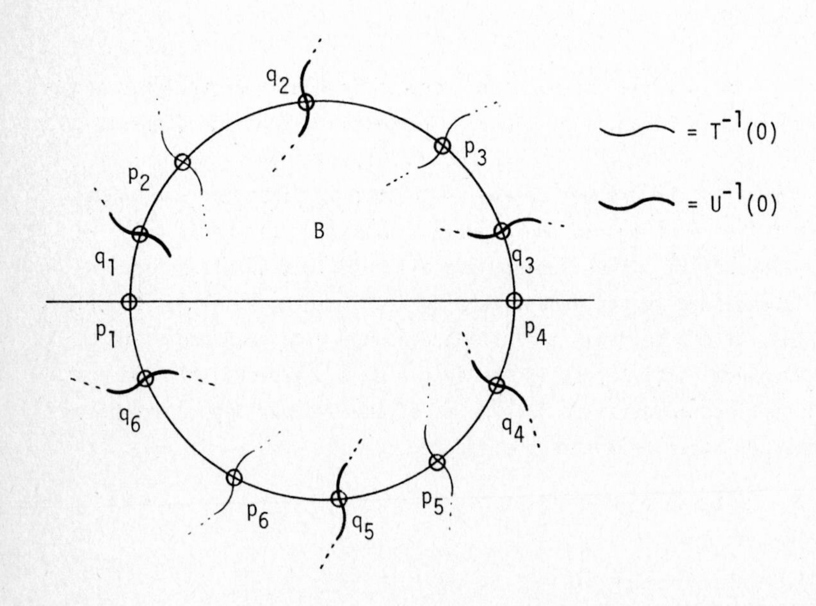

Fig. 1.

It can be shown that each point $p_j \in T^{-1}(0) \cap S$ (resp. $q_k \in U^{-1}(0) \cap S$) is connected by a component (lying in B) of $T^{-1}(0)$ (resp. $U^{-1}(0)$) with a point $p_{\hat{j}} \in T^{-1}(0) \cap S$ (resp. $q_{\hat{k}} \in U^{-1}(0) \cap S$), $j \neq \hat{j}$ (and $k \neq \hat{k}$). (This part in Gauss' proof is not quite complete.) Now a combinatorial argument yields the desired result that $T^{-1}(0) \cap U^{-1}(0) \neq \emptyset$.

The same train of thought works in a more general situation. This was observed by Ostrowski [30.b], who also filled the gap in Gauss' proof mentioned above, see [30.a].

Theorem (Ostrowski, 1934): Let $B := \{x \in R^2 : ||x|| \leq r\}$, $S := \partial B$, be a ball in R^2, and let $F = (F_1, F_2) : (B, S) \to (R^2, R^2 - \{0\})$ be a continuous map with the following property:

F_1 has only a finite number of zeros on S, say $p_1, \ldots, p_{2n}$ (which are assumed to be indexed according to a fixed orientation of S, comp. Fig. 1.), and F_1 changes its sign in each point p_j, $1 \leq j \leq 2n$. Label the points $p_1, \ldots, p_{2n}$ as follows:

$$\ell(p_j) := \begin{cases} 1 \quad (-1), & \text{if } F_2(p_j) > 0 \text{ and } j \text{ even (odd)} \\ -1 \quad (1), & \text{if } F_2(p_j) < 0 \text{ and } j \text{ even (odd)} \end{cases}$$

If $\sum_{j=1}^{2n} \ell(p_j) \neq 0$, then F has a zero.

Sketch of proof: (see [30.b]) Each p_j is connected by a component of $F^{-1}(0)$ with a point $p_{\hat{j}}$ such that $j \not\equiv \hat{j}$ mod 2. (proof !)

Since $\sum_{j=1}^{2n} \ell(p_j) \neq 0$, one concludes that F_2 has at least one zero on at least one component of $F_1^{-1}(0)$.

□

Remark: 1) In the case that F is a polynomial of degree n one has

$$|\sum_{j=1}^{2n} \ell(p_j)| = 2n ,$$ provided B is large enough; see 2.

2) Moreover, it is easy to see that

$|\sum_{j=1}^{2n} \ell(p_j)|$ describes the absolute value of the intersection number of

$F(S)$ with $0 \times R$. Thus,

$\sum_{j=1}^{2n} \ell(p_j)$ is a "discretization" of the winding number as known in

complex function theory:

$$|\sum_{j=1}^{2n} \ell(p_j)| = 2.|w(F_{|S}, 0)| \qquad (w = \text{winding number around the origin})$$

(For the relation between winding number and intersection number, see e.g. [23].)

From his correspondence with Bessel [16.f] one knows that since Dec. 1811 Gauss was acquainted with the foundations of complex function theory and that he had discovered complex logarithm. Thus one might guess that Gauss knew about the argument principle which would imply the fundamental theorem, too. In 1816 he published a short proof of the fundamental theorem using certain integration arguments [16.c]. However, in this proof complex function theory is carefully avoided; only implicitly the argument principle shines through. But it is reported [13] that Gauss has mentioned in later days (1840) that his proof of 1816 originated in his first one given in 1799, explaining thereby essentially the argument principle: thus, one has good reason to believe that Gauss knew the argument principle, see [13].

Explicitly the argument principle is first used for a proof of the fundamental theorem in Briot and Bouquet's "Theorie des fonctions elliptiques" (2nd ed., I, Paris 1873, p. 20), see [13].

[29, p.184] and [30.b] indicate that Cauchy knew the argument principle too.

<u>Remark</u>: Recently, several algorithms for the computation of the winding number and for finding zeros using the argument principle have been established. E.g. Delves and Lyness [9] have given algorithms which are based on quadrature formulas for the computation of the winding number

$$w(F_{|S}, 0) = (2\pi i)^{-1} \int_S (F'(z)/F(z))\, dz \quad ,$$

assuming F is analytic.
Another algorithm, introduced by Erdelski [12], computes the winding number of a (Lipschitz-) continuous map F exploiting essentially Ostrowski's theorem (i.e. computing intersection numbers) : F is replaced by a piecewise linear approximation ("PL-approximation") such that Ostrowski's theorem is applicable.

3. THE KRONECKER INTEGRAL

After Gauss had guaranteed the existence of a (complex) root of any polynomial the question of the number of real roots arose in a natural way. A first satisfactory answer to this problem was given by Sturm in 1829 - 1835 [41.a], [41.b]. Whereas the older methods ("Descartes-Fourier") provided only bounds for the number of real roots, Sturm created a method for finding the exact number. Sturm's theorem was generalized by a number of authors, e.g. Jacobi, Hermite and Sylvester [43]. All these generalizations were influenced by algebraic tools (until 1869). Distilling the qualitative (topological) contents of Sylvester's paper [43] Kronecker developed in 1869 [24.a] the concept of the "characteristic of a function system" which may be interpreted as an n-dimensional generalization of Sturm's theorem (see Kronecker's remark in [24.a]). For our purposes Kronecker's definition of the so-called "Kronecker characteristic" is of no relevance (the interested reader is refered to [38.a], [38.b]). We only mention that it is strongly related to Gauss' proofs [16.a], [16.d] of the fundamental theorem of algebra. For the topologist, the techniques Kronecker used are early examples of oriented intersection theory, which has been refined later in the work of Poincare and Lefschetz.

For numerical purposes we have to mention a representation of the Kronecker characteristic by an integral, given by Kronecker [24.a]. Using the "Kronecker integral" the celebrated Kronecker existence theorem can be formulated as follows (we use the notation of differential forms):

<u>Kronecker's Existence Theorem</u> (1869, [24.a]) : Let M be an $(n+1)$-dimensional, compact, oriented smooth manifold with boundary $\partial M \neq \emptyset$, and let $F : (M , \partial M) \to (R^{n+1} , R^{n+1}-\{0\})$ be a smooth mapping.

Let further $r : R^{n+1}-\{0\} \to S^n$ be the map $r(x) := x/||x||$.

Let $\sigma := \sum_{i=1}^{n+1} (-1)^{i+1} x_i \, dx_1 \wedge \ldots \wedge \hat{dx_i} \wedge \ldots \wedge dx_{n+1}|_{S^n}$

be the volume form of the unit sphere S^n. If

$$K(F) := (\text{vol } S^n)^{-1} \int_{\partial M} (r \circ F_{|\partial M})^* \sigma \neq 0 \quad ,$$

then F has at least one zero in M.

<u>Proof</u>: Use Stokes' theorem; cf. [17] .

□

Kronecker proved this theorem for $M = F_0^{-1}((-\infty , 0])$, where $F_0 : R^{n+1} \to R$ is a smooth map which has 0 as a regular value [24.a]. In this case $K(F)$ $(F = (F_1 , \ldots , F_{n+1}))$ describes the Kronecker characteristic of $(F_0 , \ldots , F_{n+1})$, provided $(F_0 , \ldots , F_{n+1})$ is a (regular) function system (cf. [24.a], [38.a], [38.b]).
Observe that in the case where M is a ball in R^2 the Kronecker integral $K(F)$ coincides exactly with the winding number; hence, Kronecker's existence theorem represents a higher dimensional analogue of the argument principle.

<u>Remark</u>: If we confine ourselves to smooth mappings $F : (M , \partial M) \to (R^{n+1} , R^{n+1}-\{0\})$, where $M \subset R^{n+1}$ is a compact, smooth (n+1)-submanifold with boundary ∂M , a full degree theory (in the sense of e.g. [5]) can be developed by setting:

$$\deg(F , M , y) := (\text{vol } S^n)^{-1} \int_{\partial M} (r \circ F_{y|\partial M})^* \sigma \in Z \qquad (F_y := F - y)$$

(partially done in [18]; cf. also [38.a]).

There are (beside the winding number) several embryonic forms of the Kronecker integral in the literature before 1869 - even in the work of Gauss. A prominent example is the Gaussian linking number of two loops in R^3 which appears in his studies on electrodynamics [16.e] .

Back to computational aspects: One of the first attempts to compute the Brouwer degree in R^n $(n > 2)$ is due to O'Neil and Thomas [28] and they use the Kronecker integral:

If M is a closed ball in R^{n+1}, say $M = \{x \in R^{n+1} : ||x|| \leq r\}$, integration in polar coordinates yields the following expression for the Kronecker integral, resp. $\deg(F, M, 0)$:

$$\deg(F, M, 0) = (\mathrm{vol}\, S^n)^{-1} \int_{-\pi/2}^{\pi/2} \cdots \int_{-\pi/2}^{\pi/2} \int_{-\pi}^{\pi} ||F \circ g(t)||^{-(n+1)} \det A(F(t))\, dt$$

$$(t = (t_1 , \ldots , t_n))$$

where

$$g(t) = (r \cos t_n \ldots \cos t_1 , \; r \cos t_n \ldots \cos t_2 \sin t_1 , \; \ldots$$

$$\ldots , \; r \cos t_n \sin t_{n-1} , \; r \sin t_n)$$

$$(|t_1| \leq \pi , \; |t_i| \leq \pi/2 , \; i = 2 , \ldots , n)$$

are polar coordinates and $A(F)$ is the matrix

$$A(F) = \begin{bmatrix} F_1 \circ g , & \dfrac{\partial(F_1 \circ g)}{\partial t_1} & , \ldots , & \dfrac{\partial(F_1 \circ g)}{\partial t_n} \\ \cdot & \cdot & \cdot & \cdot \\ \cdot & \cdot & \cdot & \cdot \\ \cdot & \cdot & \cdot & \cdot \\ F_{n+1} \circ g , & \dfrac{\partial(F_{n+1} \circ g)}{\partial t_1} & , \ldots , & \dfrac{\partial(F_{n+1} \circ g)}{\partial t_n} \end{bmatrix}$$

Applying Gauss-Legendre quadrature to this integral O'Neil and Thomas computed the degree (provided F is analytic) in several dimensions (3, 6 and 10). Note that this procedure generalizes the algorithm of Delves and Lyness [9].

About the same time another computational formula for the degree appeared - the Stenger formula [40]. Again, the Kronecker integral is an essential tool for Stenger to derive an inductive degree relation [40, ch. 4.2] on which the computational formula is based. We will discuss Stenger's formula more precisely in section 5.

4. TOWARDS DEGREE THEORY

After the pioneering work of Kronecker many papers appeared on the Kronecker characteristic, respectively the Kronecker integral. It turned out that the Kronecker characteristic could be successfully applied to several geometrical-topological questions. We mention Poincare who used the Kronecker integral for investigations into the qualitative theory of autonomous ordinary differential equations [32], and we mention v. Dyck who made important contributions to the n-dimensional Gauss-Bonnet theorem using the Kronecker integral [10] (see also [38.a], [38.b]).

A very important and still recommendable paper on the Kronecker integral, respectively degree theory, is due to Hadamard [18]. In his paper Hadamard collects the facts (known in 1910) about the Kronecker characteristic, and refines them to a degree theory (see remark in 3.). (We should mention that Hadamard defines the degree for continuous mappings using approximation steps.) Although his style is rather sketchy many properties of a degree theory (e.g. homotopy invariance) are essentially proved. Moreover, the Brouwer fixed point theorem is proved in this paper: For a continuous mapping $F : B \to B$ (B a closed ball centred at the origin) without fixed points on the boundary ∂B, one has $(x , x - F(x)) > 0$ for $x \in \partial B$. Therefore the Poincare-Bohl principle (a special case of homotopy invariance of degree) implies
$\deg(Id - F , B , 0) = \deg(Id , B , 0) = 1$, and the fixed point theorem follows.

The reader should not be confused in view of the fact that Brouwer published the fixed point theorem one year later. Hadamard knew the fixed point theorem (without proof) from a letter of Brouwer (date 4-1-1910). (For this and for other very interesting historical remarks concerning Brouwer we refer to Freudenthals excellent expositions in [14] and [15].)

The homotopy invariance of degree roots in works of Poincare [32] and Bohl [6], however, Kronecker too has shown the invariance of his characteristic under certain deformations [24.b]. The paper of Bohl is interesting also from another point of view. Bohl proved for the first time that the boundary of a cube is not a C^1-retract of the solid cube, which is equivalent to Brouwer's fixed point theorem.

Before we return to the computational aspects let us mention a curiosity about Hadamard's paper which is reported by Hopf [19]. Hadamard's paper contains in ch. 41 , 42 the famous Poincare-Hopf theorem on the singularities of a vector field - however, without proof. Thus, one might suspect that Hadamard is the author of this theorem. Actually, there was no proof at that time. The quotation of the theorem was caused by a misunderstanding between Hadamard and Brouwer: Brouwer had proved his theorem in the special case of spheres. (The interested reader may inform himself in [14] about Hadamard's role as a "midwife" in Brouwer's degree theoretic works.)

5. INTERSECTION THEORY

Recall that the winding number $w(F_{|S}, 0)$ in complex function theory can be represented as the intersection number of $F(S)$ with a half-ray, emanating from the origin (which intersects $F(S)$ transversally, see 2.). The same is true in higher dimensions for the Kronecker integral:

$K(F)$ = "intersection number" of $F(\partial M)$ with a half-ray, emanating from the origin (which intersects $F(\partial M)$ transversally) .

A proof of this relation is contained in Hadamard's argument [18] saying that the degree respectively $K(F)$ is always integer. More precisely, Hadamard developed a relation between $K(F)$ and certain (n-1)-dimensional Kronecker integrals (the sum of these (n-1)-dimensional Kronecker integrals corresponds to the above mentioned intersection number), such that $K(F)$ being an integer follows from the fact that the winding number in complex function theory is always an integer. In particular, one has

$2(n+1)\,K(F)$ = "intersection number" of $F(\partial M)$ with the coordinate axes of R^{n+1} .

Looking at the "inductive degree relation" [40, ch. 4.2] which is the basis for Stenger's degree formula, one observes that it describes the same property. Thus, roughly speaking, one can say that the degree formula of Stenger roots in the early work of Hadamard which again was discovered first by Stynes [42].

Let us sketch now how this intersection property may be exploited in a different way for numerical degree computation:

Let $P \subset R^n$ be a homogeneous n-dimensional polyhedron (i.e. there exists a finite triangulation T of P by n-simplices), and let $F : (P, \partial P) \to (R^n, R^n - \{0\})$ be a continuous mapping. Let F_T be the PL-approximation of F associated with the triangulation T of P. If the triangulation is small enough on ∂P the homotopy invariance of degree implies $\deg(F, P, 0) = \deg(F_T, P, 0)$.
To compute $\deg(F_T, P, 0)$ we use the above noted intersection property which is also true in the PL-case, cf. [2, p.467].

Theorem: Let $C_v(F ; \partial T)$ be the set of (n-1)-simplices τ on ∂P ($\tau \in T$) such that $co\{0, F_T(\tau)\}$ is an LP-basis (i.e. there exists $\varepsilon_0 > 0$ such that $\mathcal{O}(\varepsilon) := (\varepsilon, \varepsilon^2, \ldots, \varepsilon^n) \in co\{0, F_T(\tau)\}$ for all $0 \le \varepsilon \le \varepsilon_0$, cf. [3]).
The degree $\deg(F_T, P, 0)$ can be computed as follows:

$$\deg(F_T, P, 0) = \sum_{\tau \in C_v(F; \partial T)} or(F_T ; \tau)$$

where:

$$or(F_T ; \tau) := \text{sign} \det \begin{bmatrix} 1 & 1 & . & . & . & 1 \\ a & t^0 & . & . & . & t^{n-1} \end{bmatrix} \det(F_T(t^0), \ldots, F_T(t^{n-1}))$$

($co\{a, t^0, \ldots, t^{n-1}\}$ is the unique n-simplex in the triangulation T of P having $\tau := co\{t^0, \ldots, t^{n-1}\}$ as a face.)

Sketch of proof:

$$\sum_{\tau \in C_v(F; \partial T)} or(F_T ; \tau) \quad \text{is just the intersection}$$

number of $F_T(\partial P)$ with $R_+ \mathcal{O}(\varepsilon)$, for an appropriate $\varepsilon > 0$ ($= \deg(F_T, P, 0)$, see [2]).

□

Remark: Using another degree formula which will be discussed in the next section there is a simpler proof (without intersection theory), see Remark ii) and iii) in 6. Moreover, the above degree formula may also be interpreted as a "vector labelling version" of a degree formula due to Prüfer [34.b], [35], see Remark 3) in 7.

6. TOPOLOGICAL TOOLS

A completely new approach to degree theory, avoiding any analytical tool, was created around 1910 by Brouwer [14]. The methods which were used by Brouwer initiated a new epoch in algebraic topology. Let us roughly describe how Brouwer - implicitly - introduced the degree (more precisely, the local degree $deg(F, P, p)$), when he proved the "invariance of dimension" [7.a] .

Let $P \subset R^n$ be a homogeneous n-dimensional polyhedron, and assume the n-simplices of the triangulation T of P to be of positive orientation.

(i) For a PL-map $L : (P, \partial P) \to (R^n, R^n-\{p\})$ such that p is an "ordinary point" of L (i.e. $p \notin L(T^{n-1})$, where T^{n-1} is the (n-1)-skeleton of T) define the degree by setting:

$deg(L, P, p) := A - B$,

where A (resp. B) is the number of n-simplices $\sigma = < s^0, \ldots, s^n >$ of T such that $p \in L(\sigma)$ and $L(\sigma) = < L(s^0), \ldots, L(s^n) >$ has positive orientation (resp. negative orientation).

Then for two ordinary points p , p' lying in the same component of $R^n-L(\partial P)$ the identity $deg(L, P, p) = deg(L, P, p')$ holds. Moreover, for a PL-map L' near L one has $deg(L, P, p) = deg(L', P, p)$. Furthermore, the degree does not change if the triangulation of P is subdivided.

(ii) By approximating a continuous mapping $F : (P, \partial P) \to (R^n, R^n-\{p\})$ by PL-maps the concept can be extended to the continuous case.

Note that the degree Brouwer obtains in this way coincides with the degree known since Kronecker - however, Brouwer's definition is much easier to handle than the old one (described in [18]). Moreover, Brouwer seems to be the first one who recognized the full power of degree theory. This is demonstrated in the proof of the invariance of dimension [7.a] .

In view of the (counter-) examples of Cantor and Peano the invariance of dimension had become a fundamental problem in mathematics around the turn of the century. In [7.a] Brouwer gave the first correct (and very short) proof of the invariance in n dimensions (special cases, $n \leq 3$, had been proved earlier). A crucial idea in Brouwer's proof is the observation that the degree $deg(F, K, 0)$ ($0 \in int K$, K = cube) of a map F which is homotopic to the identity map equals one, and that therefore F is surjective in a whole neighbourhood of zero.

At the same time (more precisely, in the same volume of Math. Ann.) Lebesgue published another (sketch of) proof for the invariance of dimension based on a "paving principle" [25.a] . This paving principle which should become very important later in dimension theory, implies the invariance of dimension by a simple argument. However, Lebesgue's sketch of proof was "so wrong that one hardly understands how it could have ever been submitted and printed" [14, p.438] .
The first correct proof of the paving lemma was given by Brouwer in [7.c] (using degree arguments) and later by Lebesgue [25.b] .

A very simple proof of the paving lemma was published in 1928 by Sperner [39]. The core of his proof is the famous "Sperner Lemma".

Sperner Lemma (1928 [39]): Let $\sigma := co\{s^0 , \ldots , s^n\}$ be an n-simplex in R^n , and let $\ell : \sigma \to \{0 , \ldots , n\}$ be a function ("proper labelling") which satisfies
i) $\{\ell(s^0) , \ldots , \ell(s^n)\} = \{0 , \ldots , n\}$,

ii) $x \in co\{s^{i_0} , \ldots , s^{i_k}\} \Rightarrow \ell(x) \in \{\ell(s^{i_0}) , \ldots , \ell(s^{i_k})\}$ $(0 \leq k \leq n)$.

Then there is an odd number of completely labelled ("Sperner-") simplices (see e.g. [44]) for any triangulation of σ .
(Counting orientations the number of Sperner simplices equals one.)

Although Sperner's lemma has a combinatorial aspect, and can be proved by combinatorial arguments, Sperner's lemma may (or should) be seen also as a result of degree theory:
The proper labelling ℓ induces a simplicial map $L : \sigma \to \sigma$ which maps each vertex p of the triangulation of σ into $s^{\ell(p)}$. Now conditions i) and ii) mean that L is a simplicial approximation of the identity map $\sigma \to \sigma$, and degree yields the result.
(The notion of a simplicial approximation is explained in the following section.)

□

As is well known, utilizing Sperner's lemma, Knaster, Kuratowski and Mazurkiewicz gave in 1929 a simple proof of the Brouwer fixed point theorem without using degree theory explicitly [21]; (nevertheless, the "philosophy" behind their proof may (or should) be called degree theory.)

There are several "simplicial methods", see e.g. [3], [44], to compute a Brouwer fixed point, based essentially on Sperner's lemma and the Knaster-Kuratowski-Mazurkiewicz argument. That many of these algorithms work (also in a more general situation than in the case of Brouwer's fixed point theorem) can be explained by a degree formula given by Prüfer [31], [34.b] . We present the formula for the case of vector labelling, the integer labelling version will be introduced in the next section.

<u>Theorem</u> ([3]): Let $P \subset R^n$ be a homogeneous n-dimensional polyhedron, and let $F : (P , \partial P) \to (R^n , R^n - \{0\})$ be a continuous mapping. Assume the triangulation T of P to be small enough (on ∂P) such that $\deg(F , P , 0) = \deg(F_T , P , 0)$.
Let $C_V(F ; T)$ be the set of n-simplices σ in T such that $F_T(\sigma)$ is an LP-basis (i.e. $C_V(F ; T)$ is the set of completely vector labelled n-simplices of T , according to the notation of [3]).
Then the degree $\deg(F , P , 0) = \deg(F_T , P , 0)$ can be computed as follows:

$$\deg(F_T , P , 0) = \sum_{\sigma \in C_V(F ; T)} \operatorname{or}(F_T ; \sigma)$$

where

$$\operatorname{or}(F_T ; \sigma) := \operatorname{sign} \det \begin{bmatrix} 1 & \ldots & 1 \\ s^0 & \ldots & s^n \end{bmatrix} \begin{bmatrix} 1 & \ldots & 1 \\ F_T(s^0) & \ldots & F_T(s^n) \end{bmatrix}$$

for $\sigma = \operatorname{co}\{s^0 , \ldots , s^n\}$.

<u>Proof</u>: Claim that

$$\sum_{\sigma \in C_V(F ; T)} \operatorname{or}(F_T ; \sigma) \quad \text{equals} \quad \deg(F_T , P , \mathcal{O}(\varepsilon))$$

for an appropriate $\varepsilon > 0$ (such that $\mathcal{O}(\varepsilon)$ is an ordinary point of F_T) . □

Remark:

i) Note that this degree formula is close to the heart of the above described definition due to Brouwer in [7.a].

ii) Since the degree is already determined by $F_{|\partial P}$ respectively $F_{T|\partial P}$ one has $\deg(F , P , 0) = \deg(\tilde{F}_T , P , 0)$, where $\tilde{F}_T$ is the PL-map which coincides with F_T on the boundary ∂P , and which maps each vertex of the triangulation T lying in the interior of P into the origin. This yields the degree formula given in the previous section, too.

iii) The degree formula stated in the previous section can also be derived from the above one using "door in - door out" pivoting arguments, see [34.b], [35].

7. SIMPLICIAL APPROXIMATION

One might guess that Brouwer's work in degree theory was caused in the attempt of solving the invariance problem. However, this seems to be not the case; probably the invariance problem was not the germ, the study of vector fields and their singularities led Brouwer to degree theory (see [14]). An "exercise book" recently discovered by Freudenthal [14], [15], shows that the proof of the invariance of dimension was a "streamlined byproduct, a mere derivative of richer material [14, p.423]" . Shortly after [7.a] Brouwer's most famous paper "Über Abbildungen von Mannigfaltigkeiten" [7.b] appeared. There degree theory is extended to maps between connected, oriented, compact n-manifolds (with triangulation), and several applications are given, e.g. the fixed point theorem. (It should be mentioned that Brouwer's proof of the fixed point theorem is different from Hadamard's one.)

This generalization of degree is provided by a simplicial approximation theorem which states that each continuous map is homotopic to a simplicial map (provided the triangulation of the domain is fine enough).

For a simplicial map $F : M \to M'$ (M , M' are n-manifolds with triangulation, connected, oriented and compact) the degree is defined as before:

$$\deg(F) := \deg(F , M , p) \quad , \quad p \in M' \text{ ordinary point of } F$$
$$= A - B \quad ,$$

where A (respectively $\mathcal{B}$) is the number of n-simplices σ in the triangulation T of M such that $p \in F(\sigma)$ and $F(\sigma)$ is positively oriented (respectively negatively oriented).
Using the simplicial approximation theorem degree can be defined even for continuous mappings.

Let us sketch how the tool of simplicial approximation can be made constructive. The following definition is different from Brouwer's, but easier to work with; it seems to go back to Alexander [1.a], [1.b] .

Definition: Let P and P' be arbitrary polyhedra with triangulation T respectively T' , and let $F : P \to P'$ be a continuous mapping.
A simplicial map $G : P \to P'$ is called a simplicial approximation of F , iff the following condition is satisfied:

$$\text{for all} \quad p \in T^0 \; : \; F(st_T(p)) \subset st_{T'}(G(p))$$

($st(x) :=$ the set of all open simplices in the triangulation containing the vertex x) .
We will denote a simplicial approximation G of F by $F_{T,T'}$.

Remark: Observe that the PL-approximations "F_T" used in the previous section and simplicial approximations "$F_{T,T'}$" are different things; in order to define a simplicial approximation the target space has to be a simplicial complex.

In order to state our theorem we use integer labelling as studied e.g. in [3] and [34.b]. (Other labellings can be used too.)

For $0 \le j \le n$ let B^j be the following sets in R^n :

$$B^0 := \{x \in R^n \; : \; x_1 \le 0\} ,$$

$$B^i := \{x \in R^n \; : \; x_1 , \dots , x_i > 0 , x_{i+1} \le 0\} , 0 < i < n,$$

$$B^n := \{x \in R^n \; : \; x_1 , \dots , x_n > 0\} .$$

Define $\ell(x) = j \; : \Longleftrightarrow \; x \in B^j$.

Let $S_n \subset R^n$ be the n-simplex $S_n := co\{b^0, \dots, b^n\}$ with

$b^0 = (-1, 0, \dots, 0)$,

$b^i = (1, \dots, 1, -1, 0, \dots, 0)$ for $0 < i < n$,

↑ i-th component

and $b^n = (1, \dots, 1)$.

Note that one can interpret each compact polyhedron P with vertices $T^0 = \{p^0, \dots, p^n\}$ as a subpolyhedron of S_n by identifying p^j with b^j.

Theorem: Let P and P' be compact polyhedra with triangulations T respectively T', and assume P' to be a subpolyhedron of S_n. Let $F : P \to P'$ be a continuous mapping. Define a vertex map

$$\mathcal{F} : T^0 \to (T')^0 \quad , \quad \mathcal{F}(p) = b^j \; :\Leftrightarrow \; \ell(F(p)) = j \; .$$

If the triangulation T of P is small enough, the piecewise linear extension of $\mathcal{F}$ is a simplicial approximation $F_{T,T'} : P \to P'$ of F.

We omit the technical proof. □

Remark: The theorem may be seen as a generalization of Prüfer's argument in [34.a], [34.b] who was the first one to discover a relation between the integer labelling ℓ used above and simplicial maps in the context of degree theory, see also [35] and the following example (i).

Examples:

(i) (Prüfer) Let $P \subset R^n$ be a homogeneous n-dimensional polyhedron, and let $F : (P, \partial P) \to (R^n, R^n-\{0\})$ be a continuous mapping. Let $\varepsilon > 0$ be so small that

$$\hat{S}_n \cap F(\partial P) = \emptyset \quad , \quad \hat{S}_n := \varepsilon S_n \; .$$

Then one has $\deg(F, P, 0) = \deg(R \circ F, P, 0)$, where $R : R^n \to \hat{S}_n$ is the radial retraction onto $\hat{S}_n$. If the triangulation T of P is small enough, then the vertex map $\mathcal{F}$ induced by the labelling ℓ :

$$F(p) = \varepsilon b^j \; :\Longleftrightarrow \; \ell(F(p)) = \ell(R \circ F(p)) = j$$

induces a simplicial approximation $(R \circ F)_{T,T'}$ (T' = standard triangulation of $\hat{S}_n$) of $(R \circ F)$.
Since $R \circ F$ and $(R \circ F)_{T,T'}$ are homotopic as maps of pairs $(P , \partial P) \to (\hat{S}_n , \partial\hat{S}_n)$, we have

$$\deg(R \circ F , P , 0) = \deg((R \circ F)_{T,T'} , P , 0) .$$

Therefore, the degree can be computed as follows, see [31], [34.a], [34.b], [35] :

$$\deg(F , P , 0) = \sum_{\sigma \in C_\ell(F;T)} or((R \circ F)_{T,T'} ; \sigma)$$

where $or((R \circ F)_{T,T'} ; \sigma)$ may be defined as in the previous section and $C_\ell(F;T)$ is the set of completely integer labelled n-simplices of T .

Remark:

1) Examples show that for our argument the triangulation T of P (more generally $T_{|\partial P}$ of ∂P) has to be rather small. Using (slightly) different techniques Prüfer derived the above formula under weaker smallness conditions of the triangulation.
2) The first reference where degree is related to integer labelling (different to the one used here) seems to be Krasnoselski's classical monograph [22].
3) Using exactly the same argument as described in remark ii) and iii) in section 6. the above degree formula can be replaced by a formula counting certain codimension one completely integer labelled simplices lying on the boundary ∂P of P , see [34.b] and [35]. This formula can also be derived from Stenger's formula, provided the triangulation is fine enough. This was done by Kearfott [20] who used combinatorial arguments; for a geometric proof see [35].

Using the simplicial approximation theorem stated above one can easily give computational degree formulas in a more general situation, i.e. for maps between triangulated (pseudo-) manifolds. Instead of stating such formulas we give another application - a computational formula for the Lefschetz number in terms of "completely integer labelled" simplices.

(ii) Computing the Lefschetz number of a continuous map $F : P \to P$ is an "easy" task, provided P is a homogeneous n-dimensional polyhedron in R^n and $F(P) \subset \text{int } P$:
one has to compute $\deg(Id - F, P, 0)$ only; see [11, p.163], [8] .
Clearly, in this context the Lefschetz fixed point theorem is constructive, see [34.a], [45].
In the case where P is an arbitrary compact polyhedron the situation is much more difficult. In the following we sketch a computational formula for the Lefschetz number in the general case, avoiding all technical expositions:

Let P be an arbitrary compact polyhedron with triangulation T, and assume P to be a subpolyhedron of S_n. Let $F : P \to P$ be a continuous map. If the triangulation of P (one has to subdivide the triangulation in the domain of F only) is small enough - denote it by T_1 - the vertex map F induced by the labelling ℓ induces a simplicial approximation $F_{T_1,T}$ of F ; thus, the Lefschetz numbers $L(F)$ and $L(F_{T_1,T})$ coincide.

Let τ_1 be a simplex in the triangulation T_1 of P. Call τ_1 a fix-simplex [2] , iff $\tau_1 \subset F_{T_1,T}(\tau_1)$, and define

$$or(\ _1) = \begin{Bmatrix} 1 \\ -1 \end{Bmatrix}$$

according to the orientation behaviour of $F_{T_1,T|\tau_1}$.

Set for $0 \le i \le n$: $Sp(i) := A_i - B_i$,
where A_i (respectively B_i) is the number of i-simplices τ_1 in the triangulation T_1 of P such that τ_1 is a fix-simplex and $or(\tau_1) = 1$ (respectively $or(\tau_1) = -1$) .

Theorem: Under the above hypothesis the Lefschetz number L(F) of F can be computed as follows:

$$L(F) = \sum_{i=0}^{n} (-1)^i \ Sp(i) \quad .$$

Acknowledgment:
The author is indebted to Prof. Dr. H.O. Peitgen and to Dr. K. Georg for drawing his attention to the historical aspect of degree theory and for many helpful discussions.

REFERENCES

1.a. Alexander, J.W., A Proof of the Invariance of Certain Constants of Analysis Situs, Trans. Amer. Math. Soc. 16, 148-154 (1915).

1.b. Alexander, J.W., Combinatorial Analysis Situs, Trans. Amer. Math. Soc. 28, 301-329 (1926).

2. Alexandroff, P.; Hopf, H., Topologie I, Springer, Berlin 1974 (Berichtigter Reprint).

3. Allgower, E.L.; Georg, K., Simplicial and Continuation Methods for Approximating Fixed Points and Solutions to Systems of Equations, to appear in SIAM Review.

4. Amann, H., Lectures on Some Fixed Point Theorems, Monografias de Matematica, IMPA, Rio de Janeiro 1974.

5. Amann, H.; Weiss, S., On the Uniqueness of the Topological Degree, Math. Zeit. 130, 39-54 (1973).

6. Bohl, P., Über die Bewegung eines mechanischen Systems in der Nähe einer Gleichgewichtslage, J. Reine Angew. Math. 127, 179-276 (1904).

7.a. Brouwer, L.E.J., Beweis der Invarianz der Dimensionszahl, Math. Ann. 70, 161-165 (1911).

7.b. Brouwer, L.E.J., Über Abbildung von Mannigfaltigkeiten, Math. Ann. 71, 97-115 (1912).

7.c. Brouwer, L.E.J., Über den natürlichen Dimensionsbegriff, J. Reine Angew. Math. 142, 146-152 (1913).

8. Brown, R.F., The Lefschetz Fixed Point Theorem, Scott Foresman and Co., Glenview Illinois 1970.

9. Delves, L.M.; Lyness, J.N., A Numerical Method for Locating the Zeros of an Analytic Function, Math. Comp. 21, 543-560 (1967).

10. v. Dyck, W., Beiträge zur Analysis Situs I, II, Math. Ann. 32, 457-512 (1888); Math. Ann. 37, 273-316 (1890).

11. Eisenack, G.; Fenske, C., Fixpunkttheorie, Bibliogr. Institut, Mannheim 1978.

12. Erdelsky, P.J., Computing the Brouwer Degree in R^2 , Math. Comp. 27, 133-137 (1973).

13. Fraenkel, A., Der Zusammenhang zwischen dem ersten und dem dritten Gaußschen Beweis des Fundamentalsatzes der Algebra, Jahresber. der DMV 31, 234-238 (1922).

14. Freudenthal, H. (Ed.), L.E.J. Brouwer - Collected Works II, North-Holland, Amsterdam 1976.

15. Freudenthal, H., The Cradle of Modern Topology, According to Brouwer's Inedita, Hist. Math. 2, 495-502 (1975).

16. Gauss, C.F., Gesammelte Werke, Gesellschaft der Wissenschaften zu Göttingen,

16.a. Band III, 1 - 31 (1799);

16.b. Band III, 33 - 56 (1815);

16.c. Band III, 57 - 64 (1816);

16.d. Band III, 72 - 103 (1850);

16.e. Band V, 605 (1833);

16.f. Band VIII, 89 - 91 .

17. Guillemin, V.; Pollack, A., Differential Topology, Prentice Hall, New Jersey 1974.

18. Hadamard, M.J., Sur quelques applications de l'indice de Kronecker, in Introduction a la theorie des fonctions d'une variable, tome II, J. Tannery, Hermann, Paris 1910. (= Oevres, 875 - 915).

19. Hopf, H., Ein Abschnitt aus der Entwicklung der Topologie, Jahresber. der DMV 68, 182-192 (1966).

20. Kearfott, R.B., An Efficient Degree Computation Method for a Generalized Method of Bisection, submitted to Numer. Math.

21. Knaster, B.; Kuratowski, C.; Mazurkiewicz, S., Ein Beweis des Fixpunktsatzes für n-dimensionale Simplexe, Fund. Math. 14, 132-137 (1929).

22. Krasnoselskii, M.A., Topological Methods in the Theory of Nonlinear Integral Equations, Pergamon Press, 1964.

23. Krasnoselskii, M.A.; et al., Vektorfelder in der Ebene, Akademie Verlag, Berlin 1966.

24.a. Kronecker, L., Über Systeme von Functionen mehrer Variabeln I (1869), Ges. Werke, Band I, Teubner, Leipzig 1895, pp. 177-234.

24.b. Kronecker, L., Über die Charakteristik von Functionen Systemen (1878), Ges. Werke, Band II, Teubner, Leipzig 1897, pp. 71-82.

25.a. Lebesgue, H., Sur la non-applicabilite de deux domaines appartenant respectivement a des espaces a n et $n+p$ dimensions, Extrait d'une lettre a M.O. Blumenthal, Math. Ann. 70, 166-168 (1911).

25.b. Lebesgue, H., Sur les correspondances entre les points de deux espaces, Fund. Math. 2, 256-285 (1921).

26. Milnor, J.W., Topology from the Differential Viewpoint, Univ. Press of Virginia, Charlottesville 1965.

27. Nagumo, M., A Theory of Degree of Mapping Based on Infinitesimal Analysis, Amer. J. Math. 73, 485-496 (1951).

28. O'Neil, T.; Thomas, J.W., The Calculation of the Topological Degree by Quadrature, SIAM J. Num. Anal. 12, 673-680 (1975).

29. Osgood, W.F., Lehrbuch der Funktionentheorie, Teubner, Leipzig 1907.

30.a. Ostrowski, A., Über den ersten und vierten Gaußschen Beweis des Fundamentalsatzes der Algebra, in Gauss - Gesammelte Werke, Band 10_2 , Abh. 3 .

30.b. Ostrowski, A., Über Nullstellen stetiger Funktionen zweier Variabeln, J. Reine Angew. Math. 170, 83-94 (1934).

31. Peitgen, H.O.; Prüfer, M., The Leray-Schauder Continuation Method is a Constructive Element in the Numerical Study of Nonlinear Eigenvalue and Bifurcation Problems, in Functional Differential Equations and Approximation of Fixed Points, H.O. Peitgen; H.O. Walther (Eds.), Springer Lecture Notes, 1979.

32. Poincare, H., Sur les courbes definies par une equation differentielle I (1881), II (1882), III (1885), IV (1886), Oeuvres I, Gauthier-Villars, Paris.

33. Pontryagin, L.S., Smooth Manifolds and their Applications in Homotopy Theory, AMS Translations, Ser. 2, II, 1-114 (1959).

34.a. Prüfer, M., Sperner Simplices and the Topological Fixed Point Index, Universität Bonn, SFB 72, Preprint no. 134 (1977).

34.b. Prüfer, M., Simpliziale Topologie und globale Verzweigung, Dissertation, Universität Bonn, 1978.

35. Prüfer, M.; Siegberg, H.W., On Computational Aspects of Topological Degree in R^n, in Functional Differential Equations and Approximation of Fixed Points, H.O. Peitgen; H.O. Walther (Eds.), Springer Lecture Notes, 1979.

36. Sard, A., The Measure of the Critical Values of Differentiable Maps, Bull. Amer. Math. Soc. 48, 883-897 (1942).

37. Schwartz, J., Nonlinear Functional Analysis, Gordon and Breach, New York 1969.

38.a. Siegberg, H.W., Abbildungsgrade in Analysis und Topologie, Diplomarbeit, Universität Bonn, 1977.

38.b. Siegberg, H.W., Some Historical Remarks Concerning Degree Theory, in preparation.

39. Sperner, E., Neuer Beweis für die Invarianz der Dimensionszahl und des Gebietes, Abh. Math. Sem. Univ. Hamburg 6, 265-272 (1928).

40. Stenger, F., Computing the Topological Degree of a Mapping in R^n, Num. Math. 25, 23-28 (1975).

41.a. Sturm, C.F., Analyse d'un Memoire sur la resolution des equations numeriques, Bull. des Sciences de Ferussac XI, 419 (1829).

41.b. Sturm, C.F., Memoire sur la resolution des equations numeriques, Mem. des savants etrangers VI, 271 (1835).

42. Stynes, M.J., An Algorithm for the Numerical Calculation of the Degree of a Mapping, Thesis, Oregon State University, 1977.

43. Sylvester, J.J., On a Theory of the Syzygetic Relations of two Rational Integral Functions, Comprising an Application to the Theory of Sturm's Functions ... , Phil. Trans. Roy. Soc. London 143 (1853).

44. Todd, M.J., The Computation of Fixed Points and Applications, Springer Lecture Notes in Economics and Math. Systems, no. 124, 1976.

45. Vertgeim, B.A., On an Approximate Determination of the Fixed Points of Continuous Mappings, Soviet Math. Dokl. 11, 295-298 (1970).

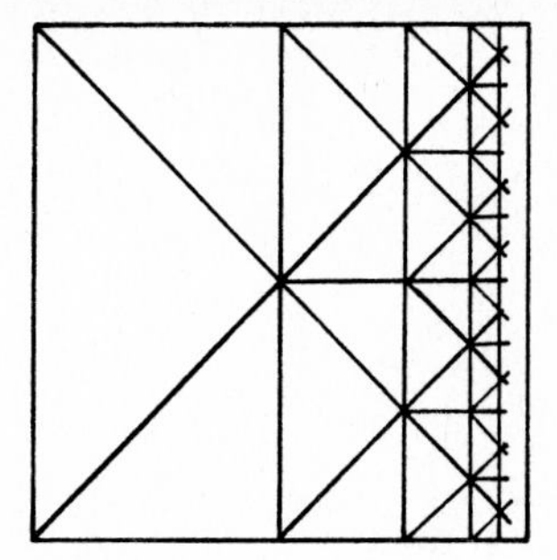

In 1979 J. Schauder would have been 80 years old. J. Schauder is perhaps best known for his fixed point theorem and his work on applications of fixed point theorems to differential equations. The following two papers are biographical in nature.

Juliusz Pawel Schauder
(1899 - 1943)

© 1980 by Walter Forster

© 1980 by Walter Forster

J. SCHAUDER - FRAGMENTS OF A PORTRAIT

Walter Forster

University of Southampton,
Southampton, England

"dreissig Jahre später"

One of J. Leray's papers [1] begins with the following note:
"À la mémoire du profond mathematicien polonais Jules Schauder, victime des massacres de 1940."
I was intrigued by this remark and wanted to find out more about it. During one lunchtime at the Durham Symposium in 1976, I happened to sit at the same table as Leray. I took the opportunity to ask him about the origins of the Leray-Schauder fixed point theorem. The following details emerged.

Leray spent the end of 1932 and the beginning of 1933 in Berlin. This was the time Hitler came to power. Leray and his young wife left Berlin the day the Reichstag burnt (27-2-1933). The next day various human rights were abolished in Germany. This was the reason for many mathematicians in famous Göttingen to leave Germany. Hans Lewy left. Nearly all the other mathematicians also went. Only Hilbert, a sick old man, stayed on.

[1] La théorie des points fixes et ses applications en analyse, Proc. Int. Congress of Math., vol II, A.M.S., New York 1950, pp. 202-208.

At that time, Juljusz Schauder, a Polish mathematician from Lwów, won a scholarship. As a Jew he was not allowed to hold a university position in Poland. He was only allowed to teach in schools. Schauder planned to use his scholarship (Rockefeller) to go to Göttingen. But Göttingen was already an empty place, and so he went to Paris. In Paris, Hans Lewy, a young mathematician, highly praised by Hilbert, Hadamard and many others, was waiting for offers from North- and South-America. He was about to go to Brazil, but changed his mind and went to some remote and unknown place called Berkeley. Thanks to him Schauder and Leray met. Leray said to Schauder: "I very much admire the topological methods you use in your paper about the relation between uniqueness and existence of second order elliptic partial differential equations [Über den Zusammenhang zwischen der Eindeutigkeit und Lösbarkeit partieller Differentialgleichungen zweiter Ordnung von elliptischen Typus, Math. Ann. 106, 661-721 (1932)], but I do not believe there is such a relation". After a few words, Schauder said: "That would be a theorem". Two days later, there was a theorem, two weeks later it was a complete paper, whose subconscious preparation had of course been going on for several years. This joint paper is the now famous paper with the Leray-Schauder fixed point theorem [Topologie et équations fonctionnelles, Ann. Ecole Normale Sup. (3), vol. 51, 45-78 (1934)]. For this paper they received the Malaxa price.

Schauder returned from Paris to Lwów in Poland. World War II started and Poland was divided between Germany and Russia. Two Russian mathematicians, S.L. Sobolev and I.G. Petrovskii, showed interest in Schauder, but Schauder with his Polish background was not at ease with the Russians. By the way, during the war, Petrovskii as a very important mathematician was removed to eastern Siberia for his own security. Then the war between Germany and the USSR began. Lwów was occupied by the German army and the Nazi police. Schauder was now married and had a daughter called Eva. The situation in Lwów got more and more dangerous. Schauder and his family went into hiding, but he was not careful enough and went to the cinema one day (He loved going to the cinema). He was recognized by a traitor and almost immediately arrested by the Gestapo. This was the last one heard of Schauder. As for the traitor, he was executed soon after the war.

Schauder's wife and his daughter hid in a cellar for a while. But the cellar was damp and both became ill. Schauder's wife brought her daughter to a convent and left her in the care of the nuns. She then gave herself up to the police and nothing more was heard of her. Schauder's daughter survived the war in a convent.

Immediately after the war, a Jewish lady, whose husband, children and relatives had been killed by the Nazis, collected all the Jewish children she could find and lived with them in a big house in the Tatra mountains. Local antisemites were still active in those mountains, and during the first winter (1945/46) they broke all the windows. Then the lady succeeded in bringing these poor children to France, and she found suitable housing for them in Meudon, a suburb of Paris. From there she attempted to find the relatives of the children. The lady knew that Eva's father was a mathematician, a teacher, she thought. It happened that one of her contacts in Switzerland read the name of Eva Schauder and remembered a scientific meeting under the chairmanship of Hadamard in Geneva. Leray and Schauder had been participants at this meeting. Thus the lady heard about Leray, when it became necessary to operate Eva (a lung condition). The lady asked Leray to intervene on behalf of Eva to improve the hospital care. Leray explained that this was not necessary. Leray and his wife visited Eva in hospital. Eva Schauder was a very bright little girl. She had learnt French within a short time. In the orphanage the children had been taught in Hebrew. Leray learnt from Poland that Schauder had a brother in Italy and relatives in the States. Leray wrote to Schauder's brother, gave the address of the orphanage and left Paris for the summer vacations. When he returned to Paris, the orphanage was no longer there. Nobody had left a forwarding address. Schauder's brother had not answered (Leray heard that he died a few years later). One has to keep in mind that for several years after the war the situation in Europe was chaotic.

With this information I started in 1976. At that time, J. Schauder's collected works had not yet been published, but I knew that there were plans to publish his collected works in Poland. J. Schauder's works were eventually published in 1978 [Oeuvres, Juliusz Pawel Schauder, PWN - Editions Scientifiques de Pologne, Warszawa 1978]. In the introduction to the collected works there is a brief biography. The information given in the biography can be supplemented with the following details and anecdotes.

J. Schauder was born in Lwów, Poland, on 21st September 1899. Hugo Steinhaus (University of Lwów) considered him to be one of the four best young mathematicians in Poland. Schauder wanted to pursue studies in hydrodynamics, potential theory and topology of function spaces under Léon Lichtenstein (University of Leipzig) and under Richard Courant (University of Göttingen). He spent the end of 1932 and the beginning of 1933 with Léon Lichtenstein (1878-1933) in Leipzig. Because of his Jewish origin, Léon Lichtenstein was to be deprived of his chair in mathematics by the Nazis as of September 1st, 1933. He knew that already in May 1933 and died on August 21st, 1933. Léon Lichtenstein was the editor-in-chief of a number of journals, and J. Schauder was at one time employed as an assistant to the editor-in-chief. From May 1933 to September 1933, J. Schauder was at the Collège de France in Paris with P. Montel and Jacques Hadamard.

Schauder and his wife met Hans Lewy in Spring 1933 in Paris. Schauder was some sort of assistant to Banach. Banach later came to Paris too, and Banach, Lewy, Schauder and others spent many evenings together. All mathematicians met at Hadamard's seminar. Seminars other than that did not exist in Paris - they were felt to be un-French. Schauder was at the time applying himself to partial differential equations. He had put his fixed point theorem [J. Schauder, Der Fixpunktsatz in Funktionalräumen, Studia Mathematica 2, 171-180 (1930)] in a "vorläufig endgültige Fassung" (according to his paper), i.e. "temporarily definitive form" when he met Leray. Like most of Schauder's papers it is written in German.
As a Rockefeller fellow Schauder received $ 200.- per month. This allowed a very comfortable existence, and everybody enjoyed the glory of Paris.

In 1933, J. Schauder and H. Lewy visited the French mathematician Georges Girard (1899-1943). Georges Girard was paralysed from his hips downward and lived in the village Bonny-sur-Loire, where he wrote, confined to his bed, a profound, but somewhat voluminous work on elliptic partial differential equations and singular integral equations. J. Schauder had been able to prove in an elegant manner some of Girard's most important results. He wanted to explain this to Girard, but the latter remained attached to his methods.

On another occasion, S. Banach, J. Schauder and H. Lewy spent an evening in the well-known cafe "Capoulde" in the Quartir Latin. They heard a black man offer peanuts in a shrill voice "for your wives, for the wives of the others". Banach beckoned him to come and asked him: "What do you think of the white race?" "Oh, sir, that is a barbaric race" the black man answered. S. Banach, J. Schauder, and H. Lewy were amused by this reply, ... without having a presentiment that this was a terrible prophecy.

Mrs. Schauder told Hans Lewy that her grandfather, being a freethinker (i.e. atheist), had been expelled from the village by the Jewish community and could never again see his family.

In 1943, during the Nazi occupation of eastern Poland, Mrs. Schauder and her daughter Eva took refuge in Warszawa in the two-room apartment which Mrs. Tarski shared with her sister and her four children. Mrs. Tarski and her children had fled there. Her husband Alfred Tarski, a mathematician and logician [see e.g. A. Tarski, A Lattice-Theoretical Fixpoint Theorem and its Applications, Pacific Journal of Math. 5, 285-309 (1955)], being of Jewish origin, had taken refuge somewhere else. The two sisters welcomed Mrs. Schauder and her daughter in the flat, and they lived there for some time. But this became more and more dangerous for everybody since the block of flats was one of the centres of the Polish resistance. Then Mrs. Schauder and her daughter went into hiding in the sewers of Warszawa, and the Tarskis lost their trace.

After the war, a notice was published in Fundamenta Mathematicae 33, p. V (1945). Among the many victims of World War II, the note mentions that J. Schauder perished in September 1943 in the hands of the Gestapo.

In autumn 1978, during my sabbatical leave, J. Leray kindly allowed me to read the letters written to him by J. Schauder. Because of their confidential nature, J. Leray regards most of these letters unsuitable for publication. These letters are highly interesting and shed light on the work and relationships of many great mathematicians.

There is e.g. the relationship with the Italian mathematician Renato Caccioppoli (1904-1959). [Renato Caccioppoli, Opere, Volume I, Volume II, Edizioni Cremonese, Roma 1963]. Caccioppoli has written extensively on

applications of fixed point theorems (in Italian). He was well known for his habit to dress sloppily, but at the same time he insisted on his assistants being dressed impeccably. R. Caccioppoli shot himself through the head in 1959. R. Caccioppoli published a result very similar to J. Schauder's result [R. Caccioppoli, Un Teorema Generale Sull'Esistenza Di Elementi Uniti In Una Trasformazione Funzionale, Rend. Acc. Naz. Lincei, s. VI, v. 11, 794-799 (1930) = Opere, Volume II, pp. 23-29], but did not give a reference to Schauder's paper. This made Schauder very angry. [see also: R. Caccioppoli, Opere, Volume I, Prefazione, p. XIX, first paragraph on top of the page]. As a matter of fact, R. Caccioppoli quoted Schauder in later papers [R. Caccioppoli, Sugli Elementi Uniti Delle Trasformazioni Funzionali: Un'Osservazione Sui Problemi Di Valori Ai Limiti, Rend. Acc. Naz. Lincei, s. VI, v. 13, 498-502 (1931) = Opere, Volume II, pp. 34-38;
R. Caccioppoli, Sugli Elementi Uniti Delle Trasformazioni Funzionali: Un Teorema Di Esistenza E Di Unicita Ed Alcune Sue Applicazioni, Rend. Sem. Mat. Padova, v. 3, 1-15 (1932) = Opere, Volume II, pp. 39-52;
R. Caccioppoli, Sulle Equazioni Ellittiche Non Lineari A Derivate Parziali, Rend. Acc. Naz. Lincei, s. VI, v. 18, 103-106 (1933) = Opere, Volume II, pp. 79-83].

J. Schauder refereed A. Tychonoff's (or A.N. Tikhonov) important fixed point paper [A. Tychonoff, Ein Fixpunktsatz, Math. Ann. 111, 766-776 (1935)]. Schauder claimed in a letter to Leray that he had mentioned the possibility of extending his own result to locally convex spaces, but J. Leray cannot find this letter. Schauder commented on Tychonoff's paper that this generalized version was of no interest to applications. Nowadays the general opinion seems to be that Tychonoff's theorem is an important generalization with many applications.

In a letter from the International Congress of Mathematicians (Oslo 1936) J. Schauder commented on his meeting with J.v. Neumann. Among other comments J. Schauder says that the spaces J.v. Neumann understands best are Hilbert spaces.

In other letters certain lines of research were suggested and ideas which had later been abandoned or did not lead to anything were discussed. J. Leray is of the opinion that a mathematician should have control

Moscow Topological Conference 1935

© 1980 by Walter Forster

of the things he wants to be published and therefore Schauder's letters should not be published. In this context J. Leray mentions Lebesgue, who made unfavourable remarks about people looking through the dustbins of great mathematicians.

At the end of these notes, I would like to quote a brief passage from "Adventures of a Mathematician" by S. Ulam. This passage describes the history of the picture of J. Schauder included in this volume.
"In connection with the Moscow topological meeting [Moscow 1935], several years after World War II, I received a letter from the French mathematician Leray, who with the Lwów mathematician Juliusz Schauder had written a celebrated paper on fixed points for transformations in function spaces and applications in the theory of differential equations. Schauder, our mutual friend, was murdered by the Nazis. Leray wanted to have a photograph of him for himself and for Schauder's daughter who survived the war and lives in Italy. But he could not find any in Poland or anywhere and he wrote to me asking whether I might have a snapshot. Some months after Johnny von Neumann's death I was looking at some of the books in his library and a group photo of the participants in the Moscow conference fell out. Schauder was there, as were Alexandroff, Lefschetz, Borsuk, and some dozen other topologists. I sent the photograph to Leray. It has since been reproduced in several publications."

ACKNOWLEDGMENTS

This paper would not have been possible without the cooperation of many people, and I would like to thank everyone who helped me in this task.
I would like to thank especially J. Leray (Paris) without whose help and constant encouragement this paper could not have been written. Furthermore, I would like to thank H. Lewy (Berkeley), A.N. Tychonoff (Moscow), M. Kline (New York), S. Ulam (Santa Fe), The Rockefeller Foundation (New York) and Members of the Polish Academy of Sciences (Warszawa).
None of the Individuals or Institutions mentioned here bear any responsibility for opinions or facts expressed in this paper. The content of this paper is solely my responsibility.

I would like to thank J. Leray for kindly providing the photographs of J. Schauder included in this paper, and I would like to thank S. Ulam for allowing me to quote a passage from his book "Adventures of a Mathematician, Charles Scribner's Sons, New York 1976".

© 1980 by Walter Forster

MY FRIEND JULIUS SCHAUDER

Jean Leray

Collège de France,
Paris, France

During the academic year 1932/33, Julius Schauder had a Rockefeller Fellowship (a very advantageous fellowship of two years' duration, which requires the recipient to stay abroad). During the five academic years 1931/36, I held a French scholarship of the Caisse des Sciences (which had just been created and was to become the C.N.R.S. [1]). It very generously made it possible for me to spend the first semester of the academic year 1932/33 with E. Schmidt and von Mises in Berlin. The day I left Berlin at the end of the term, the Reichstag was burning [2] and Hindenburg made Hitler Chancellor. The following day, while the first of Hitler's decrees abolishing various fundamental rights were being proclaimed, Lichtenstein in Leipzig kindly gave me the interview I had requested. He repeated to me the advice about reading Julius Schauder's publications which one had given me in Berlin several times. Lichtenstein, a far-sighted man, was frightened. This was going to impair his health, and he would quickly pass away.

After my return to Paris, my thesis was printed, and I could defend it. This defence satisfied me less than the examiners. It convinced me that I had to refine or change the method which allowed me to establish existence theorems without making an assumption implying uniqueness.

It so happened that both Julius Schauder and I had planned to spend the second semester of the academic year 1932/33 in Göttingen, but almost all the mathematicians had fled. Hans Lewy, for example, had taken refuge in Paris, where he was well-known and highly esteemed on account of his publications and his contributions in the seminar of Jacques Hadamard (Collège de France). Julius Schauder had also come to Paris.

One beautiful morning in spring, Hans Lewy introduced us to each other in a modest restaurant in the rue Soufflot. I immediately said to Julius Schauder: "I have read your paper on the relationship between existence and uniqueness of the solution of a nonlinear equation. I know now that existence is independent of uniqueness. I admire your topological methods. In my opinion they ought to be useful for establishing an existence theorem independent of the whole question of uniqueness and assuming only some a priori estimate." He replied - we were speaking German - : "Das wäre ein Satz". ["That would be a theorem"].

Forty-eight hours later, this theorem existed. In the beautiful Jardin du Luxembourg, the theorem was formulated and the proof worked out. It was in the same park a few days later that Julius Schauder suggested to enrich it with its best application, the existence of the solution of the Dirichlet problem for all two-dimensional convex regions, posed for the elliptic equation:

$$a(x , y , z , \frac{\partial z}{\partial x} , \frac{\partial z}{\partial y}) \frac{\partial^2 z}{\partial x^2} + 2b(\ldots) \frac{\partial^2 z}{\partial x \, \partial y} + c(\ldots) \frac{\partial^2 z}{\partial y^2} = 0 .$$

After fifteen days of intense work without a set back, meeting in the Jardin du Luxembourg, in Meudon, where I lived, and its woods, we had our paper "Topologie et équations fonctionnelles" summarily written up.

The summer vacation separated us. In November 1933, the only remaining question was the phrasing with which we wanted to dedicate this work to the memory of Léon Lichtenstein (see below the copy of page 1 of a letter of 15-11-1933.) But the editor of the Annales de L'École Normale Supérieure removed our dedication and added to our quotations on the first page of our paper a list of French publications without direct connection to our subject [3]. I was greatly annoyed at these two changes, and I apologized for them to Julius Schauder. I am no longer in possession of his reply, which showed his astonishment, resignation, and courtesy. Three hurried removals during the 1939/45 war caused the loss of more than one document.

What remains of our correspondence has dwindled to some fifteen letters written by Julius Schauder. The earliest one is the four page letter of 15-11-1933, pages one and four of which have been reproduced below to

bewiesen zu haben. Ich werde Ihnen erst im nächsten
Briefe – nachdem ich alles nachgerechnet habe –
schreiben können, ob alles richtig, oder ob ich mich
geirrt habe.

Von Zeit zur Zeit denke ich über das Problem der
freien Oberfläche bei idealen Flüssigkeiten nach.
Dies scheint mir aus diesem Grunde wichtig zu sein,
weil eigentlich die Gleichgewichtsfiguren rotierender
Flüssigkeiten nur einen Spezialfall davon darstellen.
Das bekannte Problem wird folgendermaßen formuliert:
Es sind die Gleichungen der Hydromechanik
zu lösen, wenn zur Zeit t=0 die Anfangslage
und die Anfangsgeschwindigkeit der begrenzten
Flüssigkeitsmasse und zu allen Zeiten t der
Drücke in den materiellen Punkten der freien
Oberfläche gegeben sind. Ich frage Sie nun: ist
auf diese Weise das Problem mathematisch
einwandfrei gestellt? Muss vielleicht der Druck
an der Oberfläche einer Zusatzbedingung
genügen. Ich kann sogar nicht beweisen, ob die
Lösung eindeutig ist. Für den Existenzbeweis
hilft hier die Analyzität nicht viel.

Viele Grüße von meiner Frau.

Mit vielen Grüßen an Sie
und Ihre Frau Gemahlin Ihr
J. Schauder

P.S.
Grüße an Herr Hirsch [?], de Possel,
Chevalley und alle andere

Lwów, 15.XI 933

Mein lieber Freund!

Ich habe mit der Antwort gezögert, um Ihnen
etwas mathematisch Interessantes mitteilen
zu können. Aber es ist nichts dergleichen geschehen.
Ich will zuerst Ihre Fragen beantworten.
Natürlich bin ich vollkommen damit einverstanden,
daß unsere Arbeit dem Andenken Lichtensteins
gewidmet ist. Ich glaube, daß die Widmung
wie folgt lauten soll: Dem Andenken Leon
Lichtenstein's gewidmet. Ich mache Sie darauf
aufmerksam, daß weder „Herr" und „Prof."
im Titel vorkommen soll, da diese beide
Attribute nur den Lebenden zukommen. Ich weiss
aber nicht, an welcher Stelle die Widmung
erscheinen soll. In den „Separata" auf jeden Fall,
so wie üblich, auf einem Blatte vor dem Beginn
des Textes. Aber wie soll es im Bande der Zeitschrift
selbst sein? Wissen Sie, wie man das tut?

Bei uns in Polen erscheint ein Heft einer
polnischen Zeitschrift, das ganz dem Andenken
Lichtensteins gewidmet ist. Ich schreibe
dort über seine Untersuchungen aus der
Theorie der elliptischen Differentialgleichungen,
Potentialtheorie und vielleicht auch aus
der Hydromechanik.

/.

Page 1 and page 4 of a letter (15-11-1933) written by J. Schauder to J. Leray .

render homage to Léon Lichtenstein. Pages two and three point out the necessity of rectifying an a priori estimate given by Serge Bernstein, which I succeeded to do in 1937/38. They also mention our friend Hans Lewy, to whom we were deeply grateful for taking an interest in our joint work and for making it widely known. Finally, they announce the very beginning of Julius Schauder's research into hyperbolic equations even before he had verified the results (see top of page 4).

The other letters contain:

- statements of results, one of which is particularly important and has unfortunately not been published by Julius Schauder; the possibility of extending the theory of topological degree to locally convex vector spaces (March 1935);
- sketches of some attempts which he gave up;
- announcements of various publications of his;
- an authentic report on the Moscow "Congress of Topology" in 1935, in which he gives evidence of very sound judgement and great breadth of mind, analysing with penetration the capabilities and limits of everyone, whether they were admirably great or somewhat narrow (2-12-1935). While recognizing the great importance of the definition of homotopy groups given by Hurewicz, Julius Schauder notes with regret (page 2): "Die Untersuchungen sind im allgemeinen nicht auf die Anwendungen zugeschnitten und die dort anwesenden Mathematiker interessierten sich nur für die Topologie als solche." ["In general, the studies are not geared to applications, and the mathematicians present were only interested in topology as such."]. He adds a note at the bottom of this page: "Ich habe immer unterstrichen, daß ich kein Topologe bin." ["I have always stressed that I am not a topologist."], and somewhere else (8-7-1936, page 1), "Ich bin, so wie Sie, ein Mann der Anwendungen." ["I am like you a man of applications."];
- some information on the Oslo Congress;
- comments to some of our publications, in particular to the ones following our joint paper: We generalize the definition of the topological degree given by Brouwer, without explaining the details of this definition and without proving its properties; however obvious they may be, we must check them scrupulously (23-2-1935).

Lwów,8.7. 1936.

Mein lieber Freund!

Ich bin im Momente mit der Vorbereitung meines Referates für Oslo beschäftigt. Eine Arbeit über hyperbolische Differentialgleichungen,die ich zum Drucke vorbereite,wird wahrscheinlich vor der Abreise nicht mehr fertiggestellt werden können.

Ich bin mit Ihrem Vorschlag,eine gemeinsame Arbeit zu schreiben,vollkommen einverstanden. Ich bin,so wie Sie,ein Mann der Anwendungen,für welche unsere Arbeit aus E.N.S. bereits genügt. Die Bemerkungen aber,die verschiedene Mathematiker fallen lassen,zwingen mich denselben Standpunkt wie Sie anzunehmen. Ich mache Sie aber aufmerksam,dass wir in der vorgeschlagenen Arbeit noch mehr leisten sollen. Es ist meiner Ansicht nach möglich auch nichtlineare und nichtkompakte Räume zu betrachten /sogar ohne Gruppeneigenschaft/.

Die traurige Frage ist nur: wann werde ich Zeit haben,um darüber nachzudenken? Ich bin jetzt von dem Unterricht in der Schule,der erst am 30 Juni beendet wurde,furchtbar erschöpft und müsste eigentlich /mein Arzt verlangt es/ ausruhen. Ich will aber unbedingt meine Untersuchungen über hyperbolische Differentialgleichungen zum Drucke vorbereiten. Es sind drei Arbeiten,die von: 1/ gemischten Randwertaufgaben für quasilineare Differentialgleichungen / ein allgemeineres Problem als das Cauchy'sche/ 2/beliebigen nichtlinearen Differentialgleichungen /gemischte Randwertprobleme/ und 3/ Systemen von Differentialgleichungen erster Ordnung handeln; alles in n unabhängigen Veränderlichen.Dies ist auch der Inhalt meines Vortrages in Oslo. - Und am ersten September fängt die Hölle in der Schule schon wieder an!

Zu dem allem weiss ich nicht,wie sich meine Zukunft gestalten wird. Meinem Alter nach sollte ich schon eigentlich Professor sein;in meinem Vaterlande ist aber die Lage für mich <u>persönlich</u> vollkommen hoffnungslos. Da ich schnell den Brief befördern will, so schreibe ich nichts über die Einzelheiten,die mir sowieso sehr peinlich sind.Ich müsste unbedingt 1 oder 2 Jahre frei haben,um sich mit der Mathematik beschäftigen zu können. Ich fühle, dass ich da noch etwas zu sagen habe. Vor einigen Wochen hat H.Steinhaus,einer der Professoren,in Lwów sich in meiner Angelegenheit an H.Hadamard gewendet. H.Hadamard wurde gebeten meinen Versuch,wenigstens für ein Jahr nach Nordamerika zu kommen,zu unterstützen. Mein Traum wäre,dort überhaupt bleiben zu können. Wie ich gehört habe,wurde in Princeton darüber gesprochen,mich für einige Zeit einzuladen. Es würde sich also darum handeln,die Sachen kräftig zu fördern. Ich bitte aber Sie und auch H.Hadamard,wenn Sie vielleicht mit

Page 1 of a letter (8-7-1936) written by J. Schauder to J. Leray .

"Ich mußte erklären, daß bei Fixpunkten [es] sich um eine Abbildung einer Menge auf sich selbst handelt, während dies bei topologischem Grad nicht sein muß." ["I had to explain that for fixed point theorems the mapping has to be to itself, whereas this is not necessarily the case with the topological degree."] (2-12-1935, pages 6 and 7);

- the hope that our correspondence and new meetings would enable us to collaborate again;
- some brave and clear-sighted references to the difficult, unfair, and finally dangerous character of his personal circumstances.

All these letters are very friendly, warm, and absolutely confidential. I would betray Julius Schauder's confidence if I passed them on to a third person in greater length. Furthermore, they could not be correctly interpreted unless I added a detailed commentary.

At a Colloquium in Geneva organized in 1935 by Robin Wavre and chaired by Jacques Hadamard, I had the great pleasure of seeing Julius Schauder again. In 1938, towards the end of spring, a mission to the Institute Français in Warsaw allowed us to meet in Lwów a third time. We did not, however, succeed in writing another joint paper as we had hoped.

When we met through Hans Lewy, we had been unconsciously preparing our collaboration for a long time. Our methods and intuitions complemented each other. Very quickly each of us knew the other's thought well enough to solve without discussion the many problems not yet solved.

I cannot explain the nature of our exchanges except by analysing one of them, for example, the one concerning the extension of the Jordan-Alexandroff theorem to Banach spaces:
If two closed subsets F and F' of a Banach space B are homeomorphic images of each other with $x \to x + F(x)$ such that F is compact, then $B \setminus F$ and $B \setminus F'$ have the same number of connected components.
In 1933, Julius Schauder thought that this statement was probably true, its proof, however, difficult. He told me so. In 1935, when preparing my lecture on the theory of the topological degree at the Cours Peccot (Collège de France), I calculated the degree of a composite map and concluded from it an extremely simple proof of that extension of the Jordan-Alexandroff theorem. I drafted a note for publication in the Comptes Rendus de l'Académie des Sciences. It would have been tactless

to say that what Julius Schauder had conjectured could be proved so easily. It was polite and scientifically correct to present this paper as a simple by-product of our joint article. I asked Julius Schauder to scrutinize it. His reaction proves the high quality of his character. Considering it to be quite superfluous to mention that he had conjectured what I had proved, he replied (23-2-1935, in French): "Il me fait grand plaisir que vous pouvez généraliser et compléter la théorie d'une manière si belle et élégante. Les différentes phrases où vous confrontez vos résultats avec les anciens portent témoignage de votre trop grande modestie . . . C'est vraiment étonnant qu'on peut obtenir le théorème d'Alexandroff d'une manière si élégante et facile et - si je dois être sincère - c'est même inquiétant pour moi (*); mais je ne suis pas capable de trouver des fautes." ["It gives me great pleasure that you can generalize and complete the theory in such a beautiful and elegant manner. The various sentences where you compare your results with the conventional ones bear witness of your far too great modesty . . . It is really amazing that one can obtain the Alexandroff theorem in such an elegant an easy manner, and, if I must be honest, it is even disquieting for me; but I am not able to find any mistakes."]. In the report on the Moscow "Congress of Topology", which he sent me on 2-12-1935, he wrote: "Ich habe die Bekanntschaft aller führenden Topologen gemacht und erzählte ihnen von Ihrem schönen Ergebnis (C.R. Note)" [4) , ["I met all leading topologists and told them of your beautiful result"], attributing to me a merit that was partly his own. He expresses himself with such clarity and enthusiasm that the one of these topologists whom we revere most says to him: "Ich schäme mich sehr, es für den endlich-dimensionalen Fall nicht bewiesen zu haben." ["I am ashamed not to have proved it for the finite-dimensional case."]. The aim of these quotations is obviously not to establish the modesty which Julius Schauder attributes to me, but to bear witness to his character which was highly sensible, very scrupulous, thoroughly honest, and truly unpretentious.

(*) The context shows that one must understand: "inquiétant à mon avis" ["disquieting in my opinion"].

Hel, 8. VIII 1936.

Mein lieber Freund!

Mein Osloer Aufenthalt hat – wenn man die Reise miteinrechnet – fast zwei Wochen gedauert. In Lwów erwarteten mich nach meiner Rückkehr viele dringende Geschäfte und es ist mir nur teilweise gelungen diese zu erledigen. So z.B. konnte ich nur die erste Arbeit über hyperbolische Differentialgleichungen fertigstellen und an die Redaktion der Studia Math. abgeben. Die anderen Arbeiten über diesen Gegenstand bin ich gezwungen erst später zusammenzuschreiben. Wann werde ich dies tun? Ich war furchtbar ermüdet (Schule, Oslo) und musste mich seelisch und physisch endlich ausruhen. Gegen 1.VIII bin ich aus Lwów weggefahren. Ich befinde mich jetzt am baltischen Meere an der Halbinsel Hel in der Ortschaft, die denselben Namen trägt. Zwar habe ich viele Bücher mitgenommen (ich soll im nächsten akademischen Jahre an der Universität Mechanik vortragen) doch glaube ich kaum während meines kurzen Aufenthaltes hier etwas lesen, geschweige den wissenschaftlich arbeiten zu können.

Was Ihre Suggestion anbetrifft, die Hilfe von S. Bernstein etc zu beanspruchen, so sind Sie nicht der erste, der mit diesem Vorschlag sich an mich wendet. Ich kann nur darauf folgendes antworten. So lange ich meinen Lebensunterhalt in meinem Lande ~~[illegible]~~ – wenn auch auf die möglichst ungeeigneste Weise – verdienen kann; will ich dies nicht tun. Andere Lösungen in meiner Lage wären mir lieber.

Part of a letter (8-8-1936) written by J. Schauder to J. Leray .

And yet, he had at that time a real need of scientific prestige; the anti-semitism of the Pilsudski regime, increased by Nazi contamination, forced him to work in difficult conditions. Although Privatdozent at the university, he had to teach in a grammar school for a living. He wrote to me on 4-4-1934: "Die Ferien sind die einzige Zeit, wo ich etwas freier bin." ["The holidays are the only time when I have some spare time."]. He informs me of his situation in detail in his letter of 8-7-1936, the first page of which I believe I can make known (see copy below; the end of the last sentence is "ihm über mich sprechen, um Diskretion."). I want to be discreet and will not quote the entire letter. I suggested to him to accept an offer by Serge Bernstein inviting him to come to Leningrad. I could not foresee the martyrdom this city would undergo. He replies to it on 8-8-1936, at the bottom of page one (see copy below). He knows that France is an uncertain asylum, already submerged. The U.S.A. are his only hope. On 11-1-1938, he writes me a letter whose beginning (copy below) shows despair. I would visit him in Lwów in June 1938. The future would become so dark that we would no longer write to each other.

In 1939, after the conquest of Lwów by the Soviet armies, he was visited by I. Petrowsky and S. Sobolev.

In 1941, after the German conquest, he had to go into hiding. I learned from a Polish source that he was the victim of an informer who was tried and executed immediately after the end of the war.

MADAME JULIUS SCHAUDER

The flat in Lwów, where Julius Schauder, his wife and their newly born daughter Eva lived in 1938, was modern, comfortable, and in good taste. Julius Schauder was backed up by his wife with much devotion..For example, his typed letter of 8-7-1936 contains the note: "Ich habe den Brief der Eile wegen meiner Frau zur Schreibmaschine diktiert." ["Because I am in a hurry, I dictated the letter to my wife for typing."]. From now on, his letters would often be typed.

I learned from a Polish source that under the German occupation Madame Julius Schauder hid with her daughter in a cellar where water came in and where Eva fell seriously ill. Madame Julius Schauder then

Lwów, den 11 I 938.

Mein lieber Freund!

Von meinem letzten Brief an Sie ist bereits ein halbes Jahr verstrichen. Der Grund aus welchem ich Ihnen nicht geschrieben habe ist immer derselbe: meine professionelle Beschäftigung lässt mir keine Zeit für die Mathematik übrig. Ausserdem werden Sie leicht begreifen,dass die Atmosphäre,die jetzt bei uns herrscht,mich sehr nervös macht und ich kaum meine Gedanken von den laufenden Ereignissen losreissen kann. Jede Wissenschaft aber und insbesondere die Mathematik verlangt ausser freier Zeit noch viel Ruhe und Konzentration. Die jetzigen Zeiten sind aber wirklich nicht denach.

Ich habe alle Ihre Sendungen wohl erhalten,insbesondere Ihre beiden C.R. Ko-

Part of the last letter (dated 11-1-1938) written by J. Schauder to J. Leray .

gave her daughter into the custody of a catholic convent, allowed herself to be discovered by the Nazis, and disappeared.

EVA SCHAUDER

Immediately after the war, in 1945, Madame Lene Kuchler, a woman of action and heart, whose entire family had disappeared, gathered as many Jewish children as possible, who had been entrusted to catholic convents during the occupation. She wanted to search for their families and give the children a Jewish education. The orphanage she had founded in Poland in the Tatra mountains was the target of attacks: groups of Nazis survived in the mountains. In deep winter all windows were broken. With approximately a hundred or less of these children she took refuge in France, in Bellevue near Meudon.

She looked after Eva Schauder, knew that her father was a mathematician and believed him to be professor at some grammar school. On making inquiries in February 1948, she contacted Harry Zeimer, an engineer at l'Ecole Centrale. He remembered the Colloquium in Geneva (1935), wrote to his former professor, Robin Wavre, who wrote to me. I informed H. Steinhaus (Wroclaw), who (in March 1948) informed Eva's family:
a cousin, Dr. Joseph Bratter, in Wroclaw;
other cousins in the U.S.A.;
Professor Mario Schauder, in Pisa (Italy), the brother of Julius Schauder.

Eva was ten years old, seriously ill, had undergone one operation in Poland, and had to undergo a second one at the Hopital Laennee (Paris). My wife and I visited her at the hospital, where, in a few weeks, she had learned French. She was very attached to the memory of her father, of whom I could give her some photographs. She was a very intelligent and deeply moving little girl.

In April 1948 Madame Kuchler gave me her news. After the summer vacation I telephoned her in vain. I made a trip to Bellevue and found that the orphanage was no longer there. I wrote to Mario Schauder, without receiving a reply. I wrote to him again, in vain. Then I was afraid to become indiscreet. I know from an Italian source that he died some years later. To my great regret, I never again had any other news of Eva Schauder.

EDITOR'S NOTES:

1) C.N.R.S. = Centre national de la recherche scientifique.

2) "Reichstagsbrand" : 27-2-1933.

3) J. Leray et J. Schauder;
Topologie et équations fonctionnelles,
Annales l'Ecole Normale Supérieure 51, 45-78 (1934).

The publications in question are the references (29) on page 67 of the paper cited above. The original reference (29) contained only E. Gevrey's paper. All of E. Picard's papers appearing in reference (29) were added. At the time E. Picard was Directeur of the Annales.

The dedication to Léon Lichtenstein should have read (to be inserted after the title of the paper and after the names of the authors):

"Dédié à la mémoire de Léon Lichtenstein."

(communicated by J. Leray to W. Forster).

There is another paper

J. Leray;
Discussion d'un problème de Dirichlet,
Journ. de Math. 18, 249-284 (1939),

which has a list of publications of E. Picard on the first page. E. Picard was at the time editor of the Journal de Mathématiques.

[4] J. Leray;
Topologie des espaces abstraits de M. Banach,
Comptes Rendus de l'Académie des Sciences 200, 1082-1084 (1935).

Translation:

Translation from the French original by B. Forster and W. Forster with the kind permission of J. Leray.